BASIC BIOGEOGRAPHY
SECOND EDITION

BASIC BIOGEOGRAPHY

SECOND EDITION

NIGEL PEARS
SENIOR LECTURER IN GEOGRAPHY, UNIVERSITY OF LEICESTER

Longman
Scientific &
Technical

Copublished in the United States with
John Wiley & Sons, Inc., New York

Longman Scientific & Technical,
Longman Group UK Limited
Longman House, Burnt Mill, Harlow,
Essex CM20 2JE, England
and Associated Companies throughout the world.

Copublished in the United States with
John Wiley & Sons, Inc., 605 Third Avenue, New York, NY 10158

First published 1977
Second edition 1985
Reprinted 1990, 1992, 1993

British Library Cataloguing in Publication Data
Pears, Nigel
  Basic biogeography. – 2nd ed
  1. Biogeography
  I. Title
  574.9   QH84
  ISBN 0-582-30120-3

Library of Congress Cataloging in Publication Date
Pears, Nigel.
  Basic biogeography.
  Includes bibliographies and indexes.
  1. Biogeography.   I. Title.
QH84.P4   1985   574.5   83-26728
ISBN 0-470-20561-X (USA only)

Set in 10/12 Linotron 202 Plantin Roman
Produced by Longman Singapore Publishers Pte Ltd
Printed in Singapore

CONTENTS

PREFACE TO THE SECOND EDITION

In preparing this second edition the two considerations which have guided the approach are, of course, just those that make all preparations of further editions a difficult task. The first of these is the desire to bring the material in each chapter up to date, in so far as it is practical to include the latest techniques and theories in an introductory textbook. The second is the need to contain the expansion of the text, which this implies, within reasonable bounds. The book has to be of appropriate length and complexity to serve its intended purpose, namely, for first-year University or College students. In trying to reconcile these two aims, every section has been carefully scrutinized to see if any reduction of the original text was possible. The space created has allowed the insertion of new text material and illustrations in each chapter in such a way that the overall increase in book length has not been excessive.

One problem in such an approach is to prevent the book becoming too indigestible. There has been a vast increase in the published output of ecology and biogeography since the first edition was prepared. This is reflected in the many additional references to be found in this edition which, nevertheless, are merely 'the tip of the iceberg'. However, many recent scientific papers are highly specialized and consequently cannot be included in a first-year book which is essentially introductory. For example, this certainly applies to much of the research material stemming from the International Biological Programme, 1965–75. The quantitative data from these studies have been published in the last few years in several lengthy technical reports. Likewise, an increasing awareness of the many ways in which man is disturbing the biosphere has been a feature of the last decade or so, giving rise to several important national and international statements. A growing interest in applied ecology, resource management and conservation is evident at all levels. This has produced a wealth of material, much of which is contained in a very detailed literature. The structure of the first edition, which most reviewers found an attractive feature, has been retained for this edition and a reasonable selection of these research findings has been incorporated. This has been done in such a way, I hope, as to maintain the careful construction and lucidity of the text.

I am particularly grateful for the assistance of Miss R. Rowell and Mrs K. Moore with the new line figures and Mrs Elaine Humphries with the typing of this edition. My wife Christine deserves a huge 'thank you' for keeping the smiles and the glasses of sherry flowing at critical moments.

N. V. P.
Leicester, 1983

Acknowledgements

We are grateful to the following for permission to reproduce copyright material:

Academic Press Inc & the author for our fig 5.5 from fig 1 (Pimentel 1966); Applied Science Publications, the author & The Foundation for Environmental Conservation for our fig 10.11 from fig 7 (Diamond 1975) & fig on p 257 (Helliwell 1976); Associated Book Publishers Ltd for our fig 6.1 modified from figs 2.2, 2.3, 2.4 (Kellman 1980); Botanical Society of the British Isles & the authors for our fig 8.17 from fig 2 (Brown 1974) & our Table 10.3 from Table 3 (Streeter 1974); British Society of Soil Science & the authors for our Tables 9.1 from Table 1 (Ball 1966), 9.5 (Avery 1973); Cambridge University Press for our figs 2.7 from fig 4 (Richards 1952), 8.7 from figs 26, 165 (Godwin 1975); The Ecological Society of America for our figs 4.8 from fig 2 (Whittaker 1956), & 4.9 from fig 9 (Curtis & McIntosh 1951); The Editor, Journal of Ecology for our figs 4.13 from fig 1 (Yeaton 1978), & 4.16 from figs 5, 7 (Lambert & Williams 1966); Edinburgh University Press for our fig 10.12 from fig 3 (McVean & Lockie 1969); The Editor of *Evolution* & the author for our fig 10.10 from figs 7, 8 (McArthur & Wilson 1967); Field Studies Council for our fig 9.1 from fig 3 (Burnham & Mackney 1964); Forestry Commission for our Tables 8.8, 10.1, 10.2 (Rooke 1974) HMSO; W. H. Freeman & Co for our fig 3.3 from fig p 200 (Oosting 1956), Copyright © 1956 W. H. Freeman & Co; W. H. Freeman & Co for *Scientific American* for our fig 5.4 from pp 96–7 (Gosz et al 1978), Copyright © 1978 by Scientific American Inc; Hodder & Stoughton Ltd for our fig 8.22 & Table 8.5 from fig 4 Table 4 (Ovington 1965); The Journal of Animal Ecology for our fig 5.7 from fig 1 (Lack 1945); The Journal of Applied Ecology & the authors for our figs 3.7 (Witcamp et al 1966), 6.24 from (Hopkins 1965) & Table 8.7 from (Crisp 1966); The Ecological Society of Great Britain for our figs 8.24 from fig 6 (Watt & Jones 1948), 4.15 from fig 2 (Marler & Boatman 1952), 4.12 from fig 1 (Williams & Lambert 1960), 9.12 from fig 8 (Grubb, Green & Merrifield 1969); Macmillan Publishing Co Inc for our fig 3.4 from figs 7.20, 7.21 p 193 (Brady 1974); Ordnance Survey for data in our Table 9.3; Oxford University Press for our fig 3.8 from figs 2.1, 7.4 pp 19, 82 (Trudgill 1977);

W. B. Saunders Company & the authors for our figs 3.5, 3.6, 5.1 & Table 5.1 from figs 4.2, 4.3, 3.7 & Table 7.8 (Odum 1971); Scottish Geographical magazine for our figs 9.4, 9.5, 10.2 & Table 9.2 (Burnham 1970); Sidgwick & Jackson Ltd for our fig 8.16 from fig 4.5 (Shimwell 1971); Soil Survey of England & Wales for our fig 9.6 (Mackney & Burnham 1964); Springer Verlag Heidelberg & the authors for our figs 6.17 (Tranquillini 1982), 8.25 (Perkins 1978); University of Chicago Press for our fig 2.8 from fig 2 in article 'Evidence for the Existence of Three Primary Strategies in Plants' by J. P. Grimes *ANII* 982 (1977) pp 1169–94; John Wiley & Sons Inc for our fig 2.6 & Table 2.1 from fig 3.11 & Table 3.5 pp 152, 148 (Dansereau 1957) Copyright © 1957 John Wiley & Sons Inc; John Wiley & Sons Ltd for our Tables 2.3 & 2.4 from Table 1 & Table 20 pp 8, 129 (Grime 1979) Copyright © 1958 John Wiley & Sons Ltd.

Part 1

BASIC CONSIDERATIONS

1

INTRODUCTION

DEFINITIONS

Geography students quickly become aware of the problems of defining their subject. Whereas the subject matter of some branches of the discipline can be easily stated this is not so with biogeography. Geography itself has been variously defined as the study of: areal distributions, spatial patterns, locational analysis, man-land relationships, the environmental relationships of man. Biogeography implies a linkage between Biology and Geography. It studies the distribution of biological materials over the earth's surface and the factors responsible for the observed spatial variations. This provides a spatial pattern for study as fundamental as the variations in rock type (geology), land forms (geomorphology) or atmospheric processes (climatology). We seek not simply to describe these patterns but also to explain them: the question 'Where?' must be followed by the question 'Why?'.

Biologists also ask these same questions. If past or present global distribution patterns of species (e.g. the great spread of coniferous trees in the high latitudes of the Northern Hemisphere; the floral elements comprising the vegetation of Australia; the world distribution of a particular species) are examined, then this study is what biologists call biogeography but what the geographer refers to as Plant or Animal Geography. But if the study is at the more local level (large scale) and centres on the interaction of species with their effective immediate environment, then the study becomes Plant or Animal Ecology. This can be sub-divided into *autecology* which deals with individual organisms or factors, and *synecology* which studies groups of organisms or complexes of factors.

Now the biogeographer may study the same phenomena as the ecologist but he usually places as much emphasis on the distributional aspects as on the environmental relationships in the study. Further, he will tend to stress the role of man in these patterns and processes or the importance for man of the findings in terms of past, present or future man-land relationships. Over the years it is this study which increasingly has become what the geographer regards as biogeography. Quite obviously the distinctions are not clear cut and, like ecology, biogeography draws heavily on information obtained from many sources (e.g. botany, zoology, meteorology, geomorphology, geology, archaeology, sociology). Both aim at an explanation through a synthesis of these data. Hill has examined the position of biogeography as a sub-field of geography, noting the various definitions and possible research themes. Naveh has recently reviewed changing attitudes to the role of man in ecological studies. For many years man was seen solely as an external, destructive agent. Naveh now argues strongly for a 'Landscape Ecology', which moves man to a central position in such studies, thus blurring the distinctions even more.

Whilst the contribution of the biologist to the subject is apparent, the geographical or spatial element should not be underrated. In another context, Darwin was aware of this when he declared in a letter to Joseph Dalton Hooker in 1845: '. . . that grand subject, that almost keystone of the laws of creation, Geographical Distribution.'

In the seventeenth century Francis Bacon unwittingly stated a fundamental principle for biogeography, namely, 'we cannot command Nature unless we obey her'. Only recently has the vital importance of this dictum been appreciated. With the industrial and scientific revolutions of the eighteenth, nineteenth and twentieth centuries there grew up a feeling that man was becoming somehow independent of Nature and would eventually be able to control many aspects of his environment. Now we are painfully aware that our relationship with the other parts of Nature is becoming not less but more important. Resources are not unlimited, and if we continue to destroy, pollute or overexploit our physical environment then we severely threaten our own existence. We are now the dominant species over most of the globe, either by virtue of our actual presence or as a consequence of our activities. The emphasis of biogeography on the role of man in ecological studies makes it a subject of increasing relevance for the future well-being of our population.

THE IMPORTANCE OF THE PLANT COVER

The central themes in elementary biogeography are usually illustrated by examples drawn from the plant world. There are good reasons for this emphasis on vegetation. Not all aspects can be covered in the time available at the introductory level and so concentration on selected areas and topics is necessary. Because animals are generally more mobile and elusive than plants and require more time-consuming and sophisticated techniques to study their ecology, they are less suitable for an introductory study. Moreover, vegetation makes up the most conspicuous element in the non-urban landscape. It profoundly influences processes operating at the interface between the atmosphere and the lithosphere (e.g. microclimates, soil properties). Plants are good site indicators and modify many habitat factors, creating a background or environment for the associated animal communities. But the influences are not all one way: plants and animals interact and influence each other in a most complex manner. The environment is *holocoenotic*, a term expressing the idea that the factors of the environment act collectively and simultaneously and the action of any one factor may be qualified by the other factors. The holocoenotic nature of the environment lies at the very core of all biogeographical thinking.

The study of plants provides a logical starting point for an understanding of the complexities in ecological relationships. The green plant stands as an intermediary between the inorganic and organic worlds. The basic source of energy for the biosphere is solar radiation but only plants can utilize this solar energy directly. All animals, including man, must obtain their chemical energy indirectly from solar energy through plant life. This dependency can be traced by constructing food chains for each animal species, e.g.

Sun → Cabbage → Caterpillar → Chicken → Man.

Food chains are seldom as simple as indicated but usually have numerous loops and interconnections. Animal life is thus, in this sense, parasitic on plant life. Man not only obtains his chemical energy (food) either directly from plants or indirectly as meat and other products from grazing animals but also many other necessities from the plant cover, such as wood products, fibres and drugs. It is therefore essential to know how the various types of plant cover originate, what relationships exist within the vegetation, what changes are taking place and what processes are involved.

Man's basic crops (wheat, barley, maize, etc.) have been in existence for several thousand years. Only recently have we begun a systematic study of man's relationship with these crop plants. Now, for the first time, we may soon be able, at a cost, to control environments completely to suit our crops or produce by genetical engineering tailor-made species to suit our environments. This is one of the main reasons for the conservation of all species so as to maintain the genetic variability of the biosphere. By wisely exploiting this 'gene pool' found in the wild, plant and animal breeders will be able to increase yields, improve quality, enhance environmental adaptability and improve disease and pest resistance in our crops.

Vegetation is one of man's most important resources and it is a renewable resource in a world of resource depletion. The plant cover also plays a vital role in the atmospheric balance of oxygen and carbon dioxide and it is a considerable element in the water budget of any area. But another important function of vegetation is its scenic role. Much that we value visually in the landscape is the result of a subtle blend of vegetation and topography, whether occurring naturally or as a man-made entity. Our aesthetic appreciation of these landscape elements and our ranking of them indicates a role for perception studies in biogeography as in other branches of geography.

These then are some of the main reasons why biogeography must have as a strong foundation the systematic study of the earth's plant cover.

PLANT CLASSIFICATION

Students reading biological literature frequently encounter difficulties with the wide use of Latin or Greek derivatives. At one level this centres on their use for the main sub-divisions of the Plant or Animal Kingdoms (classification) and at another it concerns the naming of individual species mentioned in particular studies (nomenclature).

The main sub-division in the Plant Kingdom is between the *Cryptogams* and *Phanerogams*. The classification is based on the morphological variations shown by plants, especially the expression of the reproductive process in morphological terms. Cryptogams are non-flowering plants and reproduction is by means of spores rather than seeds; the whole reproductive apparatus is rather inconspicuous or hidden, making up a small, less obvious part of the plant body. In contrast, Phanerogams display their reproductive mechanisms prominently, as cones or flower heads, and are seed-bearing.

Cryptogams, having less complex internal structures, are the so-called 'lower plants'. They can be further sub-divided into the following main groups:

(a) *Bacteria* – single-celled plants sometimes referred to as the Schizophyta because they reproduce by simply dividing the cell into two parts.

(b) *Algae* – seaweed is a common example but not typical of the group. Most are much smaller, often existing as microscopic forms in soil or still water.

(c) *Fungi* – mushrooms and toadstools are obvious examples but, once again, most species are much smaller and many occur microscopically in the soil.

(d) *Bryophytes* – all the mosses and liverworts belong to this group. The small capsule on a thin stalk growing up from the moss cushion and usually visible to the naked eye in many species is, in fact, the spore-containing organ of the plant.

(e) *Lichens* – they exist in several growth forms but in all cases lichens are composite plants, consisting of algae and fungi living together in a mutually beneficial (symbiotic) relationship.

The above sub-divisions are grouped together as the *non-vascular* Cryptogams: plants that have not developed a system of specialized tissue, a vascular system, for the movement of fluids (e.g. nutritional solutions) around the plant body. Instead, transport is achieved by general diffusion. The rest of the Plant Kingdom has such a system which, of course, is also a feature of higher animals and represented in man by the heart, veins and arteries.

(f) *Pteridophytes* – this group of Cryptogams are vascular and most examples are conspicuous, upstanding plants (e.g. ferns, horsetails). Many have an ancestry traceable to the Devonian Period, some 250 million years ago.

Phanerogams (or *Spermatophyta*) are of more recent origin and the evolution of structural complexity is reflected in a corresponding morphological diversity. Two main sub-divisions are recognized:

(a) *Gymnosperms* – these are seed producers in which the seed is said to be naked or not enclosed in attractive or protective tissues (cf. the cherry seed encased in a hard shell which in turn is surrounded by soft, fleshy tissue). Most are cone-bearing and the seeds are released after the cone ripens. Conifer trees are typical examples but the Maidenhair tree is a more primitive type in this group.

(b) *Angiosperms* – here the seeds are enclosed in tissues and the group is usually known as the true flowering plants. The brightly flowering herbaceous plants of hedgerows and gardens are Angiosperms and so are the common deciduous trees of Britain. They are the most recently evolved group, beginning their development during Jurassic–Cretaceous times some 100 million years ago. They now dominate vegetation in many regions, particularly those with equable climates. There are well over a quarter of a million Angiosperm species today or approximately 400 species for every one species of Gymnosperm now in existence. This is a measure of their dominance.

Angiosperms are sub-divided into the *Monocotyledons* and *Dicotyledons*: a division based mainly on differences in floral structures and the number of initial leaves (cotyledons) produced when the seed first germinates. Examples of Monocotyledons are sedges, rushes and grasses (the latter include as cereals most of man's basic food crops). The plants are usually relatively short (though palm trees are an exception). As a consequence, their flowers are small when compared with the Dicotyledons. To this second group belong the colourful herbaceous flowering plants, many ornamental species and the hardwood trees of the world. The oaks and beeches of northern temperate latitudes, the main trees of the Tropical Rain Forest, the 600 or more species of eucalypt tree found in Australia, are all Dicotyledons.

Although Phanerogams make up the most obvious vegetation in the world, an important element of Cryptogam vegetation is always present in these communities: for example, the mosses in the ground flora of beechwoods, the lichens on oak trunks, or the 'unseen' but immensely important fungi and bacteria in the top soil. In certain extreme environments Cryptogams may become the main, and sometimes the only, species present, e.g. the moss and lichen vegetation of exposed Arctic habitats.

PLANT NOMENCLATURE

The main divisions within the Plant Kingdom outlined above can be taken further to produce a descending order of plant categories (or taxa). The hierarchy, which is also used for animal classification, is shown in Fig. 1.1.

In the eighteenth century a Swedish botanist, Carl von Linné, also known as Linnaeus, advanced a scheme for naming individual plants. This enables botanists, irrespective of their own native language, to identify the specific plant referred to in any botanical study. This naturally involved the use of Latin, the only international language of the time. The Linnean System of

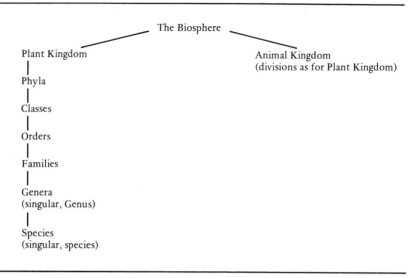

Fig. 1.1 The hierarchy of subdivisions in the biotic components of the biosphere.

nomenclature gives each plant two Latin or latinized names (a Latin binomial). The first, always starting with a capital letter, refers to the genus name. The second, in modern practice written with a small letter, refers to the species within that genus. For example, *Erica tetralix* is the species *tetralix* within the genus *Erica*. Its common name is cross-leaved heath or bog heather and the Latin *tetralix* indicates that the leaves are arranged in fours about the stem to form a cross pattern. *Erica cinerea* (bell heather), on the other hand, is a slightly different species within the same genus. By referring to plants in this way a reader anywhere in the world can identify specimens by reference to a suitable 'Flora', a descriptive key to the various plant families. The Latin binomial is often followed by the name of the authority who first scientifically described and named that particular plant, e.g. L. for Carl von Linné, R. Br. for Robert Brown, Benth. for G. Bentham. Watercress under this scheme is *Nasturtium officinale* R. Br. This increases the accuracy of the reference.

Lastly, the following abbreviations frequently appear in botanical literature and should be noted: sp = species (singular), spp = species (plural), ssp = sub-species (singular) and is sometimes also seen as var. (variety), sspp = sub-species (plural). To summarize this section the cross-leaved heath, a plant of wet upland peat bogs, can now be placed into the overall scheme of classification in use as follows:

Phylum	Spermatophyta
Sub-phylum	Angiospermae
Class	Dicotyledones
Sub-class	Metachlamydeae
	(petals united into a tube)

Order	Ericales
Family	Ericaceae
Genus	Erica
Species	*Erica tetralix* L.

DISCUSSION SECTION

The purpose of this short section (and similar sections following each chapter in this book) is to explore a few points prompted by material in the main part of each chapter. A question and answer approach is adopted, along the lines of the kind of question students in a tutorial group might ask about the content, aims and importance of a lecture previously attended. No attempt is made to be comprehensive and some comments are concerned with areas marginal to the main themes. These sections have been deliberately written in an informal style.

Practically every subject we study involves classification of one sort or another. Why do we have to bother with all these classifications?

There is a staggering diversity in the biosphere: about 1 million animal species and half a million plant species have been described and there could be a great many still undescribed. Life has existed on earth for at least 1,000 million years and the total number of species in this span may run to several hundred million. Classification is our basic method of handling this multiplicity of objects and is one of the most fundamental activities of the human mind. It underlies all forms of science and is the basis of the scientific approach. Quantification and the use of mathematics are sometimes said to be this but before that approach can be used objects must first be selected, and this presupposes a classification. Classification becomes part of an intellectual organization system, a fundamental tool of science, generated by the observer to explain certain relationships between objects being classified. It is a product of man's need to deal with his environment, and increasing knowledge frequently calls for a revision of earlier attempts. The classification of plants given here is essentially an artificial one and since it was first proposed many adjustments have taken place because species or genera were assigned to incorrect groups.

What is wrong with using the common names of plants? How important is it to know the Latin names?

This is best answered by some examples. Common names vary from region to region and often have local significance. This leads to confusion, e.g. *Eucalyptus papuana*, a native tree of Northern Australia is known as Moreton

Bay Ash in the Northern Territories (it is not an ash nor closely related to one. Presumably the white settlers had to call it something!). In nearby Queensland the same tree is variously known as White Gum, Cabbage Gum or Pudding Gum. In England, *Alliaria petiolata,* a common herb of hedgerows, wood margins, shady gardens, wall bases and beechwoods, is known as Garlic Mustard, Jack-by-the-Hedge or Hedge Garlic, and probably has other local names as well.

The only scientific way forward is to persist with the Latin binomial; with practice some names will become familiar. Most of the Latin or Greek used is essentially descriptive and an attempt to decipher it may make the name clearer and more easily remembered, e.g. in the name Cryptogam, crypto means hidden or non-apparent, originally from the Greek word *kruptos,* and gam refers to the gametes or reproductive cells, originally from the Greek word *gameo* (to wed or marry, i.e. join together). Hence an essential feature of this group, as we saw earlier, is that the reproductive apparatus does not form a conspicuous part of the plant body as in Phanerogams (derived from the Greek *phaneros,* meaning visible).

How do we find out more about a named plant species?

This is relatively easy. For most parts of the world, floras exist giving a descriptive paragraph on each local species and sometimes an illustration. For Britain, the key reference covering the vascular plants is *Flora of the British Isles* by A. R. Clapham, T. G. Tutin and E. F. Warburg (CUP 1962). A much cheaper version of this book by the same authors is the *Excursion Flora of the British Isles* (CUP (3rd edn) 1981). A companion set of illustrations will be found in any good library reference section. Also useful is *Wild Flowers of Britain and Northern Europe* by R. S. R. Fitter and A. H. Fitter (Collins 1974). Similar books exist for the lower plants: *British Mosses and Liverworts* by E. V. Watson (CUP (3rd edn) 1981) and *Introduction to British Lichens* by U. K. Duncan (Richmond Publishing Co., Richmond 1970).

If the name is not known how should we set about identifying the plant?

Most floras have a step by step key based on detailed observation and measurement of the plant morphology. Unfortunately it can take much practice and experience to work speedily through a key. Some books rely more on immediate visual recognition, with all the common plants well illustrated. Good examples are *The New Concise British Flora,* W. Keble Martin (Ebury Press 1982) and *Wild Flower Key: British Isles and Northwest Europe,* F. Rose (Warne 1981).

Help can also be obtained from local Natural History societies where experts on various local plant groups are usually willing to assist. Botanical gardens (the main one is Kew, London) will sometimes assist with identification. But such sources should not be approached unless a special study is involved for which accuracy is essential, and then only when all other avenues have been fully explored. As with the recognition of fossils in geology, the main answer is practice and patience.

Finally, it must be stressed that under the Wildlife and Countryside Act 1981, it is illegal to pick or uproot any wild plant unless permission has been given by the owner or occupier of the land. Certain species are totally protected – removal of any part is illegal.

REFERENCES

Crowson, R. A., 1970. *Classification and Biology*, Heinemann, London.

Edwards, K. C., 1964. 'The importance of biogeography', *Geogr.*, **49**, 85–97.

Eyre, S. R., 1964. 'Determinism and the ecological approach to geography', *Geogr.*. **49**, 369–76.

Gilmour, J. S. and Walters, S. M., 1964. 'Philosophy and classification', *Vistas in Botany*, **4**, 1–22.

Hill, A. R., 1975. 'Biogeography as a sub-field of geography', *Area*, **73**, 156–61.

Morgan, W. B. and Moss, R. P., 1965. 'Geography and ecology: the concept of the community and its relationship to environment', *Ann. Ass. Am. Geogr.* **55**. 339–50.

Naveh, Z., 1982. 'Landscape ecology as an emerging branch of human ecosystem science', in *Advances in Ecological Research*, **12**, (eds Macfadyen, A. and Ford, E. D.), Academic Press, London and New York, pp. 189–237.

Prescott-Allen, R. and Prescott-Allen, C., 1981. 'Wild plants and crop improvement', *World Conservation Strategy, Occasional Paper No. 1*, World Wildlife Fund, Godalming.

Simmons, I. G., 1970, 'Landuse ecology as a theme in biogeography', *Canadian Geogr.*, **14**, 309–22.

Simpson, G. C., 1952. 'How many species?', *Evolution*, **6**, 342.

Wace, N. M., 1967. 'The units and uses of biogeography', *Australian Geog. Studies*, **5**, 15–29.

Watts, D., 1978. 'The new biogeography and its niche in physical geography', *Geogr.*, **63**, 324–37.

2

INITIAL APPROACHES TO VEGETATION STUDY

INTRODUCTION

Most plants are gregarious, growing together in groups of varying size. This is because most are fixed in the soil and, in a variety of ways (by seed, vegetative reproduction from fragments, by runners, bulbs, corms, etc.), produce offspring which establish in the vicinity of the parent. This process leads to plant growth in the mass or simply vegetation.

Vegetation can usually be subdivided into groups of plants, i.e. certain combinations of species which grow together. These are not random assemblages since the same or very similar sets are often found again in similar locations. For the moment we can leave aside, until Chapter 4, the controversy surrounding the true nature of spatial arrangements in natural vegetation and accept as a starting point the existence of distinguishable units or plant communities.

The term community suggests not simply living together but also the sharing of a common background (or environment or habitat) and a degree of interdependence between members. For the initial study of a plant community we will look at some aspects of two broad approaches to the problem, the *physiognomic* (or non-floristic) and the *floristic*. But a first consideration is the sampling procedure to be used.

SAMPLING

Some communities are quite extensive and even apparently simple ones turn out to be complex on closer examination. The total species population at a site can seldom be included in the study. Thus an initial problem is to determine the size of the 'minimal area' for detailed study that will give an adequate sampling of the whole population. Ecologists commonly use quadrats to mark out such sample areas. Originally quadrats were square frames but various shapes (circles, rectangles) may be used. For a community where species are equally represented and evenly distributed (e.g. a well-kept lawn) a small quadrat would give a good picture of the overall vegetation. However, such cases are rare. Most vegetation is markedly heterogeneous, although the extent to which it is so depends on the scale of the investigation. We must therefore have quadrats of sufficient size and number to sample adequately the degree of variation present. The size and number needed are determined by constructing a *species-area curve* from the field data.

The x-axis shows the increasing size of quadrat used and the number of species recorded in each quadrat becomes the y-axis. The plotted entries invariably form a curve of the type shown in Fig. 2.1. Where the curve flattens marks the quadrat size beyond which further sampling would yield at best only a few more species. The extra effort required to include them in the survey is considered an inefficient use of the limited time usually available for field work. There are refinements for determining this point on the graph more accurately than just by eye, but these need not concern us here.

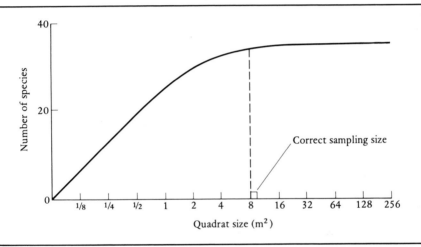

Fig. 2.1 A species-area curve for determining the minimum size (minimal area) of a representative sample from the plant community. The minimal area determined is an approximation, involving some subjective judgement.

Once quadrat size has been decided, a similar approach can be used to determine the number of these units (quadrats) which would constitute an appropriate sample. In this case the x-axis is the number of quadrats of the already determined adequate size and the y-axis is the cumulative total of species as more quadrats are used. The plotted data produce a 'law of diminishing returns' curve and the number of quadrats appropriate for the study should be obvious.

When statistical techniques are used to describe vegetation and analyse field data, the location of quadrats within the study area, strictly speaking, should be random. The data are assumed to be free of bias in a statistical treatment.

The correct method for ensuring this is to lay out a grid across the vegetation and select sampling points by reference to a table of random numbers and the grid coordinates. This is often time-consuming and, in some kinds of vegetation, difficult to carry out. A quicker method is to blindfold a person and instruct him to throw the sampling frame (or other suitable reference object) in any direction; the sampling point is then where the frame lands. But this does not give a truly random sample.

A random approach, however, can sometimes lead to large areas of vegetation being left unsampled. Several interesting, though not common, species can then fail to appear in the analysis although known to be present. Ecologists may therefore resort to a regularized pattern of plot location, arguing that systematic distribution throughout the stand gives quite satisfactory results. These practical considerations can outweigh the theoretical advantages of a random location.

PHYSIOGNOMIC METHODS

Physiognomic methods are non-floristic in the sense that they do not demand a detailed knowledge of the species composition of the vegetation (the community floristics). Rather, there is emphasis on the predominant species present and a relatively quick visual classification of these plants into an established set of main morphological categories. The criteria used are the general outward appearance of the plants in terms of growth forms and stratification (or heights reached), and their leaf patterns, colours, seasonality, density, etc.

The floristic approach demands an excellent knowledge of plant taxonomy (identification and classification), which the beginner will not have. The advantage of physiognomic methods is that they can be applied at various levels of sophistication, thus some immediate progress is possible. Even the research worker studying unfamiliar vegetation in remote regions for which accurate reference books (floras) are not available comes up against the problem of plant identification (for example, there are nearly 300,000 species of flowering plant in the world). These morphological approaches offer some way forward.

Physiognomic methods basically divide the plants present into various *life-forms* irrespective of their taxonomic position. The term life-form expresses the overall plant morphology shown by the species, its form or shape in broad terms. This idea is already present in general usage. In speaking of the needle-leaved conifer forests of high latitudes or the short, tufted grasses of the drier prairies, the form of the vegetation is referred to and not the specific plants present. Some communities consist of one main life-form (e.g. lichen and moss cushions on Arctic rock surfaces), but most have a variety of species present which exhibit several life-form categories. Two simple examples illustrate this.

In many English oakwoods there is a main tree layer (or stratum) formed by the oak (usually *Quercus robur* on the wet, heavier lowland soils or *Quercus petraea* on the lighter soils). Perhaps a few other trees such as elm or ash can

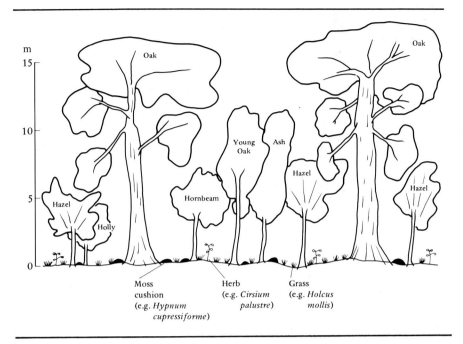

Fig. 2.2 Profile diagram of an English oakwood: the species form three main strata.

also occur in this stratum. Beneath the tall trees, a shrub layer may be present in which hazel, hornbeam, holly or other woody shrubs or small trees feature. Under this shrub stratum, a patchy ground flora of grasses, herbs and mosses frequently occurs, some of the herbs reaching several feet in height. A woodland like this has three main strata and may be visualized as in Fig. 2.2. This, of course, is a gross simplification but it does express the broad forms of the plants and the way they are arranged within the vegetation.

In some mature beechwoods on the Chiltern plateau north of High Wycombe a physiognomy less complex than the oakwood example is noted. The beech (*Fagus sylvatica*) forms a distinctive stratum but it is usually the only tall tree present. Beneath this the shrub or small tree layer is often absent or confined to a few scattered specimens of holly or yew (both evergreens). The ground flora may also be virtually absent; perhaps some scattered moss cushions and an occasional stunted patch of blackberry may be found, but large areas will be bare (Fig. 2.3).

The physiognomic nature of these two woodlands varies, but it is possible for two communities to have very similar physiognomies but not a single species in common (Figs 2.4 and 2.5). In such cases an approach which emphasizes only the species composition of the vegetation would not necessarily provide a very complete picture of the two types being compared.

Plant communities are said to possess a *structure*: the set of life-forms found and their spatial arrangement constitute the primary morphological feature of that vegetation and bestow upon it a more or less definite structure. Strictly

Fig. 2.3 Chiltern beechwood on the plateau south of Chinnor. The simplified structure reflects the dominance of the beech, *Fagus sylvatica*. The ground flora is absent over large areas.

Fig. 2.4 Birchwood (*Betula pendula*) at 320 m on Deeside, near Balmoral. Ground flora mainly of *Vaccinium* spp (whortleberry), *Calluna vulgaris* (heather) and *Deschampsia flexuosa* (wavy hair-grass). Compare this woodland structure with that shown in Fig. 2.5.

Fig. 2.5 Eucalyptus woodland (*Eucalyptus coccifera*: Tasmanian snowgum) on stony doleritic soils of the Central Plateau, Tasmania, about 10 km north of Bronte Park at 850 m. A heath-like ground flora of dwarf shrubs. Note the structural similarities with Fig. 2.4 although these two communities do not have a single species in common.

speaking, the structure of a community reflects three components: the vertical arrangement of the plants (stratification), the horizontal pattern of spatial distribution and the abundance of each species. The second and third of these components are perhaps best considered under the floristic approaches of the next section. It is the first component upon which most elementary physiognomic approaches have concentrated. At the introductory level, structure may be taken as a synonym for physiognomy. This approach is particularly suitable as an elementary technique for the biogeographer since the emphasis is on spatial distributions in the vertical plane, and it can proceed without a detailed knowledge of taxonomy.

These simple ideas have been elaborated to provide increasingly sophisticated schemes for the physiognomic description of plant communities. One of the best known is the scheme proposed by Dansereau (Table 2.1 and Fig. 2.6). It indicates just how much information can be obtained when careful observation is combined with simple measurement. British ecologists working in species-rich Tropical Rain Forest (where up to 400 different species of tall trees, 300 species of small trees or shrubs and 100 species of large woody vines may be encountered) have successfully used a structural approach for this complex vegetation. The initial survey is based on *profile diagrams* (Fig. 2.7) which are scaled drawings of actual trees, showing their positions, diameters, heights and canopy sizes along a narrow strip through the forest (conventionally plots 60 m × 8 m are used and some or all of the vegetation may be felled to facilitate accurate measurement). These long plots are known as transects

Table 2.1 Six categories of criteria to be applied to a structural description of vegetation types. See Fig. 2.6 for examples (From Dansereau, 1957)

1. Life-form			4. Function		
T	◯	trees	d	☐	deciduous
F	♀	shrubs	s	‖‖‖	semideciduous
H	▽	herbs	e	▦	evergreen
M	△	bryoids	j	▨	evergreen-succulent; or evergreen-leafless
E	☆	epiphytes			
L	⬟	lianas			

2. Size		5. Leaf shape and size		
t	tall (T: minimum 25m)	n	◯	needle or spine
	(F: 2—8 m)	g	◖	graminoid
	(H: minimum 2 m)	a	◇	medium or small
m	medium (T: 10—25 m)	h	◊	broad
	(F, H: 0·5—2 m)	v	♈	compound
	(M: minimum 10 cm)	q	⊙	thalloid
l	low (T: 8—10 m)			
	(F, H: maximum 50 cm)			
	(M: maximum 10 cm)			

3. Coverage		6. Leaf-texture		
b	barren or very sparse	f	▨	filmy
i	discontinuous	z	⊞	membraneous
p	in tufts or groups	x	▉	sclerophyll
c	continuous	k	▥	succulent; or fungoid

and their width may be so reduced as to take the form of a single line (the line transect or line intercept). All individuals in contact with this line then form the basis of the sample. Although the idea is simple, in practice difficulties arise in counting problematical intercepts, the length of line to use, and the laying out of lines in dense vegetation.

The ecologist's interest in morphological arrangements of plants is usually not an end in itself. The physiognomy may be an expression of the environ-

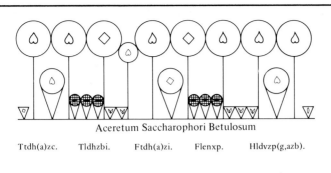

Aceretum Saccharophori Betulosum

Ttdh(a)zc.   Tldhzbi.   Ftdh(a)zi.   Flenxp.   Hldvzp(g,azb).

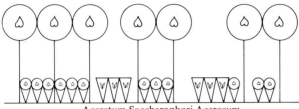

Aceretum Saccharophori Acerosum

Ttdhzc.   Fldhzp.   Hmdvzp.

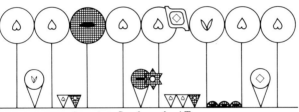

Aceretum Saccharophori Tsugosum

T t d h (v) z c (e n x b).   L t d a z b. F m d a (v) z (e n x) i.
E m e q z b. H l d h z (e a x) p. M l e n f p.

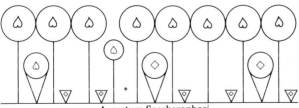

Aceretum Saccharophori

Ttdhzc.   Tldhzb.   Ftda(h)zi.   Hldazi.

Fig. 2.6 Examples of maple-dominated woodland stands in the St Lawrence Valley, Canada. The species are depicted according to the categories of criteria defined in Table 2.1 (From Dansereau, 1957.)

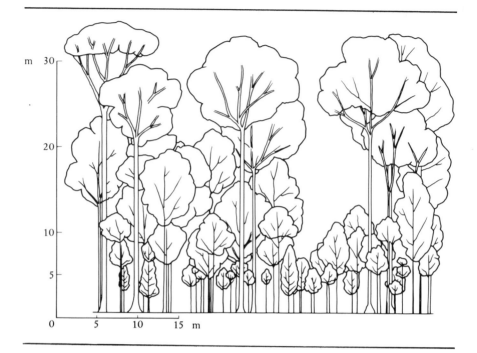

Fig. 2.7 Profile diagram of primary Mixed forest, Moraballi Creek, Guyana. The diagram represents a forest strip 41 m long and 7·6 m wide. Only trees over 4·9 m high are shown. (From Richards, 1952 after David and Richards.)

mental characteristics of the site, and study of spatial patterns might suggest areas for further fruitful investigation.

DOMINANT SPECIES

Within the structure exhibited by a plant community one species may be found which is strongly influencing the environment of the other species present. This plant is frequently referred to as the dominant. It is usually, but by no means always, the tallest species present. For example, in the English beech-wood the beech tree largely controls light intensity at lower levels (a mature beechwood in full leaf may reduce light at ground level to less than 5 per cent of that at canopy level), the amount of moisture and wind reaching lower strata, the input of organic material as leaf-fall to the topsoil and the uptake of soil nutrients and soil moisture. The beech thus exerts a strong influence on the above-ground environment, the ground-surface environment and, by virtue of its very extensive lateral spreading root system, the below-ground environment of the community.

Because a plant community shows structure or layering, it may be possible to demonstrate a dominant species for each stratum. In the oakwood mentioned above, *Quercus robur* may be regarded as the dominant in the tree stratum and the overall dominant of the community. Hazel may be the shrub stratum dominant in that it could be suppressing the growth of other shrubs and the moss species might only become established in a few patches where there are gaps in the grass cover. For a tree seedling to survive and mature into a tall tree it must be able to compete successfully at all strata levels, perhaps dominating each level in turn. Some ecologists treat each stratum in a complex vegetation as a minor community or *synusia* (defined as a social aggregate of a few closely related life-forms which have a similar ecology).

The term 'dominant' is used to express the strong influence of a particular species on community structure or on the dynamism of the community ecology. In the first case we have a physiognomic dominant and in the second an ecological dominant. *Fagus sylvatica* is, of course, an example of a species which combines both roles. If the word 'dominant' is used it should be qualified so as to avoid any confusion. However, this is not always easy to do since frequently we cannot be certain which role the dominant plays. Because of such difficulties, the term 'dominant' has fallen into disrepute for some ecologists who claim that it leads to ambiguity or is biologically meaningless or has anthropomorphic overtones (but see Discussion Section).

Dominance implies competitive ability and the competition mechanisms by which a plant achieves success may take rather subtle forms. We must not always assume that a tall species showing dominance has reached this position just by virtue of its large size. Sometimes chemical rather than physical methods are involved. An increasing number of plants are now known to release chemicals (usually through the roots but sometimes directly into the atmosphere) which are thought to control or suppress the growth of neighbouring competing species (allotoxicity). These chemicals are variously described as root exudates, antibiotics, or, more generally, *allelopathic agents*. For example, this mechanism may operate in some lichen species, in several desert plants, in the Kauri pines of New Zealand and in *Grevillea robusta* (the silky oak) of Queensland. In this last case the exudate is highly specific, killing off only the seedlings of *Grevillea robusta* itself (i.e. autotoxicity). This presumably prevents these seedlings growing up too near to the parent and thus competing with it for environmental resources.

It is very difficult to distinguish allelopathic interactions from the usual competitive interaction of plants growing together under natural field conditions. Because of this, most studies of allelopathy have relied heavily upon laboratory approaches (bioassays) where the isolation of mechanisms is much easier than in the field. However, Stowe has recently re-examined allelopathy and considered its influence on plant patterns at an old-field site in Illinois, USA. He shows that extracts from many species can cause inhibition of germination and growth in other plants when tested under laboratory conditions. Indeed, with many plants autotoxicity is as severe as allotoxicity. Nevertheless, he concludes that these inhibitory mechanisms cannot be demonstrated in the field and the distribution pattern of species does not reflect their operation. Further, he states that the experimental conditions of the laboratory approach never adequately simulate field conditions and they are clearly insufficient as a demonstration of allelopathic relationships. He does not dismiss all cases of allelopathy but only those (and there are many) which have been identified through bioassays. Whether most cases are eventually substantiated or not, these findings point to the need for caution when searching for explanations of spatial patterns in vegetation.

LIFE-FORM CLASSES

The term life-form is frequently used in a more specialized sense than the general one given above. This stems from the work of a Danish botanist, Raunkiaer. He classified plants according to the position of their regenerating parts. Depending on the relative exposure of their perennating bud to the impact of climatic extremes, they were grouped into *life-form classes* wholly irrespective of taxonomic order. Principal divisions are:

1. *Phanerophytes* (P) – bear buds high and exposed to the full force of the climate. They are numerous in warm, moist regions where no protection is required. Sub-divisions of this group are:
   (a) megaphanerophytes (Pg) – more than 30 m high
   (b) mesophanerophytes (Pm) – 8 to 30 m high
   (c) microphanerophytes (Pp) – 2 to 8 m high
   (d) nano-phanerophytes (Pn) – less than 2 m high
   (e) climbing phanerophytes (Ps) – no height restriction
2. *Chamaephytes* (Ch) – herbaceous or woody plants with buds produced close to the soil.
3. *Hemicryptophytes* (H) – buds are half buried in the top soil layer or humus.
4. *Geophytes* (G) – buds lie entirely in the soil, protected from cold or dry air.
5. *Therophytes* (Th) – annuals which produce seeds. The parent plant dies completely and seeds remain inactive until favourable conditions return.
6. *Hydrophytes* (HH) – water plants with a similar response to geophytes; the protection afforded by water is analogous to that of soil.
   (Several sub-divisions of these categories have been suggested: Raunkiaer proposed 15 sub-types of phanerophytes and Dansereau has produced a detailed classification of the hydrophytes along these lines.)

For any area a histogram can be produced showing the percentages of the total flora falling into the various life-form categories. Raunkiaer expanded these ideas with his *Normal Spectrum* of life-form categories based on sampling 1,000 species at random from the world's floral lists. The results were: 46% phanerophytes, 9% chamaephytes, 26% hemicryptophytes, 6% geophytes and 13% therophytes. This spectrum can be used as a yardstick against which to compare spectra produced from actual field data (Table 2.2). A marked variation in values for one or more life-form categories calls for an explanation. Since the classification is based essentially on plant reaction to extremes of climate, the individual spectrum or the variations when two or more spectra

Table 2.2 A comparison of life-form spectra

Region	Life-form class (%)				
	Phanero-phytes	Chamae-phytes	Hemicrypto-phytes	Geophytes	Thero-phytes
Normal (Raunkiaer)	46	9	26	6	13
Swiss Alps (Braun-Blanquet)	0	24·5	68	4	3·5
Death Valley, California (Braun-Blanquet)	26	7	18	7	42

are compared should tell us much about macroclimatic (and possibly micro-climatic) patterns at the field site. For example, we might compare by these methods the spectra for a series of sites on the same mountain, either at different altitudinal levels or on sheltered and exposed slopes at the same altitude. Raunkiaer demonstrated that for Clova Mountain in Scotland there was an increase from 7 to 27 per cent in chamaephytes when moving from zones below 300 m to those above 1,000 m.

More recently an approach to vegetation that has some similarities to that of Raunkiaer has been proposed by Grime. He suggests that the composition and structure of vegetation is largely controlled by two classes of external factors. First, there are those factors which limit the production of plant material (biomass) or habitat productivity. Examples are moisture deficiency, restricted mineral nutrients or reduced light intensity. Such factors may be regarded as causing *stress*. Secondly, there are factors which destroy or remove biomass. Examples are grazing, burning or mowing and they may be of natural origin or man-induced. These factors create an intensity of *disturbance*.

By permutating high and low values for each class of factors he arrives at four alternative *primary strategies* for plants, one of which is non-viable (Table 2.3). The evolution of plants has been associated with the emergence of these three primary strategies for survival, each of which is reflected in a distinctive set of genetic characteristics which, of course, have morphological (form) and spatial expressions in the field. A selection of these characteristics is given below.

Table 2.3 Suggested basis for the evolution of three strategies in plants (From Grime, 1979)

Intensity of disturbance	Intensity of stress	
	Low	High
Low	Competitors	Stress-tolerators
High	Ruderals	No viable strategy

(a) *Competitors.* Herbs, shrubs or trees with relatively high dense canopy of leaves and extensive lateral spread above and below ground. Relatively long life history with rapid growth rates but only a small proportion of annual production devoted to seeds. Generally large plants that tap site resources over an ever-increasing zone by root and canopy extension, leading to competitive exclusion of other specimens.

(b) *Stress-tolerators.* A wide range of growth forms from lichens to trees with a very long life history. Slow growth rates, small or leathery or needle-like leaves and a preponderance of evergreenness. Low allocation of resources to seed production.

(c) *Ruderals.* Herbs of small stature and very short life history. Rapid growth rates, flowering early and with high frequency. A large proportion of resources devoted to seed production. Weeds typically belong to this group and many members may be viewed as 'opportunists', taking advantage of disturbed sites.

Grime develops these ideas to take account of those situations which are intermediate between these three extremes of evolutionary specialization. He proposes a set of *secondary strategies* for habitats that have intermediate intensities of competition, stress and disturbance. From such considerations, he develops the model shown in Fig. 2.8.

The approach so far has been concerned with the established phase of the plant. But Grime recognizes the difficulties posed by the different phases in the life-cycle of the same organism. Juvenile and mature phases may respond in different ways to the same set of external factors. He therefore develops a scheme of five *regenerative strategies*. These are:

- Regeneration by vegetative expansion, e.g. *Prunus spinosa* – blackthorn.
- Seasonal regeneration in vegetation gaps, taking advantage of Spring or Autumn conditions, e.g. *Endymion nonscriptus* – bluebell.
- Regeneration involving a persistent seed bank (or reserve) in the soil, e.g. *Calluna vulgaris* – heather.
- Regeneration by numerous wind-dispersed seeds or spores, e.g. *Taraxacum officinale* – dandelion.
- Regeneration involving a bank of persistent seedlings, e.g. *Acer saccharum* – North American sugar maple.

By combining all these ideas, Grime produces a comprehensive system for studying the structure of vegetation and its response to the environment (but see Discussion Section). Part of the required knowledge is based on observation and simple measurement and to this extent at least the beginner can make some headway with this approach.

FLORISTIC METHODS

Floristic methods lead to a more precise description of the vegetation stand, the plot studied in the field. They are based on its detailed species composition

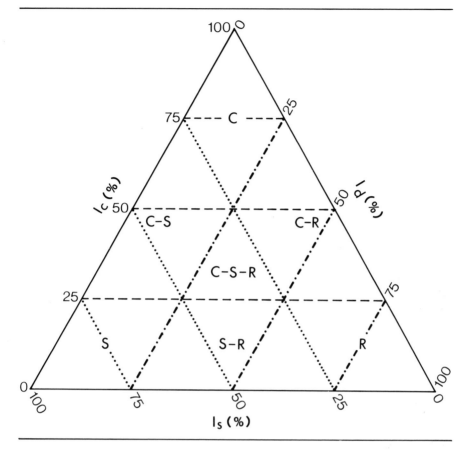

Fig. 2.8 Model describing the various equilibria between competition, stress, and disturbance in vegetation and the location of primary and secondary strategies. I_c, relative importance of competition (---); I_s, relative importance of stress (.....); I_d, relative importance of disturbance (-.-.); C = competitors, S = stress-tolerators, R = ruderals, C-R = competitive ruderals, S-R = stress-tolerant ruderals, C-S = stress-tolerant competitors, C-S-R = strategists adapted to habitats in which the level of competition is restricted by modest intensities of both stress and disturbance. (After Grime, 1979.)

and once this has been compiled several characteristics (or properties or attributes) of all the species will be recorded. Some data may be of a *qualitative* nature (characteristics which cannot be given exact values). Other records will be *quantitative* (carefully measured properties like the number of individuals per unit area).

In the first category species are assigned to classes on a series of arbitrary scales which usually run from 1 to 5. The method is subjective and there can be marked variations between workers in assigning species. To be successful the approach calls for careful observation, wide field experience and some familiarity with the community. Since results in both qualitative and quantitative approaches can be affected by the quadrat size care has to be taken in first accurately determining this. To illustrate the basic ideas behind the assignment of plants using these measures, three examples of each follow (others can be traced in the literature). The qualitative measures, presented first, are based essentially on the early work of Braun-Blanquet, who was instrumental in advancing these methods.

A. QUALITATIVE MEASURES

1. SOCIABILITY CLASSES

Class 1. Plants growing in one place, singly.
Class 2. Plants grouped or tufted.
Class 3. Plants in troops, small patches or cushions.
Class 4. Plants in small colonies, in extensive patches or forming carpets.
Class 5. Plants occurring 'in great crowds' or pure populations.

This is a simple idea showing how individuals of each species tend to be grouped in the field. But it is not without its problems (see Discussion Section).

2. VITALITY CLASSES

Class 1. Ephemeral adventives, germinate occasionally but cannot increase their area.
Class 2. Plants maintaining themselves by vegetative reproduction but not completing a full life cycle.
Class 3. Well-developed plants, regularly completing a full life cycle.

Familiarity with the behaviour of the plant throughout the year(s) is necessary before this scale can be used. (Of course this qualitative measure could be turned into a quantitative one by expressing the vitality of the plant in terms of some aspect of its morphology which has been accurately measured. The aspect selected might be leaf length or width, the number of flowers or the number of fruits produced per individual. Such an approach is known as a *performance* measure.)

3. ABUNDANCE CLASSES

Class 1. Very sparse; *Class 2.* Sparse; *Class 3.* Not numerous; *Class 4.* Numerous; *Class 5.* Very numerous.

This is just one of several schemes (all very similar) for estimating the number of individuals of each species in a community. Obviously these classes are rather vague and may cause difficulties for beginners.

B. QUANTITATIVE MEASURES

1. DENSITY

When the exact number of individuals of a species is determined for a unit area (the quadrat), then we are dealing with density, a truly quantitative measure. Density determinations can be very time-consuming, especially when plants are small and numerous. There are also problems in deciding what is an individual plant since some grow in clump form (many grasses) or spread by runners or produce several shoots from a single bulb. The method proceeds by simply counting the individuals present and relating these values to the sampling area.

2. FREQUENCY

This measures the regularity of distribution throughout the community. If we sample a moorland with 50 quadrats and *Calluna vulgaris* (common heather) occurs in every quadrat and *Erica tetralix* in only 10 then the frequency values are 100 per cent and 20 per cent respectively. It does not matter how many individuals of a species occur in each quadrat since a single occurrence carries the same weight in frequency calculations as a whole cluster of individuals. Frequency is commonly expressed as a limited number of frequency classes: *Class 1*, 1–20%; *Class 2*, 21–40%; *Class 3*, 41–60%; *Class 4*, 61–80%; *Class 5*, 81–100%.

Although this measure is simple to obtain we have to be very careful in interpreting it. The frequency values are really a reflection of the density and pattern of occurrence shown by each species. Fig. 2.9 makes these relationships clear. The three vegetation types, A to C, have the same unit area, the same number of species present per unit area (i.e. the density) and are sampled by the same quadrat size. The only variable is the arrangement of the species (i.e. the pattern) from regular to random. Frequency values vary as indicated. Frequency values would also alter if the quadrat size was significantly changed but all else in Fig. 2.9 (A–C) remained the same. Frequency results may further be influenced by the size of plants as Fig. 2.9 (D) shows where a small, compact plant exists with a large, straggling plant. The chances of the former being recorded are obviously much less than those for the larger plant.

3. COVER

These values indicate the proportion of the ground surface covered when the aerial parts of each plant are projected perpendicularly down on to it. Cover values are expressed as percentages of the total area of the sampling unit. Because plants frequently overlap, the total percentage cover for all species in a quadrat will often exceed 100 per cent. Although rather tedious to measure

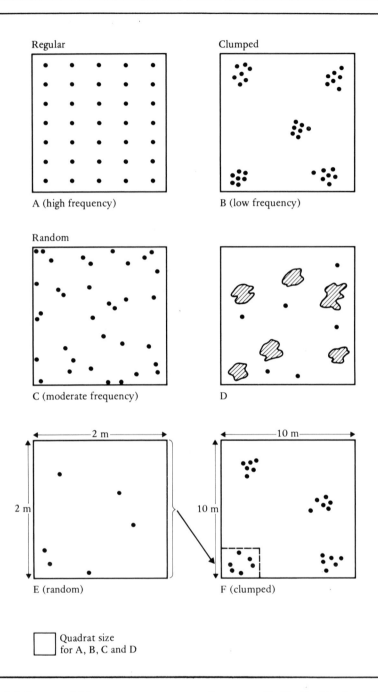

Fig. 2.9 Influence of (a) pattern on percentage frequency (A, B, C); (b) plant size on percentage frequency (D); and (c) quadrat size on distribution pattern (E, F). See text for explanation.

in the field, cover values do give a good indication of the volume of space each species occupies. This can tell us more about the likely importance of the plant in community ecology than is the case with density or frequency, which indicate numbers and distribution but in doing so give equal weight to all species recorded, e.g. six buttercups in a quadrat have the same density value as six oak trees and, if they both are regularly distributed, they could have the same frequency value. Their cover values, however, would clearly indicate the differences in their contribution to the overall vegetation. (This is not to say that the largest plants are always the most important. Beard showed that the dynamics of some savanna vegetation are controlled by the grasses in the field layer and not the taller, scattered trees.)

Cover values are often arranged into a limited number of classes as follows: *Class X* less than 1%; *Class 1*, 1–5%; *Class 2*, 6–25%; *Class 3*, 26–50%; *Class 4*, 51–75%; *Class 5*, 75–100%.

Scales can be combined and in the example given below the first four classes express the abundance and vigour of species with very low coverage values:

Abundance-vigour-coverage scale of Domin (modified, see Evans and Dahl 1955)

Class +. Occurring as a single individual with reduced vigour; no measurable cover.

Class 1. Occurring as one or two specimens of normal vigour; no measurable cover.

Class 2. Occurring as several individuals; no measurable cover.

Class 3. Occurring as numerous individuals but cover less than 4% of total area.

Class 4. Cover up to 1/10 (4–10%) of total area.

Class 5. Cover about 1/5 (11–25%) of total area.

Class 6. Cover 1/4 to 1/3 (26–33%) of total area.

Class 7. Cover 1/3 to 1/2 (34–50%) of total area.

Class 8. Cover 1/2 to 3/4 (51–75%) of total area.

Class 9. Cover 3/4 to 9/10 (76–90%) of total area.

Class 10. Cover 9/10 to complete (91–100%) of total area.

We have been describing a few measures that can be applied to the species found in a single stand. These measures lead to what ecologists call analytical phytosociology. When a number of quadrats have been analysed in this manner we can then compare and contrast the data from the stands to see to what extent the quadrats can be grouped together (synthetic phytosociology). If they are all very similar then we have one broadly uniform (homogeneous) area of vegetation. But if marked differences emerge then sub-division of the vegetation into types (associations) is indicated. The extent of sub-division depends largely on the purpose of the study: in the absurd case no two quadrats can ever be totally alike in all respects.

Various measures are used in the synthesis of quadrat data and from these we can obtain an abstract definition of the community based on the important properties of the many species which go to make up actual stands studied in the field. One example will illustrate the kind of concepts which have been developed for this purpose. *Fidelity* is a measure of the degree to which a plant is found only in a certain type of community and not in others (i.e. its relative exclusiveness). Some plants are members of several different types of vegetation; others are only ever found in a specific community. The former have low indicator value while the latter strongly characterize that community. The following fidelity classes are described by Braun-Blanquet:

Class 1. Strange species, rare or accidental intruders from another community or relics of a community that has previously occupied the same area.

Class 2. Indifferent species, without pronounced affinity for any community.

Class 3. Preferential species, present in several communities more or less abundantly, but predominantly or with better vitality in one community.

Class 4. Selective species, found most frequently in a certain community but also, though rarely, in other communities.

Class 5. Exclusive species, completely or almost completely confined to one community.

Plants of Class 1 are also known as *accidentals*. *Companions* is another name for species in Class 2, while those in Classes 3 to 5 are referred to as *characteristic species*. Again, much field experience is necessary for a successful application of these basically simple ideas. But they have proved very useful in many detailed studies, particularly those associated with European phytosociologists.

While physiognomic methods tend to emphasize the vertical arrangements within the plant community, floristic methods bring out more strongly the horizontal variations within the vegetation. The boundary between two plant communities is seldom sharply marked except where it coincides with an equally abrupt change in an important ecological factor or where the vegetation has been clearly differentiated through human activity. More commonly, there is a relatively narrow zone where species from both communities blend together to form an *ecotone* or transition zone (it may be referred to as a tension zone if the two communities are actively competing for the same territory).

DISCUSSION SECTION

Why is it important to randomly locate sampling points?

Well, for a start, not all studies use randomly located quadrats. However, particularly when statistical treatment of data occurs, random sampling should be the aim. Randomness removes personal bias. It is very easy to select subconsciously for sampling dry sites as opposed to wet muddy sites; colourful, conspicuous plants rather than small, drab species; easy access vegetation in preference to thick, tangled undergrowth. Even with a randomly located quadrat it is still quite possible to overestimate the abundance of a conspicuous species and underestimate a small plant. The vegetation in a quadrat may look very different at different seasons and so that season we select for sampling can introduce further bias.

Which is the best method to use?

There is no best approach. Every method has its own presuppositions, advantages and disadvantages. Much depends on the purpose of the study, the accuracy required and the time available. A quick reconnaissance of a large area might best be dealt with by physiognomic methods. A detailed study of floristic variations over a small area in terms of density, coverage, fidelity, etc. should best reveal subtle environmental changes. We have only dealt with a few examples here: many more can be traced in the literature, some contributed by geographers (e.g. Stamp, Küchler).

Many approaches are simple to state but very difficult to apply. For example, from the table of sociability classes previously given what value should be assigned to a tree in a forest? If we were interested in the whole forest then Class 5 would seem appropriate, but if we were studying just variations in the shrub plants and a tree was included in our sample then presumably it would come within Class 1 (or perhaps be excluded altogether to overcome the problem). This illustrates the problem of scale which enters into many aspects of vegetational analysis. A species might appear randomly distributed at one scale but show obvious clumping at another, as Fig. 2.9 (E and F) indicates.

Most methods have their critics. For example, Raunkiaer's system, first suggested in 1905 and still widely used as a descriptive tool, has recently been criticized by Schulze who writes, 'Raunkiaer's approach, namely to define plant form as a response to climate and then to use this plant form in order to define plant climates, was a basic inconsistency and the circularity of this argument has stalled comparative ecological research on different plant forms almost to the present time'. Likewise, Piggot takes exception to the choice and definition of terms used in Grime's method. He dislikes the way in which certain key-words (such as stress and strategy) have lost their precise or scientific meaning in the presentation of this approach, leading to some conceptual confusion. In a more general context, Harper has also attacked the imprecise use of certain words (e.g. adaptation, strategy and stress) in ecological writing, pointing out that some have now lost their true value in ecology and are often wholly redundant in the sense in which they are employed. Harper quotes 'the effect of temperature stress' (= the effect of temperature) as one illustration of this.

The lesson in all this is to aim for precision when using ecological terminology and to state exactly what is meant when an ordinary English word is used in a specific scientific context.

Apart from its use as a simple method of describing vegetation, has the physiognomic approach any other value?

Yes; communities with a complex structure tend to have a great influence on their own environments, particularly with respect to microclimatic modifications. Ecologists speak of the geometry of the forest and this refers to the arrangement of life-forms in the vertical plane. Study along physiognomic lines shows how this structure largely controls the temperature gradients, the humidity variations and the light intensity and wind-flow patterns of the forest. In much the same way as we speak of urban climates so we can demonstrate internal forest climates.

The study of the internal stratification of a complex plant community, such as a forest, has also revealed that the associated animal community may show similar vertical differentiation. For example, different breeds of tits utilize different levels of British pinewoods for feeding purposes. Some are essentially canopy and high branch feeders, others search for food among low vegetation or surface litter. Likewise, some insects (e.g. the mosquito) in forests daily move up and down through the profile in response to the changing daily pattern of humidity values which reflect the complex vertical structure of the forest.

You indicated some difficulties with the concept of the dominant species. Is the concept now no longer emphasized in ecological writing?

The position varies. As previously mentioned, some ecologists reject its use but

Table 2.4 The dominance index: examples illustrating its derivation (From Grime, 1979)

Species	Attributes				Dominance index (Total/2)
	(a)	(b)	(c)	(d)	
Chamaenerion angustifolium	5	5	5	2	8·5
Arrhenatherum elatius	5	4	4	3	8·0
Brachypodium pinnatum	3	4	3	5	7·5
Ranunculus repens	3	5	3	1	6·0
Helictotrichon pratense	3	2	3	2	5·0
Taraxacum officinale	3	1	4	1	4·5
Festuca ovina	2	1	3	2	4·0
Campanula rotundifolia	2	2	3	0	3·5
Arenaria serpyllifolia	1	0	4	0	2·5

Key to scoring system. (a) Maximum plant height: 1, <26 cm; 2, 26–50 cm; 3, 51–75 cm; 4, 76–100 cm; 5, >100 cm. (b) Morphology: 0, small therophytes; 1, robust therophytes; 2, perennials with compact unbranched rhizome or forming small (<10 cm diameter) tussock; 3, perennials with rhizomatous system of tussock, attaining diameter 10–25 cm; 4, perennials attaining diameter 26–100 cm; 5, perennials attaining diameter >100 cm. (c) Relative growth rate (g/g/wk): 1, <0·31; 2, 0·31–0·65; 3, 0·66–1·00; 4, 1·01–1·35; 5, >1·35. (d) Maximum accumulation of persistent (i.e. from one growing season to the next) litter produced by the species: 0, none; 1, thin discontinuous cover; 2, thin continuous cover; 3, up to 1 cm depth; 4, up to 5 cm depth; 5, >5 cm depth.

others do not query it. Grime, for example, devotes a whole chapter to it and discusses many of his ideas on plant strategies in terms of the concept. Further, he develops a *dominance index*, illustrated in Table 2.4, which is based upon four attributes, one of which is the relative rate of dry matter production measured under standardized laboratory conditions.

Greig-Smith, on the other hand, appears to adopt an intermediate position. In a recent review of the causes of pattern in vegetation, he states that every individual plant modifies its immediate habitat to a greater or lesser extent and that a tree or shrub to some degree will determine the habitat available to herbaceous species. He refers to plants having such influence as 'dominating' species and the presence of the inverted commas implies a degree of caution in using this term.

REFERENCES

Beard, J. S., 1953. 'The savanna vegetation of Northern Tropical America', *Ecol. Monogr.*, **23**, 149–215.

Braun-Blanquet, J., 1932. *Plant Sociology* (transl. G. D. Fuller and H. S. Conrad), McGraw-Hill, New York.

Cain, S. A. and Castro, G. M. de O., 1959. *Manual of Vegetation Analysis*, Harper Bros, New York.

Dansereau, P., 1957. *Biogeography: an ecological perspective*, Ronald Press, New York.

Evans, F. C. and Dahl, E., 1955. 'The vegetational structure of an abandoned field in southeastern Michigan and its relation to environmental factors', *Ecol.*, **36**, 685–705.

Greig-Smith, P., 1979, 'Pattern in vegetation', *J. Ecol.*, **67**, 755–79.

Grime, J. P., 1979. *Plant strategies and vegetation processes*, Wiley, Chichester and New York.

Harper, J. L., 1982. 'After description', in *The Plant Community as a Working Mechanism* (ed. E. I. Newman), Special publication No. 1, British Ecological Society, Blackwell, Oxford.

Kershaw, K. A., 1973. *Quantitative and Dynamic Ecology* (2nd edn), Arnold, London.

Küchler, A. W., 1967. *Vegetation Mapping*, Ronald Press, New York.

Oosting, H. J., 1956. *The Study of Plant Communities*, W. H. Freeman, San Francisco.

Phillips, E. A., 1959. *Methods of Vegetation Study*, Holt, Rinehart and Winston Inc, New York.

Piggot, C. D., 1980. 'A review of "Plant strategies and vegetation processes"', *J. Ecol.*, **68**, 704–6.

Raunkiaer, C., 1934. *The Life Forms of Plants and Statistical Plant Geography*, Clarendon Press, Oxford.

Rice, E. L., 1979. 'Allelopathy – an update', *Bot. Rev.* **45**, 15–109.

Richards, P. W., 1952. *The Tropical Rain Forest*, CUP, Cambridge.

Schulze, E. D., 1980. 'Photosynthetic production and water relations in different plant-forms as related to altitudinal gradients' in *Mountain Environments and Subalpine Tree Growth* (eds U. Benecke and M. R. Davis), Technical Paper No. 70, New Zealand Forest Service, Wellington, pp. 213–29.

Stamp, L. D., 1934. 'Vegetation formulae', *J. Ecol.*, **22**, 299–303.

Stowe, L. G., 1979. 'Allelopathy and its influence on the distribution of plants in an Illinois old-field', *J. Ecol.*, **67**, 1065–85.

3

INITIAL APPROACHES TO SOIL STUDY

DEFINITIONS

Pedology is the scientific study of soil genesis, i.e. the factors in the formation of soil, the processes involved and the classification of the resulting soil types. *Edaphology* is the scientific study of the influence of soils on living things, particularly plants, and it includes man's use of soil for plant growth. It is often concerned with applied aspects of soil study and has strong links with agriculture. In this introduction, the approach will be mainly pedological but at several points edaphological implications should become apparent.

Soil usually results from the weathering of underlying solid rock surfaces. Some soil types, however, develop from transported materials, such as glacial debris, wind-blown sands or alluvium. Others have much of their bulk composed of organic material (peaty soils). The soil is a dynamic zone within which physical, chemical and biological processes are at work. This implies frequent changes within soil which, consequently, should never be regarded as inert material.

Soil consists of matter existing in the three basic forms, namely, solid (inorganic rock particles and organic plant and animal remains), liquid (soil water and chemicals in solution) and gaseous (atmospheric gases and those released by chemical and biological activity within the soil). It results from three processes of weathering. Physical weathering is brought about by the action of frost, rain-drop pressure, differential heating and cooling of surfaces, gravity, etc. It breaks down the initial rock surface to smaller fragments. Chemical weathering produces changes in the rock surface and fragments as when, for example, some chemicals are washed out to lower levels or combined with atmospheric gases or other chemicals present. As the soil develops, a zone of chemical alteration advances through the weathering rock (or *parent soil material*). Biological weathering is caused by plants and animals (including man) and can be either a physical or chemical process. The pressures caused by roots are in the first category while changes due to chemicals released during organic decay are in the second category.

PEDOGENIC FACTORS

Soil results from the interaction between three broad sets of pedogenic factors – those related to the climatology, geology and biology of the site. Climate is usually considered to be the most important. It broadly determines the type of physical processes at work (e.g. number and severity of frosts), the rate of chemical activity (which is often temperature-dependent) and the complexity of the biotic community in the region. Since soils develop over a considerable period of time, it is not necessarily the present-day climate which has had the main influence on their pedological characteristics. For example, many Australian soils are best understood in terms of past climatic patterns.

The geology provides the 'starting material' for soil development. The physical properties and chemical composition of the parent material will be important influences on the type of soil to evolve. But the actual fertility and productivity of soil is mainly due to the biological processes at work. A fertile soil is packed with living organisms. Many of these are minute and largely unseen in a superficial examination of a sample. They constitute the soil microflora and microfauna; almost countless numbers of bacteria, fungi, algae, protozoa (such as *Amoeba*) and flatworms or nematodes. The number of bacteria in one gram of soil varies from about 1 million to 4,000 million, according to soil type, while the fungi may range from 5,000 to 1 million. The microfaunal population, often as unicellular animals, may reach nearly 2 million in a 1 gram sample. Despite these huge numbers, the microflora and microfauna seldom contribute more than 2–3 per cent of the total organic material in a sample.

Earthworms, mites, spiders, etc., although usually clearly visible, exist in numbers far in excess of what most people would expect. Russian studies, reported by Bunting, show the following earthworm densities:

2·94 million   per ha ($2\frac{1}{2}$ acres) in Russian oak forest
610,000   per ha in Russian spruce forest
880,000   per ha in a Russian wheatfield

In an old lime-rich New Zealand pasture the earthworm population was estimated as 8 million per ha and their weight was approximately equal to the weight of sheep on the pasture. Soil passing through the bodies of earthworms, as worm casts, can average 25–50 tons per ha per annum. Any large changes in these soil populations produced by land management techniques will be of prime importance in edaphological studies.

These simple ideas on pedogenesis have been expanded as the factorial (or functional) approach to soil study, the method commonly followed being one proposed by Jenny. He formulated the following equation:

Soil = f (regional climate, organisms, relief, parent material, time . . .)

Any single property of a soil is a function of the interaction between the factors listed. It is an open-ended equation since Jenny recognized that our knowledge was incomplete and other factors may have to be added. These factors may be subdivided into two broad groups: the *passive factors* (parent material, relief and time) which provide a background upon which the *active factors* (climate and organisms, seen in terms of either physical or chemical processes) operate. A pedologist studying an unfamiliar soil will usually begin by considering the operation of these various factors in turn. This will help him to decide which have been key influences in giving the sample its particular characteristics and properties. Appendix 3.1 provides a check-list of some of the questions to which a pedologist might seek answers in carrying out this approach.

The factorial approach has been developed further by several workers. For example, Jenny himself updated the concept and Runge developed an approach which stressed water availability within the soil as a factor. But it was Simonson who emphasized process within soils by proposing a generalized theory that all soils could be viewed in terms of four main processes: *additions*, *removals* (occurring as either inputs or outputs of solid, liquid or gaseous 'elements' of organic and mineral origin), *transfers*, *transformations* (movements and changes of form of these 'elements' within the soil body). Such processes proceed simultaneously in all soils. The balance within the combination of changes possible under these headings will govern the ultimate nature of the soil and differentiate one soil from another. Recent work which follows this dynamic, process-oriented approach to soil study has been summarized by Huggett.

THE STUDY OF SOIL EXPOSURES

Initial study of a soil should be carried out in its natural setting, i.e. in the field. If necessary, this field investigation can then be followed by a laboratory analysis to provide more precise information. With practice and experience many soil properties can be adequately determined in the field, reducing the need for laboratory treatment. The field study of a soil begins with the *soil profile*, the various layers seen in a vertical section extending from the ground surface to the bedrock. This arrangement of horizons is seen when either a trench is dug or samples are systematically brought to the surface with a soil auger.

In some soil profiles three main horizons can be recognized: A and B represent the true soil (or solum) and C is the weathering parent material (see Fig. 3.1). Each horizon is usually capable of subdivision and the likely sequence for a soil developing under a mid-latitudinal, maritime climate with high precipitation, the climate of the wetter parts of western Britain, is set out below. Although not all horizons need be present or so clearly demarcated at least some will be.

L – litter layer; composed of recently deposited organic material.

F – fermented layer; as for L but partly decomposed.

H – humified layer; a further stage of F. The original structure of leaves is no longer discernible and much material is now in a colloidal state (a semi-solid, non-crystalline condition with gluey properties).

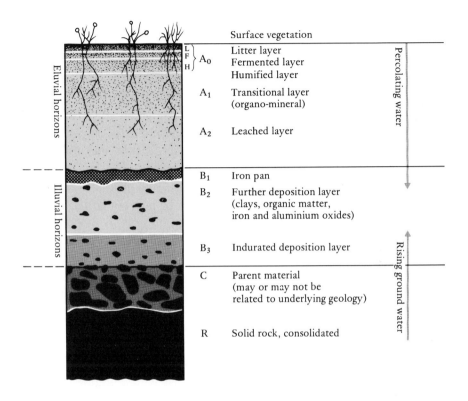

Fig. 3.1 A diagrammatic profile of a soil developed under conditions of a cool wet climate. The number of horizons and their thicknesses are only given as a guide to illustrate the degree of profile differentiation possible.

L, F and H make up the A_0 horizon. Once litter has become well decomposed it is then known as *soil humus*.

A_1 – transitional layer; first layer of the soil proper where organic material becomes well mixed in with the mineral fraction.

A_2 – leached layer; percolating water has removed some mineral and organic particles to lower levels. The zone is often grey in colour due to loss of clay, iron or aluminium and the concentration here of quartz and other resistant minerals.

Horizons A_0, A_1 and A_2 are subject to removal of organic material, fine clay particles and chemicals – a *zone of eluviation*. This loss will occur as substances either suspended or dissolved in percolating waters. The process of clay translocation as particles in suspension is known as *lessivage*. The removal of dissolved chemicals in solution is called *leaching*.

B_1 – ironpan; here some iron salts have accumulated from zones above together with a concentration of redeposited clay minerals. If present and well developed, this layer can impede drainage and plant rooting.

B_2 – deposition layer; a lower zone where more chemicals (clays, iron and aluminium oxides, organic material) are redeposited.

B_3 – transitional layer; just above the parent soil material.

Horizons B_1, B_2 and B_3 receive material from above – a *zone of illuviation* where substances are precipitated from solution or deposited from suspension. For example, salts and oxides may precipitate and clay particles may flocculate to form coatings or *cutans* on the surfaces to which they adhere.

C – parent soil material; depending upon whether the soil has developed on drift material or solid rock, this may or may not be related to the solid geology of the site.

The shorthand letter system used to denote each horizon varies from textbook to textbook but is basically similar in most cases. A comprehensive scheme of horizon nomenclature, as used in recent Memoirs of the Soil Survey (e.g. the report by Thomasson on the soils of the Melton Mowbray District), is set out below. Here horizon A_2 is designated as E.

ORGANIC AND ORGANO-MINERAL SURFACE HORIZONS

L – Undecomposed litter.

F – Partially decomposed litter.

H – Well-decomposed humus layer, low in mineral matter.

A – Mixed mineral-organic layer.

Ap – Plough layer of cultivated soils.

Ag – A horizon with rusty mottling, subject to periodic waterlogging.

SUBSURFACE HORIZONS

E – Eluvial horizon, from which clay and/or sesquioxides have been removed.

Ea – Bleached (ash-like) horizon in podzolized soils.

Eb – Brown (paler when dry) weak-structured horizon depleted of clay.

B – Altered horizon distinguished from overlying A or E horizons and underlying less altered C horizon, by colour and/or structure, or by illuvial concentrations of the following materials denoted by suffixes: t – illuvial clay; h – illuvial humus, characteristic of podzols; fe – illuvial iron, characteristic of podzols.

C – Horizon little altered, except by gleying, and either like or unlike the material in which upper horizons have developed.

Bk or Ck – Horizons containing appreciable amounts of calcium carbonate (1 per cent of the fine earth).

Bg, Cg, etc. – Mottled (gleyed) horizons subject to waterlogging; where gleying is only weakly expressed the suffix (g) is used.

A/C, B/C, etc. – Horizons of transitional or intermediate character.

(The Soil Survey of England and Wales now use an extended version of this notation which has more suffixes to increase the scheme's descriptive powers. Full details are presented by Hodgson.)

Plant roots may penetrate to the lowest layers and bring back into circulation chemicals deposited there. If, however, the water table is near the surface lower horizons will be waterlogged and this can restrict plant rooting. Seasonal variations in the position of the water table lead to a *zone of gleying* in the profile (designation symbol g). This is seen as a colour change: blue-grey characterizes the permanently saturated layers, while grey and ochreous mottling is seen in the levels subject to a fluctuating water table. Here the chemicals, particularly iron oxides, are alternately reduced under saturated conditions (to ferrous compounds) or oxidized under dry conditions (to ferric compounds).

Having established the profile sequence, the soil exposure can then be examined for other characteristics which add to the description. The previous point about the blue-grey mottling indicates *soil colour* may be a good guide to soil type. Colour variations in a profile might reflect (a) the percentage of organic material and the stage in its decomposition, (b) the drainage characteristics, (c) the degree of aeration and (d) the oxidation of certain chemicals. For example, a strong red colour often points to the breakdown of iron compounds under conditions of reasonably good drainage, while a dark soil is frequently indicative of high organic content.

Pedologists used to argue about colour interpretation since this can be a very subjective matter. But an agreed reference system solves this problem. Such a system is provided by the *Munsell Colour Chart*, a designation system specifying the relative degrees of the three simple variables of colour:

hue – caused by light of certain wavelengths.

value – relative lightness or intensity of colour.

chroma – the relative purity or saturation of a colour, being directly related to the dominance of the determining light wavelength and inversely related to greyness.

Without going into unnecessary detail, the use of the scheme can be simply illustrated. 10 YR 6/4 is a soil colour on the Munsell Chart with the hue being 10 YR, the value 6 and the chroma 4. This notation on the colour chart can be translated into an appropriate descriptive colour name. Thus, all we have to do is match our sample with the chart and record the name or notation. Samples so matched should be viewed in good natural light but not strong sunlight. As colour varies with moisture content all samples being compared should be approximately equal in this respect.

Soil texture is the percentage of sand, silt and clay in a sample. In assessing texture coarse fragments such as gravel would first be removed to leave the '*fine earths*' fraction, material which will pass through a sieve with round holes 2 mm in diameter. In the field, texture can be judged by wetting a hand sample and working it between the fingers. A soil with a high clay content will mould into a ring without fracturing. A silt will form a roll but not a ring, while a soil with high sand content will not mould into any consistent shape. With field experience, these simple tests will often be adequate for assessing texture. For a more exact analysis several laboratory procedures are available, and will be outlined shortly.

In addition to texture, soils show *structure*. The colloids of clay and organic matter bind together the other soil constituents to form aggregates. If the organic matter content is less than 3 per cent in fine textured soils (e.g. silts) then aggregate formation becomes difficult. Tisdale and Oades stress the importance of organic bindings in the formation of water-stable soil aggregates. A natural soil aggregate is known as a *ped*. Structure refers to the type of aggregate formed by soil particles within the soil and also takes into account the pores and fissures (voids) between and within these aggregates. To demonstrate this, the conventional procedure is to turn or gently throw a spadeful of soil onto a flat, hard surface. The way in which primary soil particles are arranged into secondary particles, units or peds may then be observed. A simple descriptive classification of these has been devised which is largely self-explanatory:

Class 1. Platy

Class 2. Prismatic (with level tops)
Class 3. Columnar (with rounded tops)
} Prism-like (usually seen as a subsoil manifestation)

Class 4. Blocky (angular) or nut-like (cube-like)
Class 5. Subangular blocky (rounded cubes)
} Block-like

Class 6. Granular (peds not readily porous)
Class 7. Crumb (porous peds)
} Spheroidal

Soils can be structureless (i.e. no obvious aggregation) and all states between this and a strongly structured soil can exist. Some small soil structures originate as excreta from the soil fauna (e.g. worm casts). Texture and structure influence the drainage and workability of soil and the retention of chemicals and plant nutrients. The soil *porosity* or volume of soil, as a percentage of total bulk, not occupied by soil particles is also influenced by texture and structure. The pores between peds, because they are much larger, usually have a greater influence on water drainage than those within peds.

Related to questions of structure are other terms used in describing soils. For example, *consistency* refers to the strength of peds at various moisture contents and under pressure or mechanical stress. It is a useful measure for agricultural purposes and describes the cohesion and adhesion of soil particles. *Plasticity* is a measure of the soil's ability to change shape under pressure or stress and to retain that shape when stress is removed.

Without doubt clay minerals are a most important constituent of soil. They are extremely small and have a colloidal status, i.e. they consist of small particles dispersed throughout another material, like fat particles in milk.

However, the clay minerals are not all of the same type. They have a plate-like structure and White describes three main types that are found in soils: the *kaolinite clays*, where the basic structure is a 1:1 lattice of layers (or lamellae) comprising a silica sheet $(SiO_2)_n$ bonded to a gibbsite sheet $(Al_2(OH)_6)_n$; the *hydrous mica clays*, which are the most common group and include the vermiculite and illite clays, have a gibbsite sheet between two silica sheets; the *smectite clays*, which include the montmorillonite clays, have a 2:1 lattice similar to that of the hydrous micas but with a wider range of associated elements in the interlamellar spaces. The interlamellar bonding is strongest in the kaolinite clays, weaker in the hydrous mica group and weakest in the smectites. Weak bonding makes the interlamellar surfaces more accessible to water molecules and other chemical substitutions in the lattice. Montmorillonite clay crystals have an enormous surface area (the sum of their external area and much of their internal interlamellar areas) while the tightly bonded kaolonites have only their external surface area. The relative ease with which water enters or is removed from the structure of the weaker bonded clays is expressed in the alternating swelling and shrinkage of soils rich in such clays.

The greater the surface area, the greater is the potential for *adsorption* – the attracting of substances (chemicals) to that surface. The very small size of the clays means that for the same given weight of colloidal clay and sand the former will have a surface area about 10,000 times that of the latter. It also means that a montmorillonite may have a surface area many times that of a kaolinite. Most of the chemical activity in the soil is centred on the surfaces of the clay and humus colloidal fraction and the importance of this point is discussed later in the chapter. In addition, this colloidal material confers various other properties on the soil: its plasticity, cohesion, swelling and shrinkage capacity (as water is added or removed) and flocculation state (the ability to hold particles together to give a good soil structure). But it is not just the clay content which is important; rather it is the amounts of clay, silt and sand relative to each other.

LABORATORY TREATMENTS

Useful material for the beginner may be found in Briggs and full details of the laboratory soil analyses used by the Soil Survey of England and Wales are provided by Avery and Bascomb. It would be inappropriate for this introductory account to attempt such a coverage. However, several simple laboratory treatments are set out below and it is hoped that these will provide some insight into how the pedologist adds more precise detail to his carefully observed field data.

1. MECHANICAL ANALYSIS TO DETERMINE SOIL TEXTURE

The soil can be accurately graded into size categories by passing samples through a series of sieves. The International Scale has the following size classes:

$$
\left.\begin{array}{rcl}
> 2 \text{ mm} &=& \text{gravel and very coarse sand} \\
2\text{--}0\cdot2 \text{ mm} &=& \text{coarse sand} \\
0\cdot2\text{--}0\cdot02 \text{ mm} &=& \text{fine sand} \\
0\cdot02\text{--}0\cdot002 \text{ mm} &=& \text{silt} \\
<0\cdot002 \text{ mm} &=& \text{clay}
\end{array}\right\} \text{'fine earths' fraction.}
$$

The USA Department of Agriculture uses slightly different size categories but the principle is the same. Once percentages are worked out for the 'fine earths' fraction, the soil can be named by reference to the textural triangle (Fig. 3.2). The 'fine earths' fraction is normally used in any subsequent analysis since only material less than 2 mm in diameter is considered to be chemically and physically active in the soil.

2. BOUYOUCOS METHOD

This is a relatively quick method for determining proportions of sand, silt and clay in a sample. For the very fine material, sieving does not always give accurate results and the Bouyoucos method is preferred. It proceeds as follows:

(a) First remove all coarse stones from sample which is then oven-dried.

(b) Prepare a tall measuring cylinder of distilled water to the 1,000 ml mark and add 1·0 g of deflocculating agent (e.g. sodium hexametaphosphate). This separates out the soil components, particularly the colloidal substances influencing aggregation, and causes them to become suspended in the liquid.

(c) Insert a hydrometer to measure the specific gravity of the liquid, and take a reading. Remove hydrometer.

(d) Add the soil sample, 60 g of 'fine earth' fraction, stirring vigorously for 10 minutes or so, according to soil type, to aid further dispersion of aggregates. Take a reading with the hydrometer immediately stirring ceases.

(e) After 40 seconds the sands settle out and a third hydrometer reading is taken.

(f) After 2 hours the silts settle out and a fourth hydrometer reading is taken.

(g) Since we are now left with the original mixture plus only the clay fraction in suspension, the hydrometer value for the clay component can be worked out by noting the difference between the fourth and the first readings.

In this way we can quickly compare a range of soils but all we have is a simple division of each sample into the relative proportions of sand, silt and clay. Certain precautions are necessary with this method. If the sample has

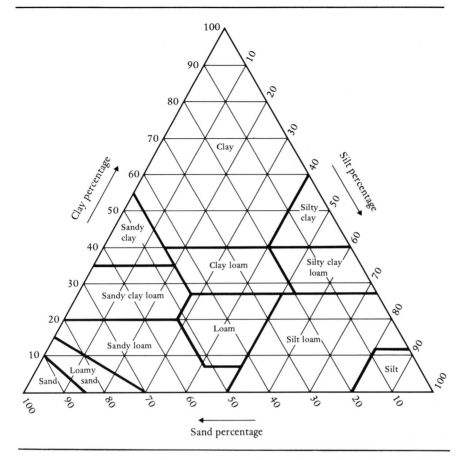

Fig. 3.2 The textural triangle: a triangular graph for determining soil texture classes by using the data from a mechanical analysis of the 'fine-earths' fraction of a soil sample.

more than 5 per cent organic content then a pre-treatment by gently boiling with dilute hydrogen peroxide is required. Also, temperature variations during the period of measurement must be allowed for, and a solution to this is to stand the measuring cylinder in a water bath set at 20 °C.

3. 'WEIGHING METHOD'

This determines water content, organic fraction and quantity of lime present in a soil. Each step is essentially based on reweighing a sample of known amount after a simple treatment. First remove very coarse hard fragments by sieving. Crush any soft large fragments by working the soil in a pestle and mortar. The next stages are as follows:

(a) Take 10 g of the soil in a porcelain basin and dry overnight (12–16 hours) in an oven at 105 °C. After cooling in a desiccator, reweigh. The weight loss indicates the water lost in drying, which is approximately the same as the water content.

(b) Another 10 g of oven-dried soil in a crucible are heated to red heat in a furnace at 850 °C for 30 minutes (2 hours at 950 °C may be needed for calcareous soils to decompose the carbonates). This destroys the organic material present. Cool in a desiccator and reweigh, calculating the percentage loss. This gives only an approximate indication of organic content since a small part of the loss will be due to dehydration of the clay fraction. Allowance for this can be made by subtracting 1/10th of the clay value (previously determined by either mechanical analysis or the Bouyoucos method).

(c) Wash with dilute hydrochloric acid a further 10 g of oven-dried soil. This will dissolve out any free lime present. Gently dry the sample and reweigh.

Quite obviously more sophisticated analysis methods than these are available to give very accurate data on all the chemicals present, including those found only in minute traces. In such work, a modern soils laboratory might use, for example: a calcimeter to determine calcium carbonate (lime) content by means of carbon dioxide release; a titration method using potassium dichromate and concentrated sulphuric acid (giving a very powerful oxidation reaction) to measure organic content; a flame photometer or an atomic absorption spectrometer to measure with great accuracy the various chemical elements (nutrients) in a sample. But enough has been said here to show how careful field observation plus a simple laboratory follow-up will reveal many of the important properties by which we describe and classify soils. However, the criteria used in soil classification are becoming much more precise and this means an increasing reliance on the exact methods of the laboratory rather than on the more subjective field observations.

SOIL ACIDITY AND BASE EXCHANGE

Most soil particles are surrounded by a thin film of water and this is particularly true of minute clay and organic colloidal particles with their large surface areas. The mineral salts dissolved in the soil water solution exist as electrically charged units or ions. The charge may be positive (cations) or negative (anions) and chemical compounds may dissolve into their respective cation and anion parts. Common salt, NaCl, may exist, for example, as Na^+ cations and Cl^- anions. The surfaces of the colloidal particles have a negative charge which consequently attract the cations. Around each such particle will be held by adsorption great clusters of cations. The strength by which these cations are held by the colloidal surface varies and a descending order of tenacity is as follows: hydrogen > calcium > magnesium > potassium > nitrogen > sodium.

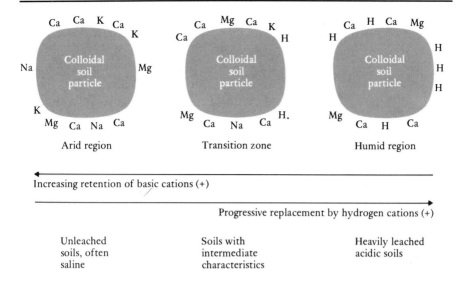

Fig. 3.3 Schematic representation of the relationship between precipitation, percentage base saturation and pH. (Modified from Oosting, 1956.)

An acid is defined as a substance which dissociates to produce hydrogen ions. Acids resulting from the decomposition of surface organic material, and those formed when chemicals in the atmosphere become dissolved in rainwater, add many hydrogen cations to the soil solution. These tend to 'dislodge' or replace some of the metal or basic cations around the colloidal particle. The basic cations so displaced (*base exchange*) then tend to be leached by percolating waters to lower soil levels or flushed out in the drainage system. The *cation exchange capacity* (CEC) is a measure of the number of cations adsorbed on the colloidal surfaces that are freely exchangeable with cations in the soil water solution. Organic particles usually have a much greater CEC than the clays. Two-way cation exchange is always occurring between the colloidal surfaces and the soil water in an attempt to achieve an equilibrium state.

Cation exchange is a common and important soil reaction. Plants obtain much of their nutrition from exchangeable bases or cations absorbed on colloidal surfaces; replacement of them by hydrogen ions will mean less available for plant use (Fig. 3.3). In wet climates, hydrogen cations are numerous and much replacement may take place. In dry climates, base cations predominate, sometimes to the extent of producing soils excessively rich in calcium or sodium. This may inhibit all but specialized species which tolerate salt concentrations.

Many plants release certain organic compounds (called *chelating agents*) to the soil which combine with some metallic cations (e.g. iron, manganese, zinc, copper) in such a way that the cations are held tightly and largely prevented from taking part in reactions with other soil constituents. They therefore remain in the soil water solution around the colloidal particles for longer periods and are not rendered insoluble through reaction with other chemicals.

In this soluble form they can be utilized as nutrients by growing plants. But it also means that these chelated cations, being soluble, move down the profile more easily in the soil water. This mobilization process is important in the loss of chemicals to lower levels and is well seen in podzolic soils, as leached layers in the profile (Fig. 3.1). Metal cations occurring in the chelated form remain in solution at much higher pH values than non-chelated (inorganic) forms. In the non-chelated form, metal cations tend to react with other soil chemicals and become readily insoluble or precipitated. This means that their availability as plant nutrients is greatly reduced. It is therefore important to measure the hydrogen cation concentration in the soil.

The concentration of hydrogen ions in a solution is measured in moles per litre. A mole is the molar weight of a substance in grams. A neutral solution has a hydrogen ion concentration of 10^{-7} g in 1 litre of solution. The *pH value* is 7. This is the negative logarithm of 10^{-7}. The pH values run in a scale from 1 to 14 as shown in Table 3.1 and each $1 \cdot 0$ movement on the scale represents a tenfold change in the hydrogen ion concentration. The extremes of this pH scale are never encountered in the field. Most soils have a pH value somewhere between 3 and 9. Even calcium-rich parent materials (Chalk, Oolitic Limestone) move towards acidity in time as percolating waters assert their influence. When a farmer limes a soil he is in effect soaking it with calcium cations to counteract the presence of excess hydrogen cations. A good supply of calcium also prevents deflocculation of the clays and organic complexes, thus preserving the structure of the soil.

Table 3.1 The pH scale: pH $= -\log [H+]$ where the concentration of hydrogen ions is in mole l^{-1}

pH	Concentration of H+ (mole l^{-1})	
1	$0 \cdot 1$	
2	$0 \cdot 01$	
3	$0 \cdot 001$	Higher concentration of
4	$0 \cdot 0001$	H+, i.e. acidity increasing
5	$0 \cdot 00001$	
6	$0 \cdot 000001$	
7	$0 \cdot 0000001$	Neutrality
8	$0 \cdot 00000001$	
↓	↓	Lower concentration of
		H+, i.e. alkalinity
14	$0 \cdot 00000000000001$	increasing

In the field, pH can be measured by using a barium sulphate soil-testing kit. A small sample of soil is placed in a glass tube with some barium sulphate and distilled water. Tap water is not used because it would influence the acidity value. A small amount of soil indicator solution is added and the tube is vigorously shaken. The mixture takes on a colour which can be matched against a standard pH colour chart. The soil indicator solution is a mixture of certain organic dyes sensitive to changes in pH. Accuracy to $0 \cdot 5$ pH unit is easily possible but the method is not suitable for highly organic soils which absorb the dye to give a false colour. More accurate values are obtained with a pH meter. A portable field version exists but assessment is often carried out in the laboratory. The instrument consists of two electrode systems: one is placed in the soil sample solution and the other into a standard solution (buffer solution) of known acidity. The differences in hydrogen cation concentration between the two solutions are shown as different electrical readings on the meter, which can be translated directly into readings on the pH scale. The method involves measuring the electrical potential developed across a membrane between the sample solution and the alkaline reference solution. The sample solution is usually one part of the soil by weight mixed with 2·5 parts of distilled water by volume. Acidity can also be roughly indicated in the field by the presence or absence of certain plants. For example, *Clematis vitalba* (traveller's joy) is invariably confined to lime-rich soils while *Vaccinium myrtillus* (whortleberry) points to a soil of moderate to strong acidity.

Hydrogen ions probably have little direct effect on plant growth. However, soil acidity does influence chemical processes important for plant nutrition. Increased acidity can directly reduce the availabilty of some soil nutrients (e.g. phosphorus) and may influence populations of soil microorganisms which are important for decomposition and release of chemicals (e.g. limitation of nitrogen-fixing bacteria). It may also lead to toxic levels of soluble alumium in the soil. Further, colloidal clay and organic complexes may become unstable at high acidities; these important constituents then being subject to removal by percolating waters. Pearsall has reviewed the ecological significance of the pH values of natural soils and concludes that they remain the most useful single measurement that can be made for ecological purposes, providing one appreciates the way in which they relate to other soil properties.

SOIL WATER

Soil water plays a vital role not only in pedological processes but also in edaphological considerations. Soil water exists in three basic forms: hygroscopic, capillary and gravitational.

Hygroscopic water adheres to soil particle surfaces as a very thin film, held by enormous pressures. Consequently, it is not available to plant roots. *Capillary* water forms thicker films and also occupies the smaller pore spaces between particles. It is water held against the force of gravity and is the only permanent form in which soil moisture exists in a liquid state. As such, it contains dissolved substances and may be regarded as the soil solution. *Gravitational* water is of a transitory nature; it is water in excess of that held as hygroscopic and capillary water and will eventually drain away. The *maximum water capacity* of a soil is the amount of water held when all pore spaces are occupied by water and the soil is then waterlogged. At the other extreme, the *wilting coefficient* is the point at which the soil ceases to supply sufficient water to the plant to maintain its turgidity or mechanical strength. At this point, permanent wilting sets in. Optimum soil water conditions for most plants will occur between these two points when the soil reaches *field capacity*. This is defined as when the amount of water retained, just after the excess has drained away by gravity, is at its peak.

The force required (a suction pressure) to withdraw a given amount of water from a given mass of soil is known as the *capillary potential* or *pF value*. The drier the soil the higher the pF value. At field capacity the pF will be about $3 \cdot 0$, at the wilting coefficient it is $4 \cdot 2$ and for an over-dry soil the value is $7 \cdot 0$ (Fig. 3.4). Many properties of the soil water are greatly influenced by soil texture. The size range of soil particles and their distribution will determine the pore space fraction of the soil. This, together with the surface areas of the soil particles, will largely control the percentages of water occurring in the three categories for plant growth, namely, unavailable (hygroscopic), available (capillary) and free draining (gravitational).

The exact measurement of the hydrological properties of a soil as it exists in the field is extremely difficult. Water enters the soil by infiltration and advances downwards by a well defined *wetting front*. The *infiltration capacity* of a soil is determined by several factors, e.g. the soil structure and texture, the time interval since the previous rainfall, the conditions of the soil – atmosphere 'interface' (i.e. the presence of surface crusts or compaction, etc.). It is usually measured by means of an infiltrometer: a simple apparatus that measures the time taken for a given volume of water from an inverted bottle reservoir to infiltrate over a known surface area of soil. Results will vary according to whether the soil is initially dry or near to or at saturation level and the infiltrometer can only provide an approximate measure.

When the soil is saturated water moves downwards under gravitational flow. This may be rapid in a sandy soil (> 100 cm per day) and slower in a clay soil (> 10 cm per day). Texture, structure and pore size distribution in a soil are vital in this respect. The force of gravity increases the tendency for soil water to move (i.e. it increases what is known as the free energy of the water). Opposed to this tendency are the *matric forces*, seen as adhesion and cohesion of water molecules to surfaces and to each other to form thin films. Water is thus retained by *tensions*, whose strength is measured on the bar or atmosphere scale. Briggs states that water molecules more than $0 \cdot 06$ mm from a soil particle surface are not held by matric forces. Hence, micropores tend to retain water but macropores facilitate free drainage. Many of the macropores in a soil are often biopores (decayed plant roots, earthworm burrows) and these permit

Fig. 3.4 (a) Volumes of solids, water and air in a well-granulated silt loam at different moisture levels. Field capacity level is reached when water has drained out of the larger pore spaces. Plants will continue to remove water until the wilting coefficient is approached. The hygroscopic coefficient represents the point when the only water left in the soil is that held tightly by the soil colloids. (b) Relationship between thickness of water films at several moisture levels and the tension with which the water is held at the liquid-air interface. The tension is shown in atmospheres. (Modified from Brady, 1974.)

(a)

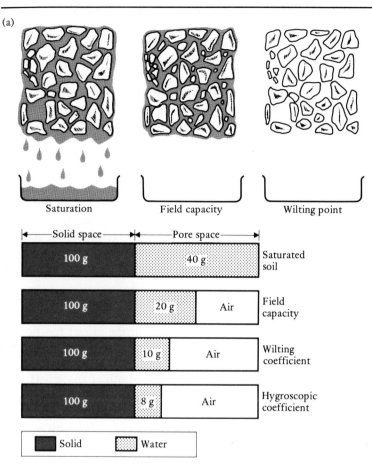

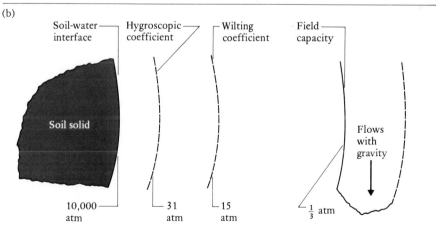

not only rapid downward water flow but also assist lateral flow within the soil body. However, even the water held by matric forces will eventually move from points of weak tension (wet areas) to points of strong tension (dry areas) as a *capillary flow*. This capillary movement is virtually zero at wilting point but is at its highest when the soil reaches field capacity.

The measurement of field capacity is particularly useful as a good guide to the potential supply of water for plant growth. It also indicates when the soil is unlikely to suffer structural damage from activities associated with cultivation. Field capacity usually corresponds to a soil moisture tension of about 0·05 bar. Briggs recommends its measurement as follows: 'to measure this parameter involves applying a force of 0·05 bar to the soil so that all the water held by tensions lower than this is withdrawn. A soil core is placed on a filter paper in a Buchner funnel and the surrounding space filled with sealing wax. The funnel can then be placed in a vacuum flask and attached to a vacuum pump. A vacuum of 0·05 bar is then developed in the vacuum flask, and the system allowed to come into equilibrium; at this point all the water held at tensions of less than 0·05 bar will have drained from the soil core. The moisture content can then be determined by oven drying.'

The soil water is the major solvent for the nutrients vital for plant growth. It is cycled through the soil-plant-atmosphere system by evapotranspiration and obviously adequate water supplies are essential for the proper functioning of ecosystems. Nevertheless, excess water leads to the exclusion of air from pore spaces and deprives plant roots and soil organic life of vital oxygen supplies. This can lead to depressed growth or rapid death of species. Hence, the balance between not enough soil water and too much is a critical consideration. This is particularly so in agricultural systems where man can improve either the water-retaining or water-shedding properties of soils by his actions.

SOIL GASES AND NUTRIENT CYCLES

The main input of soil water will be as rain. Much soil water will eventually find its way back into the atmosphere through direct evaporation from the soil surface or as plant transpiration, two important pathways in the hydrological cycle. Similarly, there is a free exchange of gases between the atmosphere and soil. Plant roots and many soil organisms require a supply of oxygen for the respiration process, which will also release carbon dioxide. To maintain this supply and to prevent the build up of carbon dioxide, a good soil will be well aerated with a total pore space often in excess of 50 per cent of its total volume. The air content of a soil is directly related to the soil water content. When water fills a pore space the air is forced out, thus poorly drained soils are always badly aerated. About 95 per cent of plant tissue is composed of carbon, hydrogen and oxygen derived mostly from air and water inputs. The remainder comes from the soil nutrients (soil solids).

Water, oxygen and carbon dioxide are exchanged between the atmosphere

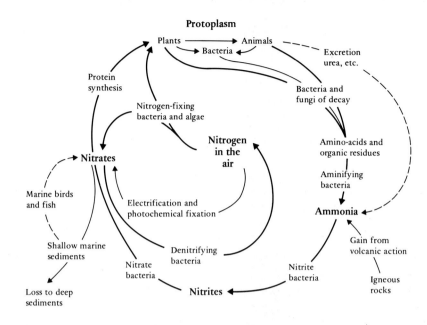

Fig. 3.5 The circulation of nitrogen between organisms and the environment as a largely self-regulating nitrogen cycle. Microorganisms play a major role: in nitrogen fixation (either as free-living organisms in the soil or in symbiotic association with leguminous plants); in the breakdown of plant and animal remains (mainly as humus) to release ammonia which, by stages, is converted back by other bacteria to nitrates; in the process of denitrifying nitrates and returning nitrogen to the main reservoir in the atmosphere. (Modified from Odum, 1971.)

and soil, with plants acting as important intermediaries in these dynamic processes. Soil dynamics can further be illustrated by reference to the cycling of chemical elements in the soil. Many important minerals other than water, oxygen and carbon dioxide are taken up by plants as nutrients. They are released back to the soil when these plants decay. For each of these elements, a *biogeochemical cycle* exists. Two examples will indicate the features of these nutrient cycles. The first is that of nitrogen (Fig. 3.5) where the atmosphere is the main reservoir or pool. Free-living bacteria in the soil and bacteria symbiotically associated with higher plants (e.g. the legumes) play a key role in this cycle. The second cycle concerns phosphorus (Fig. 3.6) where the initial source is weathered rock (a sedimentary cycle).

Similar cycles can be demonstrated for all chemical elements in the soil. Each one shows slight differences in the routes and stages followed. However, all chemical elements incorporated in organic tissues will be returned to inorganic forms at some point. Whether these forms can then be directly used by plants will depend upon where the elements are deposited. Most are returned to the soil and the green plant depends on the microbial decomposition of organic

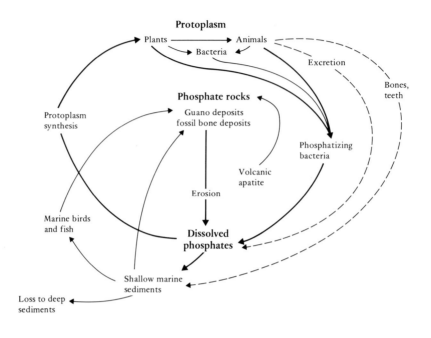

Fig. 3.6 The phosphorus cycle, where phosphatic rocks form the main reservoir. (Modified from Odum, 1971.)

material for its main supply of nutrients. Some, however, may be removed by the drainage system to deep oceanic waters, remaining there for the foreseeable future. These 'losses' must be replaced by rock weathering, which is a very slow process. But in general a rough balance exists between the uptake of nutrients by organisms and their subsequent return to the inorganic pools or reservoirs. In these processes, plants may be viewed as 'biogeochemical pumps powered by the energy of the sun.

To understand the soil we must not limit our attention to it alone. Important features in its development and properties result from interactions between the soil, the organisms in the top zone or on the surface, and the atmosphere above. These three form *the soil-plant-atmosphere continuum*. Interchanges (as

Fig. 3.7 Developments in the soil and biota of a recently planted Kudzu vine (*Pueraria lobata*) community at Copperhill, Tennessee. (*Top row*): organic matter in vegetation, litter and soil. *Middle row*: densities of biota, and oxygen uptake 1/m². *Bottom row*: carbon dioxide evolution ml/m², mean moisture contents, pH of soil (0–2·5 cm). C/N ratios of vegetation, litter and soil. Measurements are for the years 1960 (——), 1962 (– – –) and 1964 (— · —) at four distances from the original cultivated strip planted with Kudzu vine on eroded soil in 1955. By 1964 the community had spread to beyond 30 m, from the 1955 strip, but growth then became checked due to immobilization of up to 80 per cent of the cycling phosphorus and nitrogen in the litter and humus. (Modified from Witkamp *et al.*, 1966.)

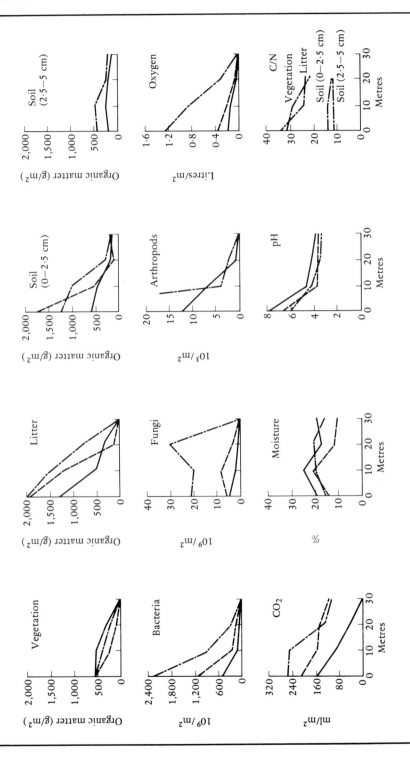

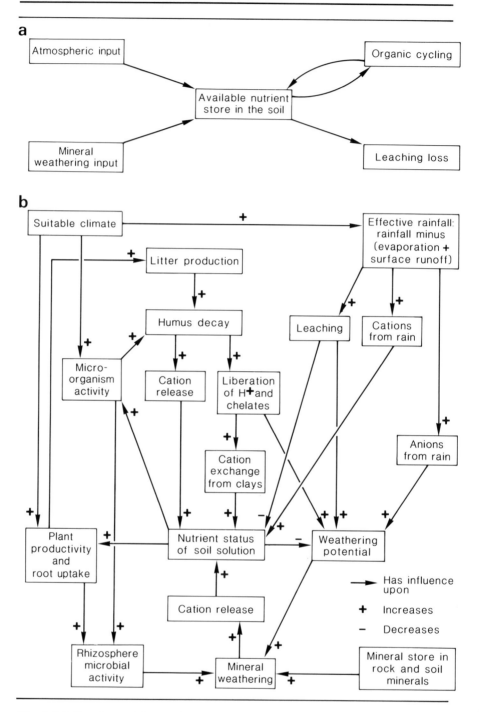

Fig. 3.8 Soil and vegetation nutrient systems: (a) A simplified model of the interaction of the five basic components of soil and vegetation nutrient systems. (b) Detailed model showing complex web of feedback relationships involved in cation balance. (Modified from Trudgill, 1977.)

inputs or removals) between these three levels are of prime consideration. In Fig. 3.7 the series of graphs show how some features of a soil change as it develops, illustrating the dynamic nature of soil maturation. These changes followed the sowing of Kudzu vine (*Pueraria lobata*); a plant used to restore soils grossly disturbed when mining operations and fumes from copper smelters destroyed the vegetation cover (see also Fig. 7.8).

Trudgill has explored the nature of soil and vegetation nutrient systems via a models approach which has five basic components (Fig. 3.8a). Under normal conditions, the leaching loss of nutrients is balanced by the input from the atmosphere and from rock weathering. Hence, in terms of the soil nutrient store, the most important consideration is often the organic cycling component, which is also the most complex. It is difficult to monitor if only because the amount of nutrients bound up in the vegetation (the biomass of stems, leaves, roots, etc.) and the speed with which it is returned to the soil store by litter fall and decay varies greatly from species to species. The main store of available plant nutrients is retained in the soil as cations adsorbed on the clay-humus complexes. Nutrients are also present as solutes in the soil water but these are more easily leached. Fig. 3.8b represents a detailed model of this cation balance.

DISCUSSION SECTION

It is not too difficult to follow the main accounts on soil formation in textbooks but many points made there are very difficult to see in the field. Quite frequently one soil exposure looks much like another to me.

This is appreciated. Soil can look very similar on first examination. But soil study calls for very careful observations, detailed measurements and as much precision in profile description as we can possibly manage. Carefully prepared field notes, set out in a standard form, are essential. Quantify the description whenever possible; a vague remark like 'very thick layer present' is rather useless. How thick is very thick? Measure it exactly. Only in this way will differences between superficially similar profiles become apparent. Soil profiles in a local study will often be very similar since one would only expect major differences when soils are compared which have developed on contrasting geologies or under different climatic regimes. Much of the soil is 'unseen' and it is therefore important to approach soil study with a list of questions to pose about each profile. Appendix A should help in this connection.

The profile is the basic unit of study in pedology. Much of soil study is concerned with differentiation of the profile into horizons. Variations in the profile call for an explanation and this starting point should lead on to an examination of the pedogenic factors. As an example of this, consider the profile differences between Figs. 3.9 and 3.10. Strata differentiation frequently reflects the important role of percolating waters.

Fig. 3.9 Alpine humus soil at 1,350 m on Mount Buffalo, Victoria. Humus-rich A horizon lying directly on largely unaltered C horizon (granitic parent material). Surface vegetation of open snow-gum woodland (*Eucalyptus pauciflora*) and alpine grasses (*Poa australis*).

Fig. 3.10 Strongly podzolized profile developed on fixed sand dunes (*Epacris* heath and scattered Eucalypts) near Eaglehawk Bay, Tasmania. Clear profile differentiation, with the deposition of iron oxides at lower levels to form an incipient, cemented, orstein horizon. The vertical channels running through this band probably represent root penetrations. Disturbed soil in the foreground.

Since much of the soil is largely unseen and time and other considerations prevent the digging of dozens of soil pits, how should we set about sampling soil?

The soil is usually examined where a significant change is suspected in at least one of the pedogenic factors, e.g. an abrupt slope change or where coniferous woodland replaces moorland. The number of sampling points within each broad subdivision of the landscape will depend, among other things, on the purpose of the study. Often the soil type examined at one site is found elsewhere in the region simply because the conditions making up that site are repeated. This observation has led to the *soil catena* concept. This may be defined as a sequence of soil types repeated throughout an area and associated with a recurring pattern of sites which have developed on a uniform lithology (geology) and also have other factors or a factor in common, e.g. type of drainage or slope facet. Milne, a soil surveyor working in Africa, was the originator of the concept and he referred to it as a 'unit of mapping convenience'. It is a useful idea for quickly linking together soils in the field (Fig. 3.11). In the United Kingdom a similar mapping unit is used known as the *soil series* (a group of similar profiles based on the same lithology). These ideas are of interest to geographers because they stress geomorphological links by essentially relating soils to landscape morphology.

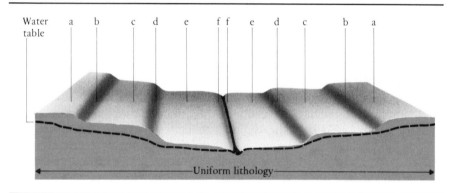

Fig. 3.11 The soils occurring at sites A to F represent a soil catena based on a repeated sequence of slope facets in the landscape. These facets are related to a series of paired river terraces, developed on a uniform lithology.

What are trace elements in the soil?

We mentioned some macronutrients required in relatively large amounts for successful plant growth. Examples were oxygen, hydrogen, carbon, nitrogen, phosphorus, calcium, potassium, magnesiun and sulphur. In addition, plants require very small amounts of other minerals such as cobalt, zinc, boron, molybdenum or copper. Only a few parts per million may be required. For example, many soils in South East Australia when cleared of scrub for pasture development were later found to be deficient in cobalt. Until this was rectified, the grass growth was severely hampered. On the other hand, too much of any element can prove harmful to organisms. This applies to micronutrients as well as macronutrients. Copper is an important trace element and the example of copper smelting in Tennessee (Fig. 3.7) is a case where too much copper has found its way into the soil leading to toxicity.

At this introductory level can much progress be made with the nutrient models suggested by Trudgill since they look quite complex when the detail of the basic components is spelt out (e.g. the cation balance model of Fig. 3.8b)?

Trudgill states that the greatest understanding of the soil-vegetation nutrient system will come when we can measure 'not only rates of processes allied to cation flow but also the amount of cations involved as well' (i.e. when values can be assigned to the boxes in the model and to the connecting arrows). Completely accurate quantification is probably impossible but Trudgill notes that 'the easiest way of tackling the problem is to start with gross basic models and then attempt to build in further details of rates, processes, and masses involved in specific environmental situations. Thus the modelling will commence with gross climatic models, building variants of geology and topography and then drainage and mesoclimate at succeeding stages.'

Appendix 3.2 presents a relatively simple computer program for a soil-vegetation nutrient model which follows this advice. It allows various inputs to the model to be changed and permits alteration of some of the environmental conditions. It is designed for use on the type of desktop computer now increasingly available in most educational establishments. The language used is a common version of BASIC. Students should have little difficulty in using this program and are encouraged to do so as an aid to further understanding of nutrient cycling. To help the beginner to understand what the program is setting out to do, marginal notes have been added to the listing in Appendix 3.2.

REFERENCES

Avery, B. W. and Bascomb, C. L. (eds), 1974. *Soil Survey Laboratory Methods*, Technical Monograph No. 6, Soil Survey of England and Wales, Harpenden.

Brady, N. C., 1974. *The Nature and Properties of Soils* (8th edn.), Macmillan, New York.

Briggs, D., 1977. *Soils*, Butterworths, London, Boston.

Bunting, B. T., 1965. *The Geography of Soil*, Hutchinson, London.

Clarke, G. R., 1957. *The Study of the Soil in the Field*, Clarendon Press, Oxford.

Fitzpatrick, E. A., 1971. *Pedology: A systematic approach to Soil Science*, Oliver and Boyd, Edinburgh.

Foth, H. D. and Turk, L. M., 1972. *Fundamentals of Soil Science*, Wiley, New York.

Hodgson, J. M. (ed.), 1974. *Soil Survey Field Handbook*, Technical Monograph No. 5, Soil Survey of England and Wales, Harpenden.

Huggett, R. J., 1982. 'Models and spatial patterns of soils', In *Principles and Applications of Soil Geography* (eds Bridges, E. M. and Davidson, D. A.), Longman, London and New York, pp. 132–70.

Jacobs, H. S. and Reed, R. M. (eds), 1964. *Soils Laboratory Exercise Source Book*, American Soc. of Agronomy, Madison.

Jenny, H., 1941. *Factors of Soil Formation*, McGraw-Hill, New York.

Jenny, H., 1961. 'Derivation of state factor equations of soils and ecosystems', *Soil Science Society of America Proceedings*, **25**, 385–8.

Jenny, H., 1980. *The Soil Resource*, Ecological Studies, **37**, Springer-Verlag, Berlin.

Mackney, D. and Burnham, C. P., 1964. *The Soils of the West Midlands*, Agricultural Research Council, Soil Survey of Great Britain, Bulletin No. 2, Harpenden.

Milne, G., 1935. 'Composite units for the mapping of complex soil associations', *Trans. 3rd Int. Congr. Soil Sci.*, **1**, 345.

Odum, E. P., 1971. *Fundamentals of Ecology* (3rd edn.), W. B. Saunders, Philadelphia: (see Ch. 2 – biogeochemical cycles).

Oosting, H. J., 1956. *The Study of Plant Communities*, W. H. Freeman, San Francisco: (see Chs. 7 and 8).

Pearsall, W. H., 1952. 'The pH of natural soils and its ecological significance', *J. Soil Sci.*, **3**(1), 41–51.

Runge, E. C. A., 1973. 'Soil development sequence and energy models', *Soil Science*, **115**, 183–93.

Russell, E. W., 1961. *Soil Conditions and Plant Growth*, Longman, London.

Simonson, R. W., 1959. 'Outline of a generalized theory of soil genesis', *Soil Science Society of America Proceedings*, **23**, 152–6.

Thomasson, A. J., 1971. *Soils of the Melton Mowbray District*, Memoirs of the Soil Survey of Great Britain, Harpenden.

Tisdale, J. M. and Oades, J. H., 1982. Organic matter and water-stable aggregates in soils, *J. Soil Sci.*, **33**, 141–63.

Trudgill, S. T., 1977. *Soil and Vegetation Systems*, Clarendon Press, Oxford.

White, R. E., 1979. *Introduction to the principles and practice of soil science*, Blackwell, Oxford.

Witkamp, M., Frank, M. L. and Shoopman, J. L., 1966. 'Accumulation and biota in a pioneer ecosystem of Kudzu vine at Copperhill, Tennessee', *J. appl. Ecol.*, **3**, 383–91.

APPENDIX 3.1: PEDOGENIC FACTORS

A check-list of some questions to consider when confronted with an unknown soil type.

This is not intended as a complete list but merely as a starting point for the consideration of pedogenic factors. Answers to some questions might be forthcoming only after years of study at the site.

PASSIVE FACTORS

1. THE PARENT SOIL MATERIAL

What is the geology of the site? How do the rocks vary in terms of:

(a) Physical properties (structure, porosity, permeability, ease of weathering)?
(b) Chemical properties (the percentage chemical composition of the rock. Do they have sufficient reserves of macronutrients and are essential trace elements present? Are the nutrients in readily available forms for uptake? Are they in easily leached forms?)

As a general rule, the geological factor increases in importance as one moves from wetter to drier climates. Another rule, generally true, is that as a soil becomes older the influence of parent soil material becomes less.

2. THE TOPOGRAPHY

What are the slope angles, slope aspects and altitudes in the area? Are the slopes stable? Is material accumulating downslope (known as colluvium) under the influence of gravity? What are the drainage characteristics of the site? Is there a zone of heavy leaching or a zone of chemical enrichment by flushing at lower levels? Where is the water table and is it subject to marked seasonal fluctuations?

While we would expect marked variation in soil in an area of pronounced relief we should remember that very small relief differences can be significant. For example, in inland New South Wales large areas which initially appear flat have a microrelief of ridges and hollows, often with no more than about a 5 cm difference between these features. This microrelief is caused by alternate expansion and shrinkage of clays as heavy downpours are quickly followed by rapid drying. This gives rise to soils locally known as *gilgai*. Soil differences are clearly detectable between the minute ridges and hollows. In the hollows more water collects and more soil nutrients are washed in. These differences can produce distinctive floras on the ridges and hollows.

3. THE TIME ELEMENT

When did the surface upon which the soil has developed become available for pedogenesis? (In much of the Northern Temperate Latitudes soils commenced development at the end of the last glacial phase. In the Tropics, however, some soils are much older.) Over what period of time has each pedogenic factor operated? What site changes have taken place, e.g. climatic shifts, man-induced changes? Are there any fossil soils present (those buried by more recent surface deposits)?

All soils should be approached from an historical viewpoint since they represent developments over a considerable time span.

ACTIVE FACTORS

1. THE CLIMATE

What are the characteristics of the regional climate in terms of:

(a) *Temperature* – seasonal variations, length of growing season, number and severity of frosts? Large temperature changes will greatly alter the rate of chemical and biological activities.
(b) *Precipitation* – seasonal variations in types and intensities of precipitation, rainfall input of dissolved industrial chemicals, salt spray, etc.?
(c) *Wind flow* – frequency of strong winds, drying effect on soils, wind-transported soil additions – road dust, sand, clay particles, etc? The microclimate at the site may differ considerably from the macroclimate of the region.

2. THE BIOLOGICAL FACTOR

After a consideration of plant and animal populations present at the site including, as far as one can, the microfauna and microflora, more specific questions can be asked about each species which may be influencing the soil type:

(a) *The flora*. Does the ecology of individual species, e.g. legumes, have a bearing on soil properties? Is the growth-form of the plant important, e.g. rooting patterns? What demands does each plant make on soil nutrients? What type of litter is produced and how easily is it decomposed?
(b) *The fauna*. What physical and chemical influences do animals have, e.g. burrowing, trampling, decay of droppings, etc? For example, pocket gophers in North America dramatically disturb the top soil but smaller, less obvious animals, such as ants, are often just as important.
(c) *Man*. What past or present land-use patterns and management techniques have operated at the site? These may be as introductions of grazing animals, moorland burning, the addition of lime or attempts at regional water table control.

We cannot always neatly separate these factors since there is frequently considerable overlap. For example, changes in soil properties on a moderately steep slope might be due to man clearing scrub and rabbits at the site and then introducing grass and cattle. This, over a number of years, could alter slope stability (as burrowing is replaced by trampling), the aeration of the profile (as burrows become infilled), the type of plant community present (due to differential grazing of the two animals), the soil microclimate (related to plant differences and soil disturbance) and the input and uptake of soil nutrients (different nutrient budgets for each biotic community). In ecological situations change in one factor will often trigger off changes in several other factors.

Many of the basic points about elementary soil study can be more easily recalled to mind by what might be called '*the three approach*': soil is matter in three states (solid, liquid, gaseous). It results from three processes of weathering (physical, chemical, biological). There are three broad pedogenic factors (climate, geology, biota). The profile can usually be divided into three horizons (A, B and C). The A horizon can sometimes be subdivided into three layers (L, F and H). Three terms useful in profile descriptions which relate to additions and subtractions in the profile are eluvial, illuvial and colluvial. Once the profile has been prepared, three other soil features can be readily studied (soil colour, texture and structure). Soil colour is the relative degree of three simple variables (hue, chroma and value). Soil texture reflects the proportions of three basic components (sand, silt and clay). These are plotted on a three-sided figure (the textural triangle). There are three types of clay mineral found in soils. Three simple laboratory methods have been described (Mechanical Analysis, Bouyoucos' and 'Weighing'). There are three basic regions on the pH scale (neutral = 7, acid = < 7, alkaline = > 7). Three methods of assessing pH are commonly used (colour charts, pH meter and indicator species). Soil water exists in three forms (hygroscopic, capillary and gravitational). For plant growth, three terms are important when considering soil water retention (Maximum Water Capacity, Wilting Coefficient and Field Capacity). In plant nutrition, the three types of biogeochemical cycle are the hydrological cycle, the atmosphere–reservoir cycles (nitrogen, oxygen) and the sedimentary–reservoir cycles (phosphorus).

The above is a gross simplification of soil complexity but it should prove useful as an aide-mémoire for beginners.

APPENDIX 3.2: 'CYCLE', A SIMPLE TEACHING MODEL OF NUTRIENT CYCLING.

INTRODUCTION

The program for this relatively simple simulation model is written in BASIC and designed to be run on a PET (Commodore) or similar microcomputer. It may look complicated but it is simple compared with the processes operating in real ecosystems. The structure of the model is along the lines suggested by Trudgill (see p. 55). All standing crop values are in kg/ha and fluxes are in kg/ha/year. Leaching losses are represented by transfer to drainage (TD).

Annotations have been added to the program listing below to explain what is happening at the various stages of the program. The beginner should study these carefully and, if necessary, pursue points further through the key references at the end of the listing. In particular, a full description is given in Haines-Young (1983), which explains the assumptions in the model and the derivation of certain parameters.

The initial data used are from the Hubbard Brook Experimental Forest ecosystem study (see Fig. 5.4) of calcium cycling. A mean annual precipitation of 1300 mm is used in the model and the mean annual temperature is set at 8 °C, to give conditions close to those found at Hubbard Brook. The model can be run under different assumptions or with different data (providing these are reasonable). For example, suggested data for different climates that can be tested are:

moist tropical	– precipitation 2300 mm, temperature 27 °C
semi-arid	– precipitation 250 mm, temperature 32 °C
moist temperate	– precipitation 1300 mm, temperature 8 °C
tundra	– precipitation 250 mm, temperature 2 °C

The model can also be run under successional conditions by setting V, L and S to zero, and M to 7,000 kg/ha as initial conditions (see key below).

The initial condition of high nutrient status is represented by V = 450 kg/ha, S = 500 kg/ha, M = 7,000 kg/ha and L = 350 kg/ha. The initial condition of low nutrient status is represented by V = 50 kg/ha, S = 50 kg/ha, M = 100 kg/ha and L = 50 kg/ha.

The model appears to be approaching equilibrium after about 100 iterations and the pattern of uptake of nutrients for the simulated system is very similar to that observed in the natural systems. Results can be displayed on the microcomputer screen either as a graph or table and are also available as a printout, if a printer is attached.

KEY

V = Vegetation store
L = Litter store
S = Soil store
M = Mineral store
PU = Potential uptake
TV = Actual uptake

TL = Litter fall
TS = Litter decay } transfers

TM = Weathering input
TD = Drainage loss
PI = Precipitation input, by rainfall
WP = Weathering potential
NP = Net above ground primary productivity

P = Precipitation (mm)
T = Temperature (°C)
E = Actual evapotranspiration
PL = ⎫ Co-efficients which
PV = ⎬ modify transfers under
C = ⎭ different climates, soil store conditions, etc.

RM = Concentration of ion in rainfall
BD = Biological demand for ion
AL = Mobility parameter

LISTING FOR 'CYCLE'

```
1 DIM V(11),L(11),S(11),M(11)
2 Y=1
```

```
5 PRINT"❑SOIL VEGETATION NUTRIENT SYSTEM"
6 PRINT"                                  "
10 REM MENU
11 PRINT"❑THE MODEL IS SET UP WITH ASSUMPTION"Y
12 PRINT"❑IF YOU WANT TO RUN THE SYSTEM WITH A"
13 PRINT"DIFFERENT ASSUMPTION THEN USE (3)BELOW"
15 PRINT"❑YOU CAN :"
16 PRINT"❑❙❙(1)RUN MODEL USING HUBBARD BROOK DATA"
17 PRINT"❙❙(2)RUN MODEL USING ALTERNATIVE DATA"
18 PRINT"❙❙(3)CHANGE ASSUMPTION"
19 PRINT"❙❙(4)DISPLAY RESULTS GRAPHICALLY"
20 PRINT"❙❙(5)DISPLAY RESULTS ON THE SCREEN"
21 PRINT"❙❙(6)OBTAIN A PRINTOUT OF THE RESULTS"
23 PRINT"❙❙(7)END SIMULATION"
25 PRINT"❑INPUT INSTRUCTION:"
26 INPUT A$
30 IF A$="1"THENGOSUB800
32 IFA$="2"THENGOSUB900
34 IFA$="3"THENGOSUB1000
36 IFA$="4"THENGOSUB2000
38 IFA$="5"THENGOSUB3000
40 IFA$="6"THENGOSUB4000
42 IFA$="7"THENGOSUB8000
45 PRINT"❑I DO NOT UNDERSTAND-TRY AGAIN!"
46 GOTO25
```

'MENU' – allows user to specify tasks e.g. run model using Hubbard Brook data; display results, etc.

Subroutines which carry out selected tasks.

```
50 REM CLIMATIC INPUT MODUEL
55 IF A$="2" THEN PRINT"❑MODEL SET UP AS FOLLOWS:
59 IF A$="2"THEN GOTO 61
60 PRINT"❑MODEL SET UP FOR HUBBARD BROOK"
61 PRINT"❑❙V="V(0)"KG/HA"
62 PRINT"L="L(0)"KG/HA"
63 PRINT"S="S(0)"KG/HA"
64 PRINT"M="M(0)"KG/HA"
65 PRINT"❑AL="AL"MG/L"
66 PRINT"RM="RM"MG/L"
67 PRINT"BD="BD
70 PRINT"❑INPUT ANNUAL PRECIPITATION (MM):"
75 INPUT P
79 PRINT"❑INPUT MEAN ANNUAL AIR TEMP.(C):"
80 INPUT T
90 X=300+25*T+0.05*T↑3
95 E=(P/(0.9+(P/X)↑2)↑0.5)
96 IF E>PTHENE=P
100 PRINT"❑ACTUAL EVAPORATION (MM):"INT(E+.5)
105 PV=0.1-(T/1000)
110 PL=T/50
111 IF T<=0 THEN PH=.02
112 PI=RM*P/100
115 Z=1.66*LOG(E)/LOG(10)-1.66
120 NP=10↑Z
125 PU=BD*NP*10
130 WP=AL+(T-8)*T/1000
140 PRINT"❑ASSUMPTION"Y
145 PRINT"❑❑ONE MOMENT WHILE I THINK"
```

see Barry (1969)

see Rosenzweig (1968)

Sets initial store sizes to specified values.

Asks for climatic data (rainfall and temperature).

Calculates actual evapotranspiration, predicts net productivity and potential nutrient uptake.

Calibrates transfer functions for a particular climatic regime.

```
150 REM MODEL
154 TT=0
155 FORI=1TO10
157 V=V(I-1):L=L(I-1):S=S(I-1):M=M(I-1)
160 FORJ=1TO10
165 IF Y=1 THEN C=1
170 IF Y=1 THEN GOTO 180
175 C=S/(S+100)
180 TD=(WP*C*(P-E))/100
190 IF S<=0 THEN TD=0
200 TV=C*PU
205 IF S<=0 THEN TV=0
211 IF (TD+TV)>=S THEN TV=S-TD
215 TL=PV*V
220 TS=PL*L
225 IF M<=0 GOTO 235
230 TM=0.01*WP*P*(M-S)/M
234 IF TM<0THENTM=0
235 IF M<=0 THEN TM=0
236 IF TM>=M THEN TM=M
237 IFS>MTHENTM=0
240 S=S-TV-TD+TM+TS+PI
245 IF S<0 THEN S=0
250 V=V+TV-TL
255 IF V<0 THEN V=0
260 L=L+TL-TS
265 IF H < 0 THEN H =0
270 M=M-TM
```

Calculates constant C under assumption 2.
Drainage loss.

Plant uptake.

Litter fall.
Litter decay.
Weathering release.

New soil store size.

New vegetation store size.

New litter store size.

New mineral store size.

By iteration, calculates transfers and new store sizes.

Mass balance equations for stores.

```
280 IF M<0 THEN M=0
284 TT=TT+1
285 NEXTJ
290 V(I)=V:L(I)=L:S(I)=S:M(I)=M
300 NEXTI
310 FOR KK=1TO150
315 PRINT"TXOK - CALCULATIONS DONE - I'LL RETURN YOU TO THE MENU
320 NEXTKK
321 PRINT"]"
325 GOTO10
```

(Stores data in array for subsequent recall).

```
800 REM HUBBARD BROOK DATA
805 V(0)=450:L(0)=350:S(0)=500:M(0)=70000
806 AL=1.62:RM=.17:BD=.01
810 GOTO50
```

see Likens *et al* (1977)

Subroutine for input of Hubbard Brook data.

```
900 REM CHANGE DATA
910 PRINT"]INPUT DATA"
911 PRINT"────────────"
915 PRINT"NINPUT V(KG/HA)"
920 INPUT V(0)
925 PRINT"NINPUT L(LG/HA)"
930 INPUT L(0)
935 PRINT"NINPUT S(KG/HA)"
940 INPUT S(0)
945 PRINT"NINPUT M(KG/HA)"
950 INPUT M(0)
960 PRINT"NINPUT AL(MG/L)"
965 INPUT AL
970 PRINT"INPUT RM(MG/L)"
975 INPUT RM
980 PRINT"INPUT BD"
985 INPUT BD
999 PRINT"]":GOTO50
```

Subroutine for input of own data.

```
1000 REM MODULE FOR CHANGE OF ASSUMPTIONS
1005 PRINT"]THE MODEL CAN BE SET UP WITH TWO      ASSUMPTIONS:"
1010 PRINT"NNN(1) THE SIZE OF THE SOIL STORE HAS  NO INFLUENCE ON UPTAKE"
1020 PRINT"NNN(2) UPTAKE IS INVERSELY PROPORTIONAL TO SIZE OF THE SOIL STORE"
1030 PRINT"NNTHE MODEL IS SET UP WITH ASSUMPTION"Y
1040 PRINT"NWHICH ASSUMPTION DO YOU WANT"
1050 INPUTY
1055 PRINT"]"
1060 GOTO10
```

Subroutine for change of assumptions:
assumption 1 = soil store does *not* influence uptake or weathering
assumption 2 = soil store does influence uptake and weathering

```
2000 REM GRAPHICAL OUTPUT
2001 GOTO9000
2010 PRINT"]"
2015 FORJ=1TO20
2020 PRINTTAB(6)"]":NEXTJ
2025 PRINT" "
2030 FORJ=1TO20:PRINTTAB(6+J)"]":NEXTJ
2035 PRINT"S"
2040 FOR D=1TO9:PRINT" ":NEXTD:PRINT"% MAX"
2045 FORD=1TO12:PRINT" ":NEXTD:PRINTTAB(16)"TIME"
2050 VX=V(0):LX=L(0):SX=S(0):MX=M(0)
2055 FORI=1TO10
2060 IFV(I)>VX THEN VX=V(I)
2065 IFL(I)>LX THEN LX=L(I)
2070 IFS(I)>SX THEN SX=S(I)
2075 IFM(I)>MX THEN MX=M(I)
2080 NEXTI
2085 FORI=1TO10
2095 IF VX=0 GOTO 2105
2100 KV=(20-INT(((V(I)/VX)*20))):GOTO2110
2105 KV=20
2110 IF LX=0 GOTO 2120
2115 KL=(20-INT(((L(I)/LX)*20))):GOTO2125
2120 KL=20
2125 IF SX=0 GOTO 2135
2130 KS=(20-INT(((S(I)/SX)*20))):GOTO2140
2135 KS=20
2140 IF MX=0 GOTO 2150
2145 KM=(20-INT(((M(I)/MX)*20))):GOTO2155
2150 KM=20
2155 REM "V"
2164 PRINT"S":IF KV<0 GOTO 2170
2165 FOR D=1TOKV:PRINT" ":NEXTD
2170 PRINTTAB(6+2*I)"V"
2175 REM "L"
2179 IF KL<=0 GOTO 2185
2180 PRINT"S":FOR D=1TOKL:PRINT" ":NEXTD
```

Subroutine for graphic display of results.

Prints graph and information.

Finds largest value of each variable.

Calculates scaling factors.

```
2185 PRINTTAB(6+2*I)"L"
2190 REM "S"
2194 IF KS<=0 GOTO2200
2195 PRINT"⌂":FORD=1TOKS:PRINT" ":NEXTD
2200 PRINTTAB(6+2*I)"S"
2205 REM "M"
2209 IF KM<=0 GOTO2215
2210 PRINT"⌂":FORD=1TOKM:PRINT" ":NEXTD
2215 PRINTTAB(6+2*I)"M"
2220 PRINT"⌂":FORD=1TO23:PRINT" ":NEXTD
2225 NEXTI
2230 PRINT"TYPE C<RET> TO CONTINUE"
2235 INPUTA$
2240 PRINT"⌂":GOTO10
```

> Plots data points on graph.

```
3000 REM DISPLAY MODULE FOR RESULTS
3010 PRINT"⌂CHANGE OVER "TT" TIME STEPS"
3011 PRINT"‾‾‾‾‾‾‾‾‾‾‾‾‾‾‾‾‾‾‾‾‾‾‾"
3015 PRINT"▓▓▓V/▓▓▓▓L▓▓▓▓S▓▓▓▓M"
3020 FOR I=1TO10
3025 V=INT(V(I)):L=INT(L(I)):S=INT(S(I)):M=INT(M(I))
3030 IFV<1000THENSV=1:IFV<100THENSV=2:IFV<10THENSV=3
3035 IFL<1000THENSL=1:IFL<100THENSL=2:IFL<10THENSL=3
3040 IFS<1000THENSS=1:IFS<100THENSS=2:IFS<10THENSS=3
3045 IFM<1000THENSM=1:IFM<100THENSM=2:IFM<10THENSM=3
3050 IFI<10THENSI=1
3055 PRINTTAB(SI)I*10
3060 PRINTTAB(5+SV)"⌂"V
3065 PRINTTAB(12+SH)"⌂"L
3070 PRINTTAB(17+SS)"⌂"S
3075 PRINTTAB(24+SM)"⌂"M
3080 SV=0:SL=0:SS=0:SM=0:SI=0
3085 NEXTI
3090 PRINT"⌂TV="INT(TV)"KG/HA"
3095 PRINT"TD="INT(TD)"KG/HA"
3100 PRINT"P="P"CMS"
3105 PRINT"T="T"C"
3110 PRINT"E="INT(E)"CMS"
3115 PRINT"⌂MODEL RAN UNDER ASSUMPTION"Y
3190 PRINT"⌂TYPE C<RET> TO CONTINUE"
3195 INPUTA$
3200 PRINT"⌂":GOTO10
```

> Subroutine for display of results on screen as table.
>
> Tabulation of data.
>
> Printing over 100 time steps.
>
> Other information.

```
4000 REM PRINTOUT MODULE
4010 OPEN3,4
4015 PRINT#3,CHR$(1)"SOIL VEGETATION SYSTEM"
4020 PRINT#3,CHR$(1)"‾‾‾‾‾‾‾‾‾‾‾‾‾‾‾‾‾‾‾‾"
4107 PRINT#3,"INITIAL CONDITIONS:"
4108 PRINT#3,"‾‾‾‾‾‾‾‾‾‾‾‾‾‾‾‾‾‾"
4120 PRINT#3,"V="V(0)"KG/HA"
4125 PRINT#3,"L="L(0)"KG/HA"
4130 PRINT#3,"S="S(0)"KG/HA"
4135 PRINT#3,"M="M(0)"KG/HA"
4137 PRINT#3," "
4140 PRINT#3,"AL="AL"MG/L"
4145 PRINT#3,"RM="RM"MG/L"
4150 PRINT#3," "
4155 PRINT#3,"MEAN ANNUAL AIR TEMP(C)="T
4160 PRINT#3,"MEAN ANNUAL PRECIPITATION(MM)="P
4165 PRINT#3,"ACTUAL EVAPORATION(MM)="INT(E)
4170 PRINT#3," "
4175 PRINT#3,"MODEL RAN UNDER ASSUMPTION"Y
4180 PRINT#3," "
4523 PRINT#3,"CHANGE OVER 100 TIME STEPS (ALL FIGURES ARE IN KG/HA)"
4524 PRINT#3,"‾‾‾‾‾‾‾‾‾‾‾‾‾‾‾‾‾‾‾‾‾‾‾‾‾‾‾‾‾‾‾‾‾‾‾‾"
4525 PRINT#3," "
4530 PRINT#3,"▓▓▓▓V/▓▓▓▓L▓▓▓▓S▓▓▓▓M"
4535 FOR I=1TO10
4540 V=INT(V(I)):L=INT(L(I)):S=INT(S(I)):M=INT(M(I))
4545 IFV<1000THENSV=1:IFV<100THENSV=2:IFV<10THENSV=3
4550 IFL<1000THENSL=1:IFL<100THENSL=2:IFL<10THENSL=3
4555 IFS<1000THENSS=1:IFS<100THENSS=2:IFS<10THENSS=3
4560 IFM<1000THENSM=1:IFM<100THENSM=2:IFM<10THENSM=3
4565 IFI<10THENSI=1
4570 PRINT#3,TAB(SI)(I*10)TAB(5+SV)VTAB(3+SH)LTAB(3+SS)STAB(3+SM)M
4600 SV=0:SL=0:SS=0:SM=0:SI=0
4605 NEXTI
4606 PRINT#3," "
4609 PRINT#3,"AT LAST TIME STEP:-"
4610 PRINT#3,"TV="INT(TV)"KG/HA"
4615 PRINT#3,"TL="INT(TL)"KG/HA"
4616 PRINT#3,"TS="INT(TS)"KG/HA"
4617 PRINT#3,"TM="INT(TM)"KG/HA"
4618 PRINT#3,"PI="INT(PI)"KG/HA"
4620 PRINT#3,"TD="INT(TD)"KG/HA"
4685 CLOSE3
4690 PRINT"⌂"
4695 GOTO10
```

> Subroutine for printing out results (see above).

```
8000 PRINT"⌂TYPE RUN TO USE THE MODEL "
```

> Final message.

```
8010 END
9000 REM NOTE ABOUT OVER-PRINTING
9010 PRINT"  A GRAPH OF THE RELATIVE CHANGES FOR EACHCOMPARTMENT WILL BE SHOWN"
9015 PRINT" EACH IS SHOWN AS THE % OF ITS MAXIMUM   VALUE"
9020 PRINT"  DUE TO OVER-PRINTING IT MAY BE DIFFICULTTO FOLLOW EACH VARIABLE."
9025 PRINT"  BUT WATCH CAREFULLY AND COMPARE IT WITH THE DATA PRINTOUT.
9030 PRINT"  TYPE C<RET> TO CONTINUE
9035 INPUTA$
9040 GOTO2010
```

Note about graphical output
called from line 2010

REFERENCES

Barry, R. G., 1969. 'Evaporation and transpiration', in *Water, Earth and Man* (ed. Chorley, R. J.), Methuen, London.

Gersmehl, P. J., 1976. 'An alternative biogeography', *Annals of Association of American Geographers*, **66**, 223–41

Golly, F. B., McGinnis, J. T., Clements, R. G., Child, G. I. and Duever, M. J., 1973. *Mineral Cycling in a Tropical Moist Forest Ecosystem*, University of Georgia Press, Athens, USA.

Haines-Young, R. H., 1983. 'Nutrient cycling and problem solving: a simple teaching model', *Journal of Geography in Higher Education*, **7**(2), 125–139.

Likens, G. E., Bormann, F. H., Pierce, R. S., Eaton, J. S. and Johnson, N.M., 1977. *Biogeochemistry of a Forested Ecosystem*, Springer-Verlag, Berlin, New York.

Rosenzweig, M. L., 1968. 'Net primary productivity of terrestrial communities: prediction from climatological data', *American Naturalist*, **102**, 67–74.

Trudgill, S. T., 1977. *Soil and Vegetation Systems*, Clarendon Press, Oxford.

ACKNOWLEDGEMENT

The author is indebted to Dr Roy Haines-Young, Department of Geography, University of Nottingham, who kindly gave his permission for 'Cycle' to be included here.

4

PLANT DYNAMICS AND THE NATURE OF VEGETATION

INTRODUCTION

Two key figures in the early development of plant ecology were H. C. Cowles and F. E. Clements. Their studies in North America led to a set of principles which had a profound influence on many ecologists in the first half of the twentieth century. Both were interested in how the species composition of vegetation changed over time. In 1899 Cowles published his studies of sand-dune vegetation around Lake Michigan, tracing the changes as a few early established plants were joined by or replaced by other species to form more complex communities. His results influenced, in particular, Clements who presented over the next few decades theories held to be generally applicable to these types of vegetation development wherever they occurred. More recently much of this work has been heavily criticized, but a review of Clements' theories forms a suitable starting point for an understanding of current views on the nature of vegetation.

As an example, the development of vegetation on an initially plant-free site, such as a recent rock fall on a scree slope in the Lake District, will be considered. The rainfall here might be high but the boulder surface retains little of this water and is essentially dry for long periods. Only plants capable of growing on dry, bare rocks become established first. This *pioneer community* mainly consists of lichens and algae. After a few years the fresh rock surface weathers slightly and loose rock fragments collect between boulders. These changed conditions may now allow other plants to join the early colonizers. Mosses and grasses may appear where a rudimentary soil is forming and water retention is better. The plants themselves cause further weathering and add organic material. More water is retained and additional supplies of nutrients become available from weathered material and organic decomposition.

These developments allow a colonization by larger plants, whose demands on water and nutrient supply can now be satisfied. Perennial shrubs (e.g. willow) and tree seedlings may follow. By this stage conditions have so altered (changes in soil, micro-climate, plant competition and animal influence) that many of the earlier species no longer find suitable habitats for their continuance at the site. Eventually a woodland could evolve and the original character of the site may not now be obvious. This will take many decades, and while it is easy to describe in general terms, it is virtually impossible for any one person to witness a full development because of the timescale involved.

A sequence of plant communities occupying the same site is known as a *plant succession* or *sere*. The sequence described here is an example of a *xerosere*, because the site is initially dry (Gr. *xero* = dry). Plants adapted to a dry site (xerophytic vegetation) have been gradually replaced by communities reflecting better water retention in the environment (mesophytic vegetation).

PRIMARY PLANT SUCCESSION

A similar sequence is seen if the developments at a very wet or waterlogged plant-free site are traced. A recently formed, shallow, freshwater pond tends to be colonized first by floating aquatic vegetation (e.g. algae). These, on dying, add organic matter to the bottom sediments. As sediment builds up and the water becomes shallower, rooted aquatics establish such as pondweeds. By trapping more sediment and adding organic matter, these assist pond infilling. Around the pond margins plants tolerant of water-logged soils will grow. These may include species of *Carex* (sedges), *Juncus* (rushes), *Typha* (reedmace) or *Phragmites communis* (the common reed).

These parts of the original site at least are now too dry for true aquatic vegetation and so here the pioneer species disappear. Shrubs colonize the margins as the infilling process continues (e.g. willow, alder), and tree species (e.g. oak, birch) may follow. In this case a *hydrosere* has been outlined. Hydrophytic vegetation, adapted to wet conditions, has been replaced by mesophytic vegetation, as the moisture regime becomes less extreme.

Both the xerosere and the hydrosere are examples of *primary seres* (or priseres) since both start from a virgin site with no previous vegetation. Xeroseres are subdivided according to the nature of the initial surface. *Lithoseres* develop on large rock surfaces such as scree boulders, while *psammoseres* colonize small rock particles, particularly the grains of sand-dunes. Hydroseres are similarly subdivided into *haloseres* (if the initial water is saline), while for the fresh or non-saline case, as outlined here, the original term *hydrosere* is retained (also known as hydrarch succession in North America). See Figs. 4.1 to 4.4.

Examples of the same type of sere will not be identical. Psammoseres, for example, may vary due to changes in either coastal exposure or in the chemical substrate itself where calcareous sand is replaced by siliceous sand. These local conditions might cause certain stages to be missed or unduly prolonged. However, there is sufficient agreement for most psammoseres to fit the generalized Clementsian statement describing their development in broad terms. Likewise, there is good agreement between the stages in a halosere wherever it occurs. That found in the River Ribble estuary in Lancashire has certain

Fig. 4.1 The Pass of Ryvoan, Cairngorms, Scotland. A lithosere of lichens, heather and Scots pine colonizing along drainage lines on a granitic scree slope at 400 m.

Fig. 4.2 Psammosere on fixed dunes along the Firth of Tay, Scotland. Marram grass (*Ammophila arenaria*) is the main species in the mid- and foreground. In the distance sea buckthorn (*Hippophaë rhamnoides*) forms a shrub stage in front of planted pines. Disturbed ground is being colonized by mosses, lichen, young marram grass, and 'opportunistic' weed species (ruderals) such as thistle and ragwort.

Fig. 4.3 Hydrosere sequence on a small lake near Edmonton, Alberta, in the Aspen Forest Belt. Floating aquatics (as a scum on the surface), rooted aquatics, rushes, sedges and reeds are all present. The climax forest forms the background.

Fig. 4.4 Halosere zonation at Warneet Creek, near Melbourne. Mangrove trees (*Avicennia officinalis*) fringe the open water (seedlings in the foreground and pneumatophores or breathing roots visible in the bare mud areas). Landward of this, a low carpet of *Suaeda maritima* gives way to dense growth of the bush *Salicornia arbuscula* (formerly *Arthrocnemum arbuscula*): both species belonging to the salt-tolerant Chenopodiaceae.

stages almost identical with those seen at Warneet Creek, south-east of
Melbourne, although the two sites are 20,000 km apart (e.g. those stages
dominated by: *Suaeda* spp (seablite), *Triglochin* spp (arrow grass) and *Pucci-
nellia* spp (sea grass); by *Juncus maritimus* (sea rush); by *Phragmites communis*
(common reed) and *Cladium* spp (twig rush)). Naturally, there are some
important differences: Victoria, enjoying a much warmer climate, has a
mangrove (*Avicennia officinalis*) as one of its pioneer species (Fig. 4.4).

The habitat modifications in a primary sere may be due to the vegetation
itself altering conditions (an *autogenic* succession). But these changes could also
be due to 'external' factors beyond the control of the vegetation. A climatic
change or a fall in sea-level may be responsible (an *allogenic* succession). But
the timescale for most successions is long and so it was thought that both types
of change feature in most seral developments (but see p. 77 for further
discussion of this point). Yarranton and Morrison describe a good example of
an autogenic succession, the psammosere of the Grand Bend sand-dune system
(Ontario) of Lake Huron. Superficially, the succession proceeds from an open
grassland to grassland dominated by juniper clumps. Finally, a more or less
closed oak-pine forest evolves. But the key role is played by the juniper
(*Juniperus virginiana*) in altering the environment so as to lead to replacement
of the colonizing species (Group I) by persistent species (Group II). In an early
stage several grass species establish beneath the juniper scrub and are then
followed by several shrubs (spp of *Prunus, Rhus,* etc.) and young oak seedlings.
The juniper acts as a centre for establishment and a nucleus for the growth
of persistent species. As this *nucleation process* occurs, the colonizers are eliminated. With the development of oaks the junipers become subordinate to them
and are also eventually eliminated.

The oak seedlings grow from acorn hoards (caches) buried by chipmunks.
Now the chipmunks are mainly resident where a humus layer is present on the
sand. The humus layer first develops under *Juniperus virginiana*. The oaks
therefore grow up at these sites because either there is a concentration of acorns
buried there or the soil conditions are better for germination under juniper
bushes (similar work has shown that on Connecticut sand plains oaks only grow
from acorn hoards buried by squirrels and acorns not so buried usually fail to
germinate). The development and distribution of oaks are therefore related to
chipmunks which in turn are related to the occurrence of juniper clumps.

If plant succession is a progressive phenomenon then we must ask where it
will lead and what form of vegetation will eventually evolve at the site.
According to Clements, this final plant community depends on the regional
climate since he regarded climate as the key factor in the environment, largely
controlling the operation of many other factors. The end-stage will be that
community which is in closest equilibrium with the climate (the *climax*
vegetation). Clements preferred the term *climatic climax*, because he believed
the community would bear a close relationship to the regional climate. We can
generalize these views along the lines proposed by Tansley, as shown in
Fig. 4.5.

Where the climate is not too extreme the climatic climax is dominated by
trees. In extreme environments it may be characterized by cryptogam veg-

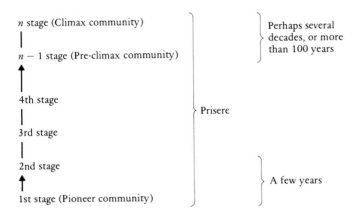

Fig. 4.5 The stages in a primary sere.

tation: the lichen and moss communities of the severe Arctic are a good example.

SECONDARY PLANT SUCCESSION

If there is no gross disturbance of a virgin site for a relatively long period then primary succession takes place. But obviously much can happen to interfere with seral developments. Man clearing forests, draining swamps, setting fire to scrub or introducing grazing animals, or relatively sudden natural disasters such as floods, volcanic eruptions or locust swarms may cause some, but perhaps not all, vegetation on the site to be destroyed. A new seral development takes place once the disturbance is reduced. Since it does not occur on a completely plant-free site this type of succession is known as a *subsere* or *secondary sere*.

Vegetation prevented from progressing to the climax community and held in a state below this end-stage is referred to as *subclimax* vegetation. Through frequent burning, and sometimes grazing, many lowland heaths in Britain, the pine-barrens of eastern USA and large tracts of Australian scrubland have this status. Indeed, there is very little true climax vegetation left anywhere in the world today.

If a disturbance occurs persistently a type of vegetation may arise which is unlike any of the expected stages in the normal prisere leading to the climax community. This disturbance (e.g. heavy grazing, frequent burning) is then said to have deflected the normal course of the prisere. The community which arises persists for as long as this factor operates simply because it is the community which is best adapted to these new environmental conditions. This deflected climax is termed the *plagioclimax* (Gr. *plagio* = oblique or deflected), or in North America the *disclimax*. It may become established by a short series of intermediate communities reflecting this progressive deflection (a *plagiosere*).

Should the factor cease to operate at some later stage then site conditions will change. The habitat may now be no longer suitable for the maintenance of plagioclimax vegetation. This vegetation then gives way to better adapted plants and reverts in a series of stages, by means of a subsere, to a type of community more normally expected in the prisere development. These trends may be expressed along the lines suggested by Tansley, as shown in Fig. 4.6.

Several simple examples of plagioclimax communities may be given. In one sense every wheatfield is a plagioclimax vegetation. It is not the natural vegetation of the area and has come about through the persistent efforts of the farmer. If these cease, the field is quickly invaded with weeds and later by shrubs and tree seedlings as the site reverts to its natural vegetation cover (Fig. 4.7).

In the Fens a stage in the hydrosere is dominated by saw sedge (*Cladium mariscus*). For centuries man stepped in at this stage and harvested saw sedge for local thatching purposes and use as litter in cattle sheds. This allowed purple moor-grass (*Molinia caerulea*) to become established as a plagioclimax, although it is not a plant featuring in the normal hydrosere stages. Recent decline in the use of saw sedge for thatching has led to some replacement of purple moor-grass as reversion to more normal stages of the hydrosere takes place.

The original altidudinal zonation of vegetation in the New Guinea Highlands was as follows:

> 900–1,800 m – oak forest (spp of *Quercus* and *Castanopsis*).
> 1,800–2,750 m – mixed forest (spp of Podocarpaceae and Lauraceae), or pure stands of southern beech (*Nothofagus*).
> 2,750–3,050 m – true montane forest (cloud forest).
> Above 3,050 m – montane shrubs and alpine grasslands.

For centuries shifting agriculture was practised on these slopes. Robbins shows that when cultivation plots are abandoned they are first invaded by tall sword-grass (*Miscanthus floridulus*). Shrubs and secondary woodland follow as the subsere develops towards the original climax forest. Where human interference is repeated too frequently the tall sword-grass stage is replaced by short-grass communities. These become stabilized as a plagioclimax. The first short-grass species appear as patches, forming a mosaic within the tall sword-grass sward. These later give way to other short-grass species (spp of *Themeda*, *Sorghum*) which gain control of the site. In the normal sere on these mountain slopes short-grass plants are not seen. They are immigrant species from adjacent lowlands. By recognizing the true status of these various seral communities those areas under greatest human pressure in the Highlands can be identified.

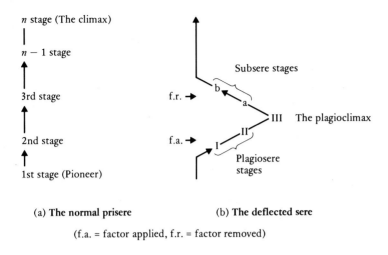

n stage (The climax)

n − 1 stage

3rd stage

2nd stage

1st stage (Pioneer)

Subsere stages

f.r. →   b

a

III   The plagioclimax

f.a. →   II

I

Plagiosere
stages

(a) **The normal prisere**   (b) **The deflected sere**

(f.a. = factor applied, f.r. = factor removed)

Fig. 4.6 The development of a plagioclimax compared with that of a normal primary sere.

Fig. 4.7 Secondary succession in an abandoned pasture, foothills of Mount Buffalo, Victoria. Bracken and *Acacia decurrens* (green wattle, a weed species) are well established and a few seedlings of the red stringy-bark (*Eucalyptus macrorrhyncha*), the tall Eucalypt in the background, are beginning to appear among the rank grasses.

MONOCLIMAX AND POLYCLIMAX THEORY

Clements regarded climate as the master factor in the plant environment. Although several types of plant succession may occur in an area (xeroseres, hydroseres, subseres) they would all tend to converge towards a final form, *the climax*. Clements postulated only one potential climax for each climatic region: the most mesophytic community which could exist under the particular climatic regime. Given a stable climate and sufficient time, initial site characteristics of soil, physiography, etc. are modified: marked differences between two neighbouring sites become less and less important. Eventually, both sites bear very similar vegetation.

Alternative views to the *monoclimax theory* of Clements soon appeared in America. At the time they received little attention due to the great impact there of his work. Nichols recognized both physiographic and climatic climaxes while Gleason argued strongly for the individualistic concept. Gleason claimed each stand of vegetation was different in some respects from any other; it was thus impractical to consider them in terms of an all-embracing monoclimax theory.

Tansley, the dominant figure in British ecology during the interwar period, accepted much of the Clementsian concept but stressed the equal importance of other factors. The *polyclimax theory* stems from such reasoning, i.e. within a climatic region several climax vegetations would develop because the climatic factor cannot wholly reduce to a subordinate level the effects of other factors. A similar view, held by a minority of ecologists, had likewise arisen in America. Subsequently, British ecologists compromised between the two theories. They regarded one climax formation in each climatic region (that found on the most widespread combination of soil and topography) as the climatic climax and others in the same area as edaphic or physiographic climaxes.

Aubréville produced data from West African tropical rain forest which conflicted strongly with a basic assumption of the monoclimax theory: climatic climax vegetation should be stable and self-perpetuating. He found examples where the present forest dominants were not producing seedlings. In fact, the seedlings in the ground flora belonged to a group of trees different from those forming the canopy. His *mosaic* or *cyclic theory* of regeneration argued that the present trees in a locality would be succeeded by a different combination of trees. The vegetation is, therefore, constant neither in space nor in time. A cyclic process is envisaged with no one permutation of trees ever forming a permanently stable community. That this may be so in some cases is clear enough and Janzen has put forward a possible mechanism in the effect of seed predators who concentrate their activities beneath the canopy of specific rain-forest trees just where the fall of seed is most dense. However, recent work has not substantiated a general rule that rainforest trees do not regenerate under their own canopy and Hubbell has concluded that a clear relationship between seed predation and the spacing of forest trees cannot always be demonstrated.

Richards, working in the primary rain forest at Moraballi Creek (Guyana), recognized five forest types: the Mora, the Morabukea, the Mixed, the Green-heart and the Wallaba. All were found under the same climatic regime as stable types that did not replace each other. A strong correlation between forest and soil type was evident. Soils varied from wet fine silt through loams to strongly leached sands. He described the Moraballi forest as a catena of climaxes, which could not be harmonized with the monoclimax view (see Fig. 2.7 for an example).

Criticism of Clementsian views has not arisen solely from tropical studies. Jones suggested a mechanism similar to the mosaic regeneration theory for Northern Temperate Zone virgin forests. Frequently, there was no regen-eration taking place under the dominant tree species. It was suggested that the Clementsian climax was like a phantom – always one step ahead of reality.

In 1953 Whittaker presented an appraisal of climax theory, assembling much evidence against both monoclimax and polyclimax interpretations. He listed 35 climax terms which have arisen over the years. 'Such a multiplicity of terms, many of them clearly exceptions to the concept as originally formulated, may imply that the concept is being stretched this way and that to cover evidence for which it is not actually adequate. If an ideal, the climax, must be so modified in application, it may be suspected that the ideal is at fault.' His criti-cism was directed mainly at the Clementsian concept. The polyclimax theory corresponded better to actual vegetation diversity but was likewise rejected by Whittaker.

In place of these two theories Whittaker advocated a third, the *climax pattern hypothesis*: a pattern of plant populations, variously related to one another, corresponding to the pattern of environmental gradients. As environmental factors change so the balance among plant populations would shift. A climax community is one which has reached a steady-state of productivity, structure and population. It is in relative equilibrium with its environment. To replace the term climatic climax, Whittaker suggested (*prevailing climax*: those dominant populations or growth-forms most evident in the climax pattern and occupying the majority of sites studied (a definable abstraction from the data thought to represent 'average' conditions in the area). Because so many factors combine to determine the nature of the population (ecological factors, species genetics, etc.) there can be no absolute climax for any area. Whittaker claimed that climax composition had meaning only relative to position along environ-mental gradients and to other factors. He regarded the series – monoclimax, polyclimax, climax pattern – as one of increasing closeness of fit to the actual pattern of vegetation (i.e. decreasing degree of abstraction).

AUTOGENIC OR ALLOGENIC SUCCESSIONS?

The classic Clementsian successional theory laid great emphasis on the community control of succession (i.e. autogenic). Allogenic processes may contribute to change but these were generally regarded as 'exceptional' events. In other words, long periods of environmental stability were envisaged during which a self-organizing community developed. Some recent presentations of succession also stress gradual, autogenic change. Margalef, for example, views it as being directed towards 'information' accumulation: the increasing diversity and structural complexity of the developing sere is seen as a trend of increasing 'information' content in the community. Odum also sees succession as largely autogenic, leading to increased stability and a high level of internal control mechanisms. Whittaker reaches similar conclusions. In essence, these workers mainly attribute community development to the evolving 'internal' properties of the community.

Drury and Nisbet have challenged what they regard as this over-emphasis on autogenic change. They do not deny autogenic change but a large number of studies have now clearly demonstrated the importance of repeated 'external' disturbances in plant dynamics (severe winds, fires, erosion). In many, but not all, North American forests external allogenic factors have regularly caused instability in what were formerly regarded as largely undisturbed sites. This disturbance is now recognized as quite normal. Heinselman, for example, has demonstrated a regular natural fire rotation of about 100 years in presettlement forests of northwestern Minnesota. Sprugel has shown that high-elevation balsam fir (*Abies balsamea*) forests in northeast America are subject to regular wind destruction. A wave-like pattern of gaps moves progressively and regularly through the forest with trees dying on the leeward side. On average, each stand of trees lasts about 60 years before another wave comes through and repeats the process. These disturbances (either as minor or major catastrophes) produce a mosaic of vegetation patches of different ages and composition. Drury and Nisbet postulate that change is not directed towards a stable end-point (the climax). They point out that gradients of exposure, soil type, moisture and other stresses exist continuously in every geographic region and these permit specialization by different species to different points on these stress gradients. At one extreme sites are occupied by stress-tolerant plants that grow and reproduce rapidly (and maybe fade rapidly to be replaced by others). In more favoured, sheltered sites the vegetation becomes dominated by plants that grow as large as possible and live as long as possible. (A view of plants that has similarities to Grime's classification, see Chapter 2.) Changing vegetation is not necessarily an orderly sequence to the climax but mainly reflects the differential behaviour of species, in terms of growth and survival, in an environment showing many spatial and temporal variations. In support of this, there is increasing evidence that many species in secondary successions were present at the outset as buried roots or seeds or arrived at the site in the first few years. That they did not 'appear' as mature plants until much later merely reflects their growth strategies and forms or that they needed some further environmental change to trigger off their germination and establishment. This contrasts with the Clementsian view that attributes successive changes in plants to successive influxes of colonizing species from outside the site. (See also Discussion section.) Some botanists now argue that in order to understand successional development in a plant community it is essential to include studies of the buried 'seed bank' already present in the soil.

GRADIENT ANALYSIS AND THE CONTINUUM

Whittaker's review of climax theory (see page 77) led him to suggest that the method for analysing natural communities should be in terms of population distribution along environmental gradients (*gradient analysis*). It recognizes the point made earlier by Gleason on the individualistic nature of the plant community. Linked with this is the view of natural plant cover as basically continuous along continuous gradients: distinct divisions do not exist unless there are pronounced environmental discontinuities. From these views evolved the concept of the *continuum* (see Fig. 4.8).

A study by Whittaker of forest vegetation in the Southern Appalachians revealed a complex continuum of populations; this was also found when Wisconsin forests and prairie were examined by Curtis and McIntosh. For each vegetation stand, Curtis and McIntosh calculated a value for their *vegetation continuum index*. This is the product of two other indices used in the Wisconsin studies: the importance value index (I.V.I.) and the climax adaption number

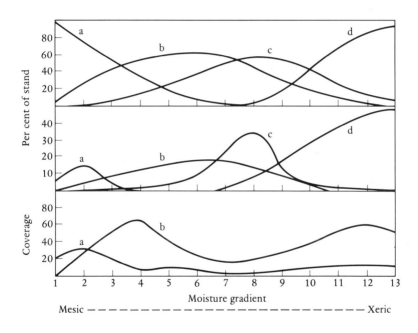

Fig. 4.8 Distribution of plant populations along the moisture gradient, 460–760 m, in the Great Smoky Mountains. The arrangement of component populations illustrates a continuum. Any subdivision into discrete communities will be arbitrary. *Top* – curves for tree classes: a, mesic; b, submesic; c, subxeric; d, xeric. *Middle* – curves for tree species: a, *Betula alleghentensis*, b, *Cornus florida*, c, *Quercus prinus*, d, *Pinus virginiana*. *Bottom* – curves for undergrowth coverages: a, herbs; b, shrubs. (From Whittaker, 1956.)

(C.A.N.). The former, with a constant value of 300 for each stand, is a measure of relative density, relative dominance (determined by measuring the basal area of the species, i.e. the ground area occupied by the stem) and relative frequency of a species in a stand. The climax adaption number is a value on the scale 1 to 10 assigned to each species to indicate its relative position along the succession from pioneer to climax status. Thus the maximum value of the vegetation continuum index for any stand is 3,000 (I.V.I. = 300, C.A.N. = 10) and the minimum is 300 (I.V.I. = 300, C.A.N. = 1).

Species A in the stand might have an I.V.I. value of 20% relative density + 30% relative dominance + 20% relative frequency, i.e. 70 out of a maximum possible of 300. Its C.A.N. value might be 5. Therefore, for species A the continuum index is 70 × 5 or 350. When this is added to the values for all other species in that stand we arrive at the stand value, which will lie between a minimum of 300 and a maximum of 3,000. Much effort is expended on these measurements in the field but once obtained the next stages are relatively simple. Vegetation stands are plotted with the continuum index as the x-axis and an environmental gradient (e.g. calcium content of the soil) as the y-axis (see Fig. 4.9). According to advocates of this method, the smooth curves of variability produced fail to demonstrate the existence of distinct communities but show the continuous nature of natural vegetation along a continuous environmental gradient.

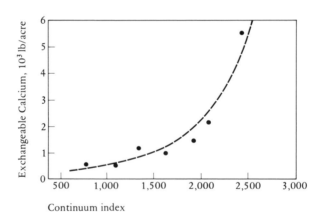

Fig. 4.9 Correlation between exchangeable calcium in the soil and the continuum index for stands of upland hardwood forest in Wisconsin. (From Curtis and McIntosh, 1951.)

THE NATURE OF NATURAL VEGETATION

The challenge to traditional concepts by Whittaker, Curtis and McIntosh and others has stimulated much further research. Their approaches produce data which can be subjected to statistical techniques and these have been increasingly applied to vegetation studies over the last three decades. Of particular importance are multivariate methods, the study of relationships between sets of interdependent variables.

Many recent methods stem from a recognition of the subjective nature of earlier work and the consequent desire for objectivity. Only a few of these methods can be outlined in an introductory text: further details may be found in Kershaw. Briefly, the statistical treatments follow one or other of two schools of thought on the nature of natural vegetation. First, those who believe natural vegetation is virtually continuous with no discrete units, i.e. any classification, if attempted, is clearly understood to be on an arbitrary basis. Secondly, those who are convinced distinct plant communities can be recognized and hence classified.

ORDINATION

The term *'ordination'* describes the treatment of stands by those who adopt the first view. Ordination results in each stand being placed in relation to one or more axes: its position relative to the axes conveys as much information as possible about its composition. The properties of each stand (i.e. the data on the species in that stand) determine its position in relation to the axes.

Ordinations are of two main types: *direct*, where the axes are based on obvious environmental gradients, and *indirect*, where the approach is based on floristics alone and the axes are mathematical constructs extracted from a matrix of similarity scores derived from comparing all the species or quadrats. The direct approach is simple in concept and execution and never poses any difficulties with interpretation of results since the environmental relationships are obvious (Figs. 4.8 and 4.9).

Of the indirect ordinations that of Bray and Curtis is known as *polar ordination*, a multivariate method of three-dimensional ordering. In outline, this method takes each sample and compares it in turn with every other sample in the analysis by means of a coefficient of similarity and a similarity score is obtained. The two most dissimilar samples are chosen as end points for the first axis of the ordination. All other samples are then located along this axis – distances apart being proportional to their similarity scores. Another axis can be obtained by selecting another pair of samples as end points, these being the next pair to emerge as least similar to each other when plotting on the first axis is complete. All samples can then be plotted according to this new framework. By repeating the procedure a third axis is added to give a three-dimensional set of coordinates. In theory, many more than three axes could be created but the coordinate framework resulting then becomes impossible to display graphically and lies beyond visualization in what is known as hyperspace. However, in many cases a large proportion of the variation in the data can be expressed by reference to a few axes.

To overcome the problem of many axes, several elaborate multivariate tech-

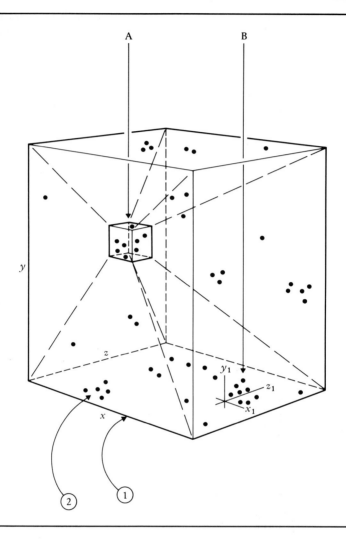

Fig. 4.10 Geometrical model of a vegetation space containing 52 records (stands). 1 – 3-dimensional abstract vegetation space: each dimension represents an element (e.g. proportion of a certain species) in the analysis (x, y, z axes) 2 – A cluster within the cloud of points (stands) occupying vegetation space. A – the results of a classification approach (here attempted after ordination) in which similar individuals are grouped and considered as a single cell or unit. B – the results of an ordination approach in which similar stands nevertheless retain their unique properties and thus no information is lost (x_1, y_1, z_1 axes). (N.B. Abstract space has no connection with real space from which the records were initially collected.)

niques have been developed, a main one being *principal components analysis*. In essence, this method mathematically reduces the many original coordinates to new, transformed variables (the principal components) so that two or three new axes account for a large proportion of the variance in the data. The new framework, within which the samples can be plotted relative to each other, is created by replacing a large number of correlated variables by a few significant uncorrelated variables (the principal components). The calculations for ordination methods are performed by computer programs.

In general, either the attributes (species) of each individual (vegetation stand) or the distribution of each species throughout the stands may be studied in ordination. Relationships can be expressed in geometric form: the individuals are visualized as points in an abstract space or geometric model and the coordinates determining their position in relation to the axes are obtained from their attribute values. Fig. 4.10 presents a model of an abstract vegetation space in which each axis represents one element of the analysis (e.g. the scale of values recorded for some property of species P – perhaps its proportional representation – in all the stands).

In direct ordinations and in polar ordination interpretation is seldom a problem and the display of results makes ecological sense. In the more advanced mathematical methods, such as principal components analysis, the final axes are based purely on mathematical considerations and may be ecologically difficult to interpret or uninterpretable. Beals has concluded that principal components analysis makes unreal assumptions about ecological data and he argues strongly that the less sophisticated Bray and Curtis approach gives better results. Whittaker has also compared several methods and reaches similar conclusions. But Hill points out that there are field situations where environment gradients are not obvious and then the more floristic-based and mathematical indirect approaches are the only types of ordination possible. His approach to this is the method called 'reciprocal averaging' which includes some features of Whittaker's gradient analysis and has a similar rationale. It also has similarities with principal components analysis but without some of its problems of interpretation.

Some consider ordination is a better initial approach to vegetation studies than classification. They argue that many properties of the stands can be utilized to produce the final ordering whereas most classifications are based on a single criterion or just a few. The latter method is therefore characterized by more information loss than the former. Daubenmire, however, criticizes the continuum concept and the mathematical manipulation of data which often accompanies its use. Lambert and Dale also refute some of the claims of ordination and advance a case for the superiority of classificatory methods.

CLASSIFICATION

THE NODUM

Vegetation classifications may follow the more usual hierarchical form or they may be non-hierarchical. An example of a non-hierarchical approach is the method developed by Poore and used extensively by McVean and Ratcliffe.

In their work, plant communities in the Scottish Highlands were regarded as forming a continuum. They observed a gradual spatial change in the vegetation and even where fairly sharp discontinuities occurred locally, they could usually find a complete series of intermediates to link any two related types, if a wider area was examined. Their essential problem was to select vegetational 'reference points', sufficient in kind and number for an adequate representation of the total range of variation found in the vegetation. Poore expressed this concept of reference points in a field of more or less continuous variation by the term *nodum*, an abstract vegetation unit of any category (see Fig. 4.11).

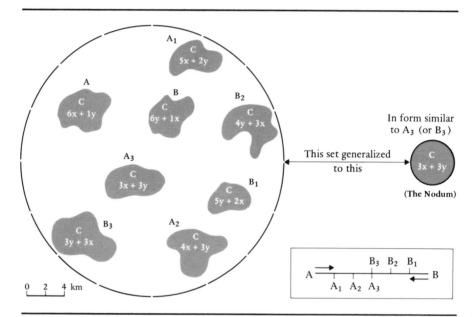

Fig. 4.11 A greatly simplified presentation of the Nodum concept (see text). Eight woodland stands have been examined in the field. Each has oak as a common element (C). Variations occur between stands in terms of the presence or absence of six ground flora species (all denoted as X) and six shrub species (all denoted as Y). The first stands to be examined, A and B, were quite unlike each other and traditionally would be classified as distinct communities. Examination of a wider area, however, shows a series of 'linking' communities (A_1, A_2, A_3 and B_1, B_2, B_3) and $A_3 = B_3$. The woodland types change gradually from A to B, i.e. a continuum situation exists.

Two concepts of vegetation are involved in this method: the 'real' vegetation (i.e. particular stands analysed in the field) and the nodum (i.e. an abstract definition stemming from the common features of a related group of these stands). Watt describes the method by using the analogy of the net where the knots represent the noda or points of reference selected from the field data. The connecting stands are the transitional vegetation between noda, which are interpreted in terms of the noda. The idea of an abstract definition stemming from the common features of a related group of objects is not new, of course. It is used frequently as, for example, when one speaks of the average

Englishman. If we examine many various alpine moss communities and erect a scheme of several noda to represent the variation observed, the set of ecological conditions related to each nodum can then be sought. In the field we may later encounter other moss stands which do not quite fit any of our existing reference groups. To cover this group it may be necessary to erect another nodum. We would ask why is this group different and what is causing these observable differences?

Although the vegetation is regarded as continuous it is classified. But no nodum is considered to be more important than any other (i.e. a non-hierarchical system). It is not strictly an objective method since the noda are subjectively determined and are solely defined by reference to floristic composition of the stands. In Fig. 4.11 only eight stands are shown and only presence or absence of species has been considered, which is a gross simplification. Many stands, many species and several properties of each species may be involved in an actual field analysis.

ASSOCIATION–ANALYSIS

In the late 1950s and early 1960s a group of botanists (Williams, Lambert and Dale), originally centred on Southampton University, developed a series of methods for the hierarchical classification of vegetation. These have progressively become more statistically complex. However, their early approaches form the basis for some of the later methods. A presentation of this early material in outline form is appropriate for this introductory text.

The aim of their first method, *normal association–analysis*, is to subdivide a plant population so that all association disappears. The basic raw data are the presence or absence of species in a relatively large number of randomly located quadrats used to sample the vegetation. For each species we record the number of quadrats in which it occurs and the number in which it is absent. These values are expressed as frequencies. The next stage compares each species with every other species in turn and measures the extent to which these pairings of species are associated. The measure used is the χ^2 (chi squared) statistic (details may be found in any elementary statistics textbook). The χ^2 values obtained for each species are then added together to give a total χ^2 value for that species.

The species with the highest χ^2 value (highest degree of association) is used as the basic dividing point, i.e. the whole population of stands (quadrats) is subdivided into two broad groups – those in which this species occurs and those without it. The two groups of stands so formed are next considered and the procedure repeated for each one, i.e. within each group we find the species which now has the highest degree of association. When this is determined it forms the basis for further subdivision. The procedure is repeated until adequate subdivision of the whole population has been made (Fig. 4.12). Interesting subdivisions occur early in the procedure and these call for ecological interpretations. The tedious calculations involved in the method are best handled with the aid of a computer program.

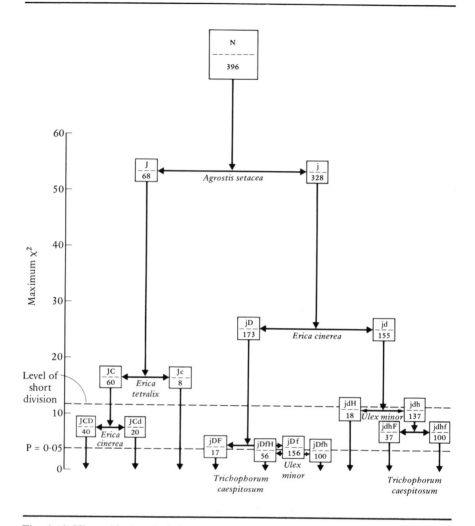

Fig. 4.12 Hierarchical analysis in a community (Matley Ridge, New Forest) containing ten species: normal association – analysis of data from 396 quadrats. Species presence shown by capital letters, absence by small letters. (From Williams and Lambert, 1960.)

Inverse association–analysis is another method developed by the 'Southampton botanists'. In principle it is very similar to normal association–analysis and proceeds by similar stages. Subsequently, both these methods were combined to produce a third method, *nodal analysis* (which should not be confused with the noda described by Poore).

Fig. 4.13 Markway, an area of New Forest heathland where heath slopes down to valley bog and different portions have been subjected to different management treatments by the Forestry Commission. A: the normal information-analysis; B: map of the area; C: map of final groupings from the information-analysis. (Modified from Lambert and Williams, 1966.)

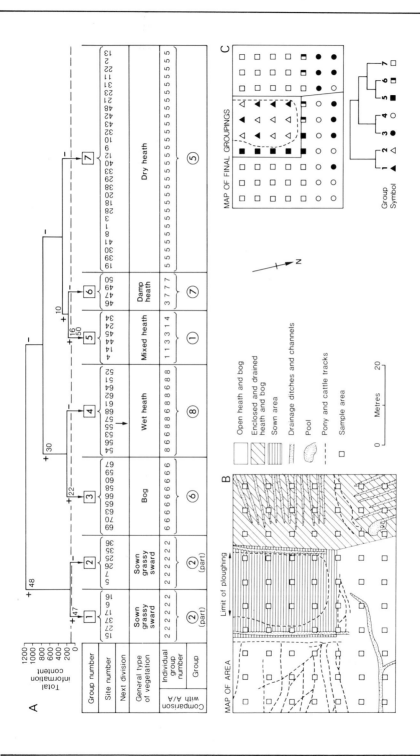

Association–analysis and its derivatives are divisive methods. In contrast, Williams, Lambert and Lance have developed agglomerative methods based on several criteria. An early approach along these lines is *information–analysis* in which single stands or quadrats are fused into larger and larger groups based on the overall similarity of the groups. If we consider two groups of quadrats then they are said to contain 'information', this being a reflection or measure of the differences in species composition between these two groups. The greater the similarity in composition the smaller will be the increment of 'information' gained when the two groups are joined together. Two identical quadrats have no 'information'. The similarity of the groups is measured by the information statistic, I. The method begins by combining together those single quadrats that result in the least increase in total information content, i.e. those that are most similar. The procedure is repeated with the fused groups that have resulted. They, too, are then compared with each other in terms of the information statistic and further fusions result until all the quadrats have been joined in a hierarchy to form the whole population (Fig. 4.13). Full explanations of these more advanced methods are beyond the scope of this text and the beginner should concentrate on the views about the nature of vegetation implicit in these methods rather than the technical details.

ORDINATION OR CLASSIFICATION

As these two sets of methods developed there grew up considerable controversy as to which was the most appropriate for vegetation studies. More recently, workers have realized that advocacy of the continuum concept does not necessarily preclude a classification of vegetation (as Poore's method demonstrated). Both classification and ordination are 'structuring' techniques, designed to obtain a structure simpler than that of the original raw data. It is now clear that the two techniques need not be mutually exclusive: classified units can be ordinated and ordination units subsequently classified. Gittins, for example, has compared ordination and association–analysis techniques for the same data from a limestone grassland in North Wales. He finds stand ordination and normal association–analysis lead to broadly identical interpretation of the data, but quadrat size and species abundance apparently affect the sensitivity of inverse association–analysis.

Anderson argues that the ultimate structural complexity of a vegetation stand depends on the part played respectively by probability (chance phenomena, such as the initial availability of species) and by selective mechanisms. One of the most important selective mechanisms is human interference with the plant cover. Where this has been pronounced, as in Europe, distinct communities are likely to exist. If such mechanisms have played a minor role, as in the more recently settled North America, the vegetation is much less clearly differentiated. In the first type of area vegetation lends itself to classification, in the second type to ordination. Thus, according to Anderson, either analytical approach may be appropriate.

DISCUSSION SECTION

Are there any types of plant succession not covered by the theories of Clements?

Clements described successions essentially linear in form but cyclic successions have since been shown to be reasonably common. Several examples should make this clear. On windswept ridges in the Scottish Highlands at about 800 m, Watt noticed that the first plant to colonize bare, mineral grit was the low-growing *Arctostaphylos uva-ursi* (bearberry). Next *Calluna vulgaris* (heather) invades these small patches, growing up over the mat of bearberry. The direction of growth is largely controlled by the strong prevailing winds. Finally, the lichen, *Cladonia rangifera* (reindeer moss), establishes itself in the patch. But by this time the other two species have begun to die off. When the lichen dies the whole cycle may start again. So we have a succession of the form:

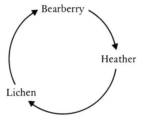

The cycle exhibits an upgrade stage, as plants become established, and a downgrade stage, as plants decline (see Fig. 4.14).

Cyclic succession can also be caused by biotic factors. Between 500–1,000 m on the basalt cone mountain of Pico Island in the Azores, Marler and Boatman described a cyclic sere related to the intensity of grazing of animals introduced by man. The biotic climax vegetation of these slopes is a mixed woodland and the main stages in the sere are outlined in Fig. 4.15. Yeaton has postulated a natural cyclic sere for a plant community in the Big Bend National Park, Texas, a northern extension of the Chihuahuan Desert where the rainfall is *c.* 250 mm and concentrated in July and August. The sere involves two main plants and two associated groups of animals: details may be followed in Figs. 4.16 and 4.17.

It should also be noted that in some extreme habitats (e.g. severe parts of the Arctic or hot deserts) only a few species are capable of growth. Here the pioneer community is also the climax, so no succession takes place.

I am not sure how many of Clements' ideas on plant dynamics are still held.

The main controversy has centred on his monoclimax views. This theory has been largely rejected now. The long periods of environmental stability required by monoclimax theory seldom occur. Even in the tropics, long thought to be areas of major stability, there is now mounting evidence of a history of natural

Fig. 4.14 A cyclic succession similar to that described by Watt. A wind-exposed surface at 1,800 m on Mount Bogong, Victoria. *Epacris petrophila* (under the rucksack) colonizes the stony alpine soils and *Poa australis* (alpine tussock grass), together with *Celmisia longifolia* (alpine daisy, beyond the rucksack), then grow in stone-free areas where an organo-mineral soil accumulates. Erosion of these fine textured soils leads to the reappearance of *Epacris*, as in the right foreground.

disturbance and instability, recently reviewed by Street. During the last glaciations of the high latitudes the tropical rainforest was apparently restricted to much smaller areas or refugia. Flenley states 'the evidence of abundant change in the equatorial vegetation is one more nail in the coffin of the climax theory'. Garwood *et al.* have also shown that earthquake-caused landslides can maintain considerable portions of some tropical rainforests in successional stages. Many of these points are well covered in the review by White.

More workers accept the basic ideas behind linear successions but even here some evidence is at variance. In 1970 Walker studied hydrosere successions around the edges of 66 British lakes by examining the sequences of plants preserved in the infilling sediments. These should provide a good record of the hydrosere succession. According to succession theory, the communities should appear in a predictable order at each site. This was not so. There was no orderly sequence and no way of predicting the community type which would next follow the vegetation already growing at any site. Walker concluded that the infill in most ponds was material brought in from outside, by erosion and run-off, and not deposited organic matter from the plants themselves. The climax for these sites appeared to be not the woodland growing on adjacent drier terrains but peat bog. This was another unexpected finding.

Colinvaux has presented a detailed examination of many ideas associated with succession theory. Among these are the following:

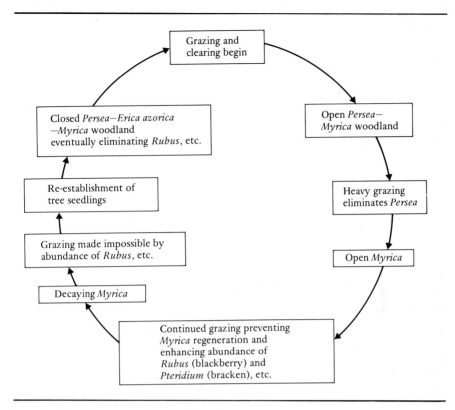

Fig. 4.15 Cyclic sere on Mount Pico (500–1,000 m), the Azores, in which animal grazing plays a key role. *Persia azorica* is a small tree occurring in clumps with *Erica azorica*, a heath. *Myrica faya* is a large, well-developed tree. (Modified from Marler and Boatman, 1952.)

1. A plant succession is an organizational process leading to a climax which has self-organizing properties, i.e. almost a superorganism in Clements' view.
2. The climax community has greatest efficiency of energy conversion.
3. All successions show increasingly efficient cycling of nutrients.
4. Climax communities show the most complex physical structures (physiognomy) and maximum biomass (total organic matter)

He concludes that we must now either reject or severely qualify many of these views.

   In support of this conclusion some successions may be *retrogressive* rather than progressive. In the Borneo swamps studied by Anderson, thick peat accumulates in the later successional stages. Nutrients are leached away and plant roots lose contact with the mineral soil. As a result, the vegetation becomes scrubby, structure is reduced and the canopy cover is less complete. The detailed studies of succession on the Storbreen glacier foreland in Norway by Matthews clearly demonstrate that successions may also be *divergent* and

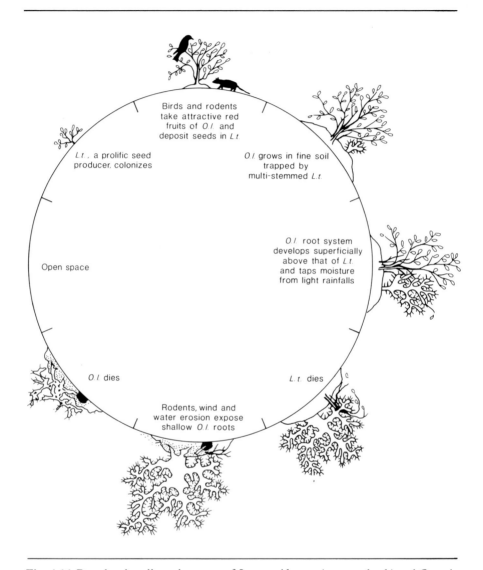

Fig. 4.16 Postulated cyclic replacement of *Larrea tridentata* (creosote bush) and *Opuntia leptocaulis* (Christmas tree cholla, a cactus) in the northern Chihuahuan Desert, Texas. (Modified from Yeaton, 1978.)

these results do not support the commonly-accepted Clementsian hypothesis that successions tend to converge. Horn has summed up a current view by saying that 'the only sweeping generalization that can safely be made about succession is that it shows a bewildering variety of patterns'. This conclusion is echoed by Whitmore: 'unambiguous generalizations about succession are difficult to make'.

Fig. 4.17 Cyclic replacement of *Larrea tridentata* and *Opuntia leptocaulis* (see Fig. 4.16): the stage when *Opuntia* has taken over from *Larrea* (now dead) but is now itself succumbing to rodent burrowing and soil removal to expose rooting system. (Photograph kindly supplied by Dr R. I. Yeaton.)

The clear ideas of Clements seem to have been replaced by many uncertainties. Are we now any nearer to some broad concepts which apply to all communities and their development?

Active debates and controversies are a sign of a vigorous science. Science progresses by someone being wrong and someone challenging accepted views until he, in turn, is shown to be wrong. This is certainly true of ecology. Developments here over the last few decades have much in common with those in geomorphology with which you may be more familiar. Consider the following parallels:

1. (a) The researches of F. E. Clements and W. M. Davis span approximately the same time period.
(b) Both developed their theories in the temperate zone of North America and were instrumental in establishing their respective fields of study as disciplines.
(c) Each came to regard his concept as representing the 'normal' situation despite recent glacial devastation in this region and the probable imbalance of vegetation and landscapes with present climatic conditions.
(d) In their theories each envisaged an 'end-form' to development (the climatic climax and the peneplain) though these differed greatly: the first characterized by maximum complexity in the community, the second representing a state of increasing uniformity and simplicity in the landscape.

(e) The phrase 'structure, process and stage', the crux of the Davisian view of landscape, reasonably epitomizes Clements' approach to vegetation studies.

2. Important early criticism came from researchers outside the 'classic' regions. Studies in tropical and semi-arid zones figured prominently in the early rise of alternative views. In Britain, Tansley accepted the Clementsian concept but introduced important modifications: the Davisian views of Wooldridge had great influence on the early course of British geomorphology.

3. In vegetation analysis and geomorphology an increasing number of statistical methods have recently been applied. Much of the traditional controversy (e.g. monoclimax v. polyclimax; Davis v. Penck) has been replaced. The 'new' geomorphology is a general systems approach. The rise of the 'new' ecology, centred on the ecosystem, a type of general system, is another, close parallel.

It is the ecosystem concept which appears to offer the best prospect for a body of general theories applicable to all communities. We shall examine this concept in the next chapter.

In view of the uncertainties expressed about the nature of natural vegetation, how should we regard atlas maps showing global distributions of vegetation?

They are gross simplifications and must be recognized as such. They are liable to give wrong impressions. Little climax vegetation now remains, most is subclimax or plagioclimax. Man is altering vegetation so quickly it is debatable whether any worldwide classification, which would be subject to constant adjustment, is worth the effort. Even within a broadly uniform climatic belt we should expect marked vegetational differences reflecting soil and topographical influence.

Some atlas maps which purport to show the relationship between climatic regions and broad vegetation types (known in the Clementsian terminology as *plant formations*) are based on a circular argument. The vegetation is used to define the climate and the climate is used to fix the boundaries of the formation. The correlation is shown to be more exact than it really is and this can lead to errors of crude determinism. Frequently we have little climatic data and few detailed vegetation studies for assessing accurately the degree of agreement between the two elements.

REFERENCES

Anderson, D. J., 1965. 'Classification and ordination in vegetation science: controversy over a non-existent problem?', *J. Ecol.*, **53**, 521–6.

Anderson, J. A. R., 1964. 'The structure and development of the peat swamps of Sarawak and Brunei', *J. Trop. Geog.*, **18**, 7–16.

Aubréville, A., 1938. 'La forêt coloniale: les forêts de l'Afrique occidentale française', *Ann. Acad. Sci. Colon., Paris*, **9**, 1–245.

Beals, E. W., 1973. 'Ordination: mathematical elegance and ecological naïveté', *J. Ecol.*, **61**, 23–35.

Bray, J. R. and Curtis, J. T., 1957. 'An ordination of the upland forest communities of southern Wisconsin', *Ecol. Monogr.*, **27**, 325–49.

Clements, F. E., 1916. 'Plant succession: an analysis of the development of vegetation', *Carnegie Inst. Wash. Publ.*, **242**, 1–512.

Clements, F. E., 1936. 'Nature and structure of the climax', *J. Ecol.*, **24**, 252–84.

Colinvaux, P. A., 1973. *Introduction to Ecology*, Wiley, New York (see Chapter 40).

Cowles, H. C., 1899. 'The ecological relations of the vegetation of the sand dunes of Lake Michigan', *Bot. Gaz.*, **27**, 95–117, 167–202, 281–308, 361–91.

Curtis, J. T., 1955. 'A prairie continuum in Wisconsin', *Ecol.*, **36**, 558–66.

Curtis, J. T. and McIntosh, R. P., 1951. 'An upland forest continuum in the prairie-forest border region of Wisconsin', *Ecol.*, **32**, 476–96.

Daubenmire, R. F., 1960. 'Some major problems in vegetation classification', *Silva fenn.*, **105**, 24.

Drury, W. H. and Nisbet, I. C. T., 1973. 'Succession', *J. Arnold Arboretum*, **54**, 331–68.

Flenley, J. R., 1979. *The Equatorial Rain Forest: a Geological History*, Butterworth, London.

Garwood, N. C., Janos, D. P. and Brokaw, N., 1979. 'Earthquake-caused landslides: a major disturbance to tropical forests', *Science*, **205**, 997–9.

Gittins, R., 1965. 'Multivariate approaches to a limestone grassland community. I. A stand ordination. II. A direct species ordination. III. A comparative study of ordination and association-analysis', *J. Ecol.*, **53**, 385–426.

Heinselman, M. L., 1973. 'Fire in the virgin forests of the Boundary Waters Canoe Area, Minnesota', *Quat. Res.*, **3**, 329–82.

Hill, M. O., 1973. 'Reciprocal averaging: an eigenvector method of ordination', *J. Ecol.*, **61**, 237–49.

Horn, H. S., 1981. 'Succession', in *Theoretical Ecology. Principles and Applications*, (2nd edn.) (ed. May, R. M.), Blackwell, Oxford and Boston, pp. 253–71.

Hubbell, S. P., 1980. 'Seed predation and the coexistence of tree species in tropical forests', *Oikos*, **35**, 214–29.

Janzen, D. H., 1971. 'Seed predation by animals', *Ann. Rev. Ecol. and Systematics*, **2**, 465–92.

Jones, E. W., 1945. 'The structure and reproduction of the virgin forests of the North Temperate Zone', *New Phytol.*, **44**, 130–48.

Kershaw, K. A., 1973. *Quantitative and Dynamic Plant Ecology*, Arnold, London.

Lambert, L. M. and Dale, M. B., 1964. 'The uses of statistics in phytosociology', *Adv. ecol. Res.*, **2**, 59–99.

Lambert, J. M. and Williams, W. T., 1962. 'Multivariate methods in plant ecology. IV. Nodal analysis', *J. Ecol.*, **50**, 775–802.

Lambert, J. M. and Williams, W. T., 1966. 'Multivariate methods in plant ecology; VI. Comparison of Information analysis and Association Analysis', *J. Ecol.*, **54**, 635–65.

Margalef, R., 1968. *Perspectives in Ecological Theory*, University of Chicago Press, Chicago.

Marler, P. and Boatman, D. J., 1952. 'An analysis of the vegetation of the northern slopes of Pico – the Azores', *J. Ecol.*, **40**, 143–55.

Matthews, J. A., 1979. 'A study of the variability of some successional and climax plant assemblage-types using multiple discriminant analysis', *J. Ecol.*, **67**, 255–71.

McIntosh, R. P., 1967. 'The continuum concept of vegetation', *Bot. Rev.*, **33**, 130–87.

McVean, D. N. and Ratcliffe, D. A., 1962. *Plant communities of the Scottish Highlands*, Monographs of the Nature Conservancy No. 1, London.

Odum, E. P., 1969. 'The strategy of ecosystem development', *Science*, **164**, 262–70.

Poore, M. E. D., 1955. 'The use of phytosociological methods in ecological investigations. II Practical issues involved in an attempt to apply the Braun-Blanquet system. III Practical applications', *J. Ecol.*, **43**, 606–51.

Richards, P. W., 1952. *The Tropical Rain Forest*, CUP, Cambridge.

Robbins, R. G., 1963. 'Correlation of plant patterns and population migration into the Australian New Guinea Highlands', in *Plants and the migration of Pacific Peoples: a symposium* (ed. Barrau, J.), Tenth Pacific Science Congress, 1961, Bishop Museum Press, Honolulu.

Sprugel, D. G., 1976. 'Dynamic structure of wave–regenerated *Abies balsamea* forests in the North-Eastern United States', *J. Ecol.*, **64**, 889–911.

Street, F. A., 1981. 'Tropical palaeoenvironments', *Progress in Physical Geography*, **5** (2), 157–85.

Tansley, A. G., 1949. *The British Isles and their vegetation*, Vol. 1, CUP, Cambridge.

Walker, D., 1970. 'Direction and rate in some British Postglacial hydroseres', in *The Vegetational History of the British Isles* (eds Walker, D. and West, R.), CUP, Cambridge, 177–239.

Watt, A. S., 1947. 'Pattern and process in the plant community', *J. Ecol.*, **35**, 1–22.

Watt, A. S., 1963. 'A review of "Plant communities of the Scottish Highlands"', *Geogr. J.*, **129**, 205.

White, P. S., 1979. 'Pattern process and natural disturbance in vegetation', *Bot. Rev.*, **45**, 229–99.

Whitmore, T. C., 1982. 'On pattern and process in forests', in *The Plant Community as a Working Mechanism* (ed. Newman, E. I.), Special Publication No. 1, British Ecological Society, Blackwell, Oxford and Boston, pp. 45–59.

Whittaker, R. H., 1953. 'A consideration of climax theory: the climax as a population and pattern', *Ecol. Mongr.*, **23**, 41–78.

Whittaker, R. H., 1956. 'The vegetation of the Great Smoky Mountains', *Ecol. Mongr.*, **26**, 1–80.

Whittaker, R. H., 1967. 'Gradient analysis of vegetation', *Biol. Rev.*, **42**, 207–64.

Whittaker, R. H., 1975. *Communities and Ecosystems* (2nd edn.), Macmillan, New York.

Williams, W. T. and Lambert, J. M. 1959. 'Multivariate methods in plant ecology. I. Association-analysis in plant communities', *J. Ecol.*, **47**, 83–101.

Williams, W. T. and Lambert, J. M., 1960. 'Multivariate methods in plant ecology. II. The use of an electronic digital computer for association-analysis', *J. Ecol.*, **48**, 689–710.

Williams, W. T. and Lambert, J. M., 1961. 'Multivariate methods in plant ecology. III. Inverse association-analysis', *J. Ecol.*, **49**, 717–29.

Williams, W. T., Lambert, J. M. and Lance, G. N., 1966. 'Multivariate methods in plant ecology. V. Similarity analyses and information analysis', *J. Ecol.*, **54**, 427–45.

Yarranton, G. A. and Morrison, R. G., 1974. 'Spatial dynamics of a primary succession: nucleation', *J. Ecol.*, **62**, 417–28.

Yeaton, R. I., 1978. 'A cyclic relationship between *Larrea tridentata* and *Opuntia leptocaulis* in the northern Chihuahuan Desert', *J. Ecol.*, **66**, 651–6.

5

ECOSYSTEMS

INTRODUCTION

Previous chapters have provided a glimpse into the complexity of the biosphere. As study progresses many apparently simple questions about plants, animals or the soil become exceedingly difficult to answer. Ecologists have sought concepts capable of handling the degree of complexity they observe in the field. They have long recognized the *holocoenotic* nature of the environment. But, as in all sciences, much of the twentieth century has seen the rise of specialization within ecology and the consequent study in detail of minute sections of the biosphere. In recent decades, however, an attempt has been made to study ecology within a single conceptual framework. This framework is provided by the *ecosystem concept.*

Hall and Fagen define a *system* as a set of objects together with the relationships between those objects and between the attributes or properties of the objects. The objects are the components of the system and relationships hold the system together. A central heating system satisfies this definition. It consists of a set of objects (boiler, pipes, radiators) and relationships between these objects (perhaps expressed as movements or changes in a heating medium such as oil or water), each object having certain attributes (boiler capacity, pipe diameter, radiator surface area, etc.).

Isolated systems are those in which no matter or energy is exchanged between system and environment. A closed system is open for energy exchange but closed to matter exchange. In an open system both matter and energy are exchanged. Most organic systems are therefore open systems, exchanging materials and energy with their environment. Ecosystems are simply ecological systems of this open form.

The ecosystem concept, variously described, appeared under different names in nineteenth-century scientific literature. But the term itself is usually attributed to Tansley in 1935: 'a particular category of physical systems, consisting of organisms and inorganic components in a relatively stable equilibrium, open and of various kinds and sizes'. Fosberg provides a more recent definition: 'an ecosystem is a functioning, interacting system composed of one or more living organisms and their effective environment, both physical and biological'. (Perhaps 'chemical' should be added to this definition to make it absolutely clear.) The concept can be applied at any scale: a small pond or the whole latitudinal range of the Tropical Rain Forest or the entire earth may be viewed as an ecosystem. The main value of the ecosystem approach is as an integrating concept, linking together plant and animal communities and the soil as components of one system. It thus stresses the holocoenotic view of the environment.

A distinction is usually made between natural and man-made ecosystems and this approach is followed here. However, some ecologists, van der Maarel for example, argue that this distinction is not always valid in the inhabited world where man's influence has long since pervaded most natural ecosystems.

THE STRUCTURE OF ECOSYSTEMS

In all ecosystems the living organisms interact with the non-living substances (inorganic and dead organic) to produce an exchange of materials. In functional terms the ecosystem has two biotic components. Odum describes these as:

1. The *autotrophic* (self-feeding) component, which is concerned with the fixation of light energy from the sun and the use of simple inorganic materials obtained mainly from the soil (lichens are an exception because they may obtain most of their inorganic materials directly from the atmosphere, dissolved in rainfall). This energy source and material source are used to manufacture complex organic substances in the cells of the autotrophs.
2. The *heterotrophic* (other-feeding) component, which is essentially concerned with using, rearranging and decomposing complex organic substances made available by the autotrophic component.

Ecosystems may also be viewed in terms of four basic elements which they all possess:

(a) *Abiotic* substances, the basic inorganic and dead organic compounds of the habitat or site. These are concentrated in the soil.
(b) The *producers*, which are mainly autotrophic green plants. They stand as intermediaries between the inorganic and organic worlds by virtue of their ability to carry out photosynthesis, mainly in their green leaves, and to absorb inorganic materials through their rooting system.
(c) The *consumers* or heterotrophic organisms. These are mainly animals, including man, which have an intake of organic material as food which is provided in the first instance by autotrophs.
(d) The *decomposers*, which are mainly micro-organisms. Although they consume as part of the process of decomposition, their main contribution is to release from complex organic substances simple ones for re-use by the

producers. Most of these heterotrophic organisms occur as the soil micro-flora and microfauna (bacteria, fungi, nematodes, etc.). The mesofauna and macrofauna of the soil are also important in this respect (insect larvae, millipedes, earthworms, etc.).

Odum declares the ecosystem to be the basic functional unit in ecology since it includes both the organic and inorganic parts of the biosphere: these influence each other and together provide the conditions necessary for the maintenance of life.

ENERGY PATTERNS

An important way of examining the ecosystem is in terms of its energy-flow pattern. For most ecosystems there is an input of energy to the system in the form of solar radiation. In some ecosystems, however, there can be another external energy input in the form of organisms (living, dead or decomposing) brought in from adjacent ecosystems, e.g. organic material washed into a pond or debris carried by tides into an estuary. Only a small proportion of the available radiant energy is fixed by green plants. Energy can exist in many forms and is defined as the ability to do work. Of the radiant energy absorbed by the green plant, most is transformed into heat energy and eventually lost from the plant and from the ecosystem. It is dispersed in time to the atmosphere. A very small proportion of radiant energy is transformed into potential or food energy and retained in the plant tissue, 'locked in' with the inorganic elements absorbed by the roots to form new protoplasm. The potential chemical energy of these organic tissues then becomes the source of energy for plant-eating animals (*herbivores*). These organisms, in turn, create new tissues as they grow and these provide energy for meat-eating animals (*carnivores*). Animals which eat either plants or other animals (*omnivores*) thus have two sources of energy. Man is omnivorous and has also developed other sources of energy to do his work (e.g. fossil fuels, atomic energy, hydroelectrical energy).

The energy pattern of an ecosystem obeys the *first and second laws of thermodynamics*. The first law is that energy is not created or destroyed but it may be transformed from one type into another. This happens when, for example, the short-wave radiant energy of the sun striking a car roof is partly changed to heat energy as the roof warms up. This heat energy will itself change as the roof cools. Then most of it will become long-wave radiant energy and be transferred to the atmosphere above. This law means that for the ecosystem as a whole energy inflow to the system will be balanced by energy outflow.

The second law of thermodynamics may be stated as follows: at each energy transfer point within the system there is a dispersion of energy into an unavailable form. This dispersed energy is in the form of heat given up to the atmosphere. It cannot be directly utilized again by any of the organisms present. Each component in the ecosystem (green plant, herbivore, carnivore or decom-

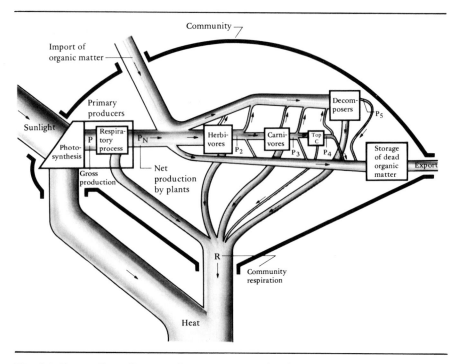

Fig. 5.1 Energy-flow diagram of an ecosystem showing successive fixation at each trophic level and transfer by components. Community respiration is the sum total of all respiratory losses at each transfer. P = gross primary production, P_N = net production and P_2, P_3, P_4 and P_5 = secondary production at the trophic level indicated: these terms are defined further on page 116. (From Odum 1959, after H. T. Odum.)

poser) uses part of its energy intake to carry out functions such as growing, reproducing, searching for grass, chasing prey. Such work requires energy which is released by the process of respiration (the chemical breakdown of food in the body). Respiration releases heat which is then generally dispersed. The sum total of this heat released by the whole ecosystem is the *community respiration*.

These relationships are shown in Fig. 5.1 where the progressive dispersal of energy through the ecosystem is represented by a narrowing of the energy flow bands along the food chain. This pattern illustrates the operation of the second law. By the time the top carnivore in the system (often man) is reached only a minute proportion of the energy originally fixed by the green plants is then available. Each energy transfer point in the system is usually referred to as a *trophic level*.

The number of trophic levels in a community is limited to four or five at most; the number depends upon the availability and stability of resources and the efficiency of their exploitation at each trophic level. The green plants represent the first trophic level, the herbivores the second level, the carnivores the third level and carnivores which eat other carnivores constitute a fourth level (the top or secondary carnivores). A food chain such as this is known as the *predator chain*. Another type is the *parasitic chain* where smaller organisms feed on larger. In the *saprophytic chain* microorganisms feed directly on dead organic material. Packham and Harding refer to this dead organic material in woodland ecosystems as the *necromass* and provide a full account of the numerous organisms (*detritivores*) active in its breakdown. The decomposers often represent a major energy pathway within the ecosystem, as Phillipson demonstrates for the Serengeti Plains. Here, no less than 68 per cent of the net primary production of 300 $g/m^2/year$ is broken down by decomposers.

In most ecosystems several types of food chain interlock to form complex food webs. Organisms may then spend time in several trophic levels, perhaps at different stages in their life cycle being, for example, a herbivorous juvenile that develops into a carnivorous adult or vice versa. In these trophic relationships it should not be assumed that only the heterotrophs benefit. There is increasing evidence that herbivore grazing does not always lead to a decline in plant yield. Owen and Wiegert speak of a 'mutualistic' relationship between plants and herbivores, where compensating mechanisms sometimes operate and moderate grazing actually promotes plant growth. Dyer's findings that the saliva of some grazing animals contains a chemical which stimulates plant growth support these ideas. It is also now well established that some plant seeds need to pass through the digestive tract of herbivores before they will successfully germinate. Temple provides an interesting example of this obligate mutualism involving the extinct Dodo!

The operation of the second law of thermodynamics means there is a greater amount of energy for conversion into *biomass* (total organic production measured as the living weight on the site) or available to perform work (dissipated by respiration) for an organism in a community where the food chain is relatively short. The ratio of gross photosynthesis to community respiration (the *P/R ratio*) is high in a pioneer ecosystem, but as succession proceeds towards the climax community it becomes low, approaching a value of unity. While it is high, biomass must accumulate in the ecosystem.

Figure 5.2 shows the food-web pattern for a coral reef ecosystem in the Marshal Islands. Here the primary producers are the plankton. Five trophic levels are involved, the fourth consumer level (quaternary) being roving carnivores such as large fish or sharks which move between various ecosystems. Relationships in this figure are shown in a *qualitative* manner. Figs. 5.3 and 5.4 show attempts to *quantify* the relationships. In the first case a relatively simple ecosystem has been studied where the herbivores (mice) are preyed upon by carnivores (weasels). This case nicely illustrates the minute portion of incoming solar radiation actually fixed by the producers (the grass). The attempt by Golley to quantify the flows in this ecosystem is not completely successful. No precise data are available for the very important decomposer element found largely in the soil. From what was said in Chapter 3 about the minute size of these organisms and their enormous populations, this is hardly surprising. The second example, published nearly twenty years later, represents the energy flow within the Hubbard Brook Experimental Forest of New Hampshire and is based on data from a long-term study of this temperate hardwood community. Considerable progress towards understanding the energetic

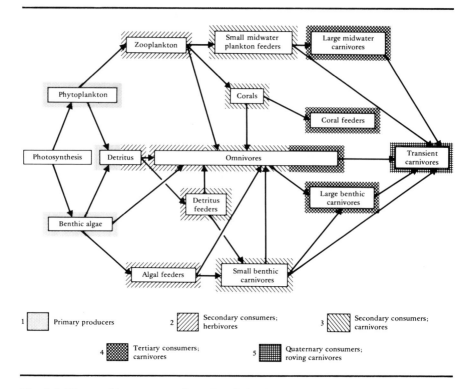

Fig. 5.2 The trophic structure of coral reefs in the Marshall Islands shown in a qualitative form. (From Stoddart, 1965, after Hiatt and Strasburg.)

and functional relations is evident but details on the consumer compartments and the factors regulating energy flow within the system are only just beginning to be elucidated.

Figure 5.5 illustrates further the immense difficulties of expressing ecosystem relationships in quantitative terms. Here the inter-relationships of just 40 of the known 210 species of insects associated with the wild cabbage plant (*Brassica oleracea*) are portrayed. Some insects are herbivorous, others parasitic or predaceous on each other. Attempts to quantify this complexity, if at all possible, would be a major research project. And we should remember that this might be just one plant species (plus its associated insect population) of the many which could be present in a complex ecosystem. Table 5.1 shows the types of population interaction possible for a simple two-species situation.

Further, while the ecosystem has been considered in terms of an energy throughput, it is also characterized by a *cycling of chemical nutrients* (as outlined

Fig. 5.3 The energy flow and trophic structure of an abandoned field ecosystem in Michigan, USA shown in a quantitative form. The field was left more or less undisturbed from 1942. The energy flow through the food chain is for the period May 1956 to May 1957 and the solar input is for the 1956 growing season. (Based on data in Golley, 1960.)

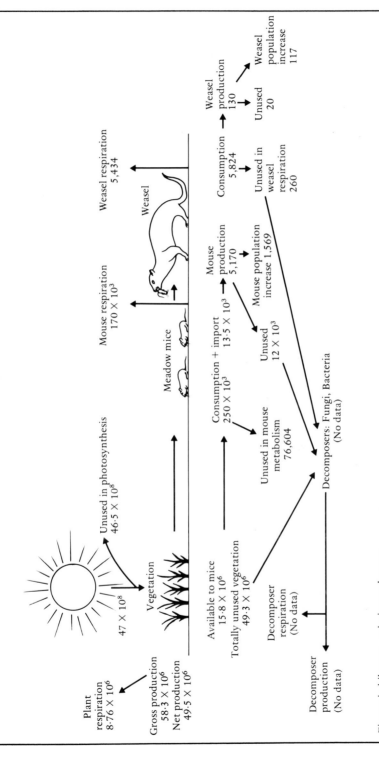

Figures in kilogram—calories per hectare

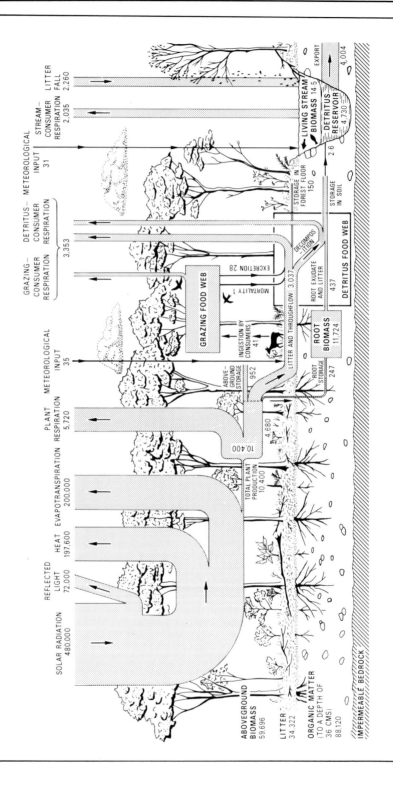

Fig. 5.4 Energy flow in the Hubbard Brook Experimental Hardwood Forest in New Hampshire, United States. The numerals are inputs and outputs of energy in units of kilo-calories per square metre per year, with relative energies indicated by width of the tubes (the width of the total plant production is arbitrary expanded at the bracket). The site has a well-developed, second-growth forest of sugar maple, beech and yellow birch which has been undisturbed by fire or cutting since 1919. It has a total area of 132,300 m², well-defined topographic boundaries (ridge divides), and is underlain by impermeable bedrock. (Modified from Gosz et al., 1978.)

in Chapter 3; see also examples in Chapter 8). Inputs of water and gases also occur, adding to the wealth of movements, exchange and transformations present in even the simplest of ecological systems. These biogeochemical cycles require an energy source and so the processes are energy-consuming operations. The study of these cycles should not be a separate exercise; they are part of the general theme of energy flow. As Reichle et al. point out, all ecosystems have evolved mechanisms that secure an energy base in the presence of a fluctuating environment. Radiant energy cannot be stored directly but its products arising from photosynthesis can and all ecosystems have a reservoir of energy and nutrients in the form of the living biomass and the organic detritus. This store is large but has a variable response time for release

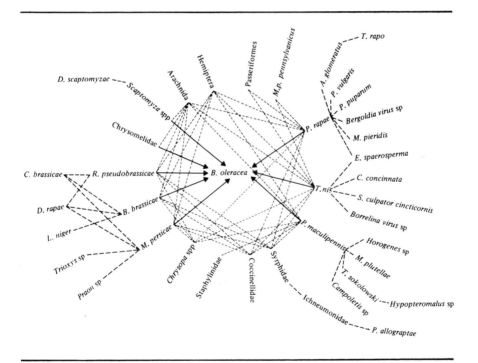

Fig. 5.5 The relationships between a number of the more abundant species associated with the wild cabbage plant, Brassica oleracea: herbivorous (———), parasitic (– – –), and predaceous (- - - -). (Modified from Pimentel, 1966.)

Table 5.1 Analysis of two-species population interactions (Modified from Odum, 1971)

Type of interaction*	Species† 1	Species† 2	General nature of interaction
1. Neutralism	0	0	Neither population affects the other
2. Competition: direct interference type	−	−	Direct inhibition of each species by the other
3. Competition: resource use type	−	−	Indirect inhibition when common resource is in short supply
4. Amensalism	−	0	Population 1 inhibited, 2 not affected
5. Parasitism	+	−	Population 1, the parasite, generally smaller than 2, the host
6. Predation	+	−	Population 1, the predator, generally larger than 2, the prey
7. Commensalism	+	0	Population 1, the commensal, benefits while 2, the host, is not affected
8. Protocooperation	+	+	Interaction favourable to both but not obligatory
9. Mutualism	+	+	Interaction favourable to both and obligatory

* Types 2 to 4 can be classed as 'negative interactions', types 7 to 9 as 'positive inter-actions', and 5 and 6 as both.
† 0 Indicates no significant interaction; + indicates growth, survival, or other population attribute benefited (positive term added to growth equation); − indicates population growth or other attribute inhibited (negative term added to growth equation).

and reuse. The turnover rate for some nutrients might be many decades when they form part of the living biomass (trunks, cones, etc.), but once they arrive in the litter layer then turnover may be accomplished within a few years. This is well seen in natural tropical forests where Went and Stark demonstrate the vital importance of endotrophic mycorrhizae (fungi that penetrate living cells) which are largely responsible for rapidly passing nutrients from decay processes direct to tree roots. Thus few nutrients leak away from the root zone into the lower soil or regional drainage water. For many heterotrophs also, the dead or decomposing organic store in the soil represents a vital source of energy and nutrients and in many ecosystems more energy is stored in or moves through the below-ground part of the system (the detritus food chain) than in the more obvious above-ground grazing food chains.

HOMEOSTATIC MECHANISMS

Most of the time we tend to take nature for granted and only notice natural ecosystems when something goes wrong with them or when there are sudden changes in their functioning. This is especially so when these changes take the form of a drastic increase or decrease in the population of one of the component species. An increase, for example, might raise the species to pest status. It is then in direct competition for some available resources in the ecosystem which man also has a special interest in. The forms these ecosystem disturbances take will be seen in some detail in Chapter 7. However, this reasoning implies that for the most part many ecosystems achieve some form of balance with apparently nothing dramatic happening in them. This allows us to accept them as 'normal' or uneventful. From a theoretical study, Yodzis claims strong support for the view that the equilibrium idea (a conception of community dynamics in which the community resides in a neighbourhood of equilibrium) is appropriate for many real communities.

To exist in this balanced state numerous checks operate within a natural ecosystem. When these are of a self-regulating nature they are known as *homeostatic mechanisms*. In the previous example of a central heating system, the thermostat is a homeostatic mechanism that maintains the temperature within an acceptable range. Natural ecosystems are characterized by the presence of many such self-regulation mechanisms. Most species have an enormous reproductive potential and yet their numbers remain relatively constant in many well-established communities. (During the early stages of succession the number of individuals of a successful species must obviously increase greatly but these populations tend to stabilize quite quickly.) If, for example, an initial breeding population of insects experiences favourable climatic conditions which allows the population to increase significantly, then many individual insects in this higher density population may have difficulty in obtaining enough food. Competition may cause starvation, thus bringing the population back towards its original size. Similarly, a severe winter could greatly reduce this insect population. For those remaining and breeding, food supplies will be more than adequate in the spring and summer, because of reduced competition. Most members will survive and breed well, leading to a swing back in population numbers to the original size.

How successfully these stabilizing mechanisms work will partly depend on how severe and prolonged are the environmental changes involved. If they are particularly marked then the population might become extinct (as the fossil record shows). Homeostasis operates not only at population level but also to maintain the internal environment of the individual in the face of changes in its external environment. The most obvious example is our own body temperature which is regulated around 37 °C. As the external temperature varies we react by shivering or perspiring, increasing or reducing activity, shedding clothes or improving our insulation; these are just some of the ways we attempt to restore the balance.

Homeostasis can be demonstrated in all organisms. Odum believes community succession operates to increase homeostasis, providing more protection for the members of the community from external change. As an undisturbed ecosystem matures, there is increased symbiosis between species and many specialized population interactions arise (see Table 5.1). High species diversity is another characteristic of a mature ecosystem. According to Odum, all these trends lead to more stability. For many years it was a central tenet of ecological theory that ecosystem diversity and complexity enhance the stability of populations. Checks and counter-checks operate between species to produce this stability. But this hypothesis has not yet been fully substantiated by field data. Indeed several studies show the position not to be quite as straightforward as theory would suggest. One problem in this field is that of semantics; stability and related terms have been defined in several different ways by ecologists. *Stability* of a system usually refers to the tendency for it to remain near an equilibrium point. *Resilience* describes the robustness of an ecosystem, its ability to maintain itself when faced with a disturbance. Orians introduces other useful terms such as *elasticity*, the speed of return to the system's former state following disturbance, and *amplitude*, the range over which the ecosystem is stable. The ease with which a system can be disrupted is referred to as its *fragility*.

Relatively simple ecosystems often show rapid population cycles and instability. A much quoted case is the predator–prey relationship seen in some Arctic communities, e.g. the lynx–snowshoe hare relationship in northern Canada. But Keith shows that hare population fluctuations are due to a number of reasons other than simply lynx predation, which is just one factor of many. On the other hand, complex and diverse Tropical Rain Forest ecosystems do not seem to experience such drastic, naturally occurring population disturbances. Ricklefs notes, however, that when disturbances do occur they spread through these communities much more slowly than in the simpler Arctic ecosystems. Nevertheless, they may trigger off a complex series of changes throughout several trophic levels. These responses are not so apparent but they are spread over a long period. In contrast, simple Arctic systems may experience drastic fluctuations because the influences are rapid and direct. But these disturbances are much more short-lived and balance may be just as quickly restored. The extent to which the influence of time-lags in tropical ecosystems outweighs or is outweighed by the dampening influence of complexity has not yet been determined. However, there is increasing evidence for the fragility of complex ecosystems. As May concludes: 'complex ecosystems (such as the tropical rainforest or Lake Baikal, with their many species and rich interaction structure) are in general dynamically fragile'. They may be well adapted to persist in the relatively predictable environments in which they have evolved and can be resilient under large natural disturbances, such as violent tropical storms, with which they coevolved, but vulnerable to recent minor perturbations of a novel kind inflicted by man (e.g. road constructions). The ecosystems of temperate latitudes seem much more robust in this respect.

Murdock has reviewed claims that species diversity leads to stability and rejects much of the evidence as presented. He argues that the stability of natural communities is to be explained in terms of the kinds of species present and their interactions, particularly related to their long-shared coevolutionary history. In his view ecologists have seized upon the wrong variable (the number of species or simple diversity) in explaining the stability properties of ecosystems. Murdock and Oaten suggest that the major aspect of diversity leading to stability might be spatial heterogeneity rather than species diversity. This applies to model and simplified ecosystems (such as laboratory examples and agriculture) and they think it may also apply to natural ecosystems. But Ricklefs notes that: 'stability is the culmination of all ecological inter-relationships, . . . the union of all lower-order properties of community, population and organism, . . . the science of ecology is not yet mature enough to mold its diverse knowledge and concepts into a unified theory of stability'. These comments indicate our lack of detailed knowledge about the workings of homeostatic mechanisms and point to a need for more research into field situations.

Where the tendency towards change is dampened down then *negative feedback* operates in the system. Feedback is the control of inputs as a function of output and in the case of negative feedback an increase (or decrease) in an initial value causes eventually a decrease (or increase) in another value. Predator–prey relationships contain a strong element of negative feedback. *Positive feedback*, on the other hand, operates to emphasize the direction and rate of change rather than to stop it. Both types of feedback operate in natural ecosystems but it is negative feedback mechanisms which attempt to remove disturbance.

POPULATION CONTROLS IN ECOSYSTEMS

Homeostatic mechanisms may determine the population level of a species in an ecosystem. However, the study of population dynamics is often much more complex than that and in this introductory text only a few broad points will be made to indicate the general set of population controls operating on ecosystems (a good account of both animal and plant population ecology is provided by Begon and Mortimer).

The population of a species will represent the difference between the *birth rate* and the *mortality rate*. But what determines these rates is quite complex. A first consideration is the age structure of the population and its sex ratio. A rapidly expanding population has by far the largest number of individuals in the young age group, either capable of breeding or soon to be so. This will give a pyramidal age structure. A diminishing population will show, in the extreme case, an inverse-pyramid structure, with few young individuals developing and most of the population in the post-reproductive age group. The reproductive potential of many species is theoretically very high indeed. If all individuals born survive to breed then in a relatively few generations the number of individuals in a species population can soon reach astronomical levels. Obviously this does not happen, except perhaps for a short while with rapid outbreaks of insect pests or in 'special' cases. The herring gull in Britain is one such case. The population is currently doubling every six years due to new additional food sources (rubbish tips, ploughed farm land) and habitats (reservoirs) and lack of predators. In most ecosystems, checks come into play immediately. The biotic potential force will be opposed by the environmental resistance force, according to Chapman.

The growth curve of an expanding population will usually take one or other of two basic forms, the S-curve or the J-curve. For any new species entering a developing ecosystem (e.g. during plant succession) there may be an initial period of rapid population growth. With the S-curve, the rise eventually tapers off as checks begin to operate. The curve reaches a 'plateau' and a balance is achieved between the biotic potential and the environmental resistance. This settled population level is sometimes referred to as the *carrying capacity* of the site for that species. Its value is not constant; it will vary as environmental factors change. With the J-curve there is the same initial rapid population increase but levelling off is not seen. Instead, the species meets sudden environmental resistance and population numbers plunge toward zero. For some insects and for annual plants this will be a yearly occurrence. A small residual population of insects or seeds overwinters and then initiates the rapid rise once more in the following year (Fig. 5.6).

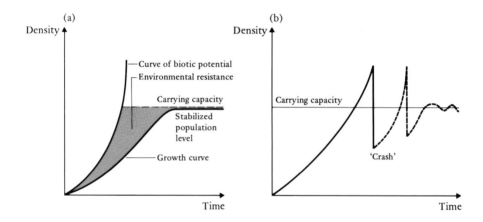

Fig. 5.6 Theoretical population curves for a species: (*a*) The S-curve, showing relationships between the biotic potential, the growth curve and environmental resistance. Some environmental resistance will operate immediately so that the growth curve is never as steep as the biotic potential would suggest. Environmental resistance becomes increasingly significant, eventually determining the carrying capacity of the site. (*b*) The J-curve, showing a population that has overshot the carrying capacity and then 'crashed'. This pattern may occur several times, eventually settling down as minor oscillations about the carrying capacity. (Based on curves in Boughey, 1968.)

Another way of looking at population curves is in terms of the strategies adopted by species to survive and prosper in their particular habitats. These strategies have been expressed by MacArthur and Wilson in terms of the *r-K continuum*. This reflects the relationship between the generation time of a species and the period over which the habitat of that species remains stable. The r-strategists are typically short-lived, opportunistic species, well adapted to exploit temporary habitats. They are small-sized, have high fecundity, mature rapidly and adopt a 'boom and crash' pattern of population growth (r = rapid). They invest little in 'defensive' structures and are vulnerable to attacks or suppression by other species, according to Southwood. However, they have a high rate of dispersal and are good colonizers of sites characteristic of the early stages of succession. Weed and pest species are good examples. In contrast, the K-strategists are species of stable habitats and are usually large-sized and long-lived. Their fecundity is lower and the development of the individual is slow, giving rise to fairly constant population densities (more or less at carrying capacity or K). Investment in 'defensive' structures is high and dispersal rate is low. Large predators and 'climax' trees are typical examples. These ideas have some similarities with the plant strategies of Grime outlined in Chapter 2. Although a useful descriptive concept, some species exhibit features of both r- and K- strategies and the r-K continuum is not without its critics (e.g. Stearns).

Breeding ability can be influenced by many factors. Some factors will operate irrespective of the species population density (*density-independent*), while with others the density of the population is all important (*density-dependent*). In the former case, reproductive capacities may change as physical or chemical environmental factors alter. Birch has shown that a female rice weevil produces about three times as many offspring at 23 °C than at 33·5 °C. Killing frosts, droughts, declines in food supply or chemical deficiencies in soils may all operate as density-independent population controls. Some density-independent controls take the form of abiotic influences while others are biotic in form.

Much population regulation is achieved through biological relationships. These controls may be *interspecific* or *intraspecific*, and they are frequently density-dependent in form. Table 5.1 lists the main forms of interspecific relationship and a moment's consideration of each should make it obvious how they may operate as checks on population growth. Perhaps the best known of these is the *predator–prey relationship*. A classic case of predator control of population numbers is seen in the case of the Kaibab deer. The Kaibab Plateau in North Arizona was declared a game reserve in 1906 and some 6,000 preda-tors (wolves, mountain lions, bobcats and coyotes) of the Mule Deer popu-lation in the reserve were exterminated over the next 25 years. This caused the deer population, estimated at 4,000 in 1906, to rise rapidly to 70,000 in 1923. The vegetation became badly overgrazed and food shortages reached critical levels. In 1925 60 per cent of the herd died and by 1939 the population dwin-dled through starvation to only 10,000. It was obvious that predator control had been keeping the herbivorous prey population at a reasonable level for the food resources in this ecosystem. In Britain also the deer populations are now without natural predators and in 1980 the Forestry Commission had to kill more than 18,000 deer (mainly roe deer) to safeguard their plantations.

Natural predators not only control prey populations but in some cases may actually increase the diversity of species in a habitat, as the studies of Paine have shown.

Intraspecific relationships are often density-dependent. Individuals of the same species are in keen, direct competition with each other for the same resource in the ecosystem. As the population becomes more crowded, the competitive stresses set up for food, mates, nesting sites, building materials or territorial space can adversely affect survival rates, breeding patterns and the quality of the offspring produced. This has been demonstrated in the laboratory and in the field for mice and vole populations. Population control often seems to relate to the social hierarchy developed in these communities, the *'pecking order'*. Large numbers of some animals, such as rabbits, mice, voles, have died from what appears to be 'shock disease', a severe physiological stress brought about by overcrowding. This causes the population to go into rapid decline. The phenomenon can occur fairly regularly: each rapid build-up being followed by a population 'crash' to give a cyclic population pattern for the species.

While crowding in a community is usually the critical condition for the operation of these population controls, Darling reminds us that the reverse can sometimes be the case. He found that some sea-birds fail to nest unless the adults in the colony reach a certain minimum density.

Population numbers may also change as a result of immigration, emigration or migration of species. These movements can be a response to a density-independent factor (climatic seasonality) or a density-dependent situation (as 'excess' population is forced to seek new territories). For many species these movements are highly hazardous and animals seeking new territories can be very vulnerable to predator attack. Some migrations cover enormous distances: the Arctic tern makes an annual round trip of 36,000 miles. These events are very effective in considerably reducing numbers.

Many of these factors in population dynamics have been expressed in mathematical equations (the early population studies of Lotka, Volterra and Gause). But most of these relate to laboratory studies of very simple communities. Much progress has been made but it remains to be seen whether the equations derived really apply to a very complex community in the field. One thing is certain, no single factor produces the population structures we see. They result from the operation of many factors, some acting to push up the population at the same time as others are tending to pull down the numbers (see also Table 5.2). Quite often the controls, whether they are density-dependent or density-independent, will benefit the population as a whole since only the fittest survive to breed. Work on Scottish grouse moors by Jenkins and co-workers shows that the young grouse picked off by eagles, hawks and foxes are most likely to be just those which have been unable to find a suitable territory quickly. Essentially, they form the 'surplus' population made up of less successful individuals. Many of the mechanisms outlined here will trigger off a response in the ecosystem (by means of a feedback loop) so that the population strives once more to achieve a balance and becomes self-regulating. It must also be remembered, however, that the abundance of a species is regulated by factors both internal and external to that species. Ayala points out that the genetic constitution of a population is of major importance in this respect and has often been ignored in ecological studies: 'the rate of evolution of a population becoming adapted to a new environment is positively correlated with the initial amount of genetic variability in the population'.

Table 5.2 Characteristics of animal species that may lower their survival rate and thereby greatly influence their population numbers (Modified from Ehrenfeld, 1970)

Endangered	Safe
Individuals of large size (cougar)	Individuals of small size (wildcat)
Predator (hawk)	Grazer, scavenger, insectivore, etc. (vulture)
Narrow habitat tolerance (orangutan)	Wide habitat tolerance (chimpanzee)
Valuable fur, hide, oil, etc. (chinchilla)	Not a source of natural products and not exploited for research or pet purposes (gray squirrel)
Hunted for the market or hunted for sport where there is no effective game management (passenger pigeon)	Commonly hunted for sport in game management areas (mourning dove)
Has a restricted distribution: island, desert watercourse, bog, etc. (Bahamas parrot)	Has broad distribution (yellow-headed parrot)
Lives largely in international waters, or migrates across international boundaries (green sea turtle)	Has populations that remain largely within the territory(ies) of a specific country(ies) (loggerhead sea turtle)
Intolerant of the presence of man (grizzly bear)	Tolerant of man (black bear)
Species reproduction in one or two vast aggregates (West Indian flamingo)	Reproduction by solitary pairs or in many small or medium sized aggregates (bitterns)
Long gestation period; one or two young per litter and/or maternal care (giant panda)	Short gestation period; more than two young per litter, and/or young become independent early and mature quickly (raccoon)
Has behavioural idiosyncrasies that are nonadaptive today (redheaded woodpecker: flies in front of cars)	Has behavioural patterns that are particularly adaptive today (burrowing owl: highly tolerant of noise and low-flying aircraft; lives near the runways of airports)

THE ECOLOGICAL NICHE

An important concept which sheds light on the population structure of an ecosystem is the *ecological niche*. Odum defines this in functional terms as the status of an organism within its ecosystem. This status results from the species adaptations, physiological responses, and behaviour. Whittaker extends this definition by describing each species as having its own place in vertical and horizontal space, in time and in manner of relating to other species in the system. He thus adds the spatial and temporal dimensions to the functional view of the niche. Recent work on niche theory has been reviewed by Pianka.

Mathematical formulae on population interactions between two species were derived by Lotka and Volterra in the 1920s and later confirmed experimentally by Gause. They indicated that when two species are in direct competition for a common resource existing in limited supply only one species will eventually survive. Competition will lead to the exclusion of the other species (this is known as *Gause's principle* or the *competitive exclusion principle* of Hardin). It is thus impossible for two species to continue to occupy exactly the same ecological niche or co-exist when having identical environmental requirements. Thus two species may show either competition between each other or co-existence with each other according to whether the ecological niches that each is attempting to occupy are the same, very similar or quite different. Competitive relationships are often the most obvious in a community but cooperation in nature is equally important (see the several examples outlined in Table 5.1).

Specialized behaviour by a species will allow it to use certain resources not available to non-specialized species, thus avoiding direct competition. Sometimes two species will use the same resource at different times of the day or year. Most ecosystems contain a blend of *non-specialist* and *specialist* species and this greatly extends the number of ecological niches available (*niche diversification*). Sometimes two species which appear to be identical in most aspects of their behaviour will differ slightly in one and this may permit their co-existence. A good example is provided by Lack who studied the feeding habits of two similar sea-birds around the British coast, the cormorant (*Phalacrocorax carbo*) and the shag (*P. aristotelis*), which appear to have almost identical ecologies or considerable *niche overlap*. Both utilize the same cliffs for nesting, but there are small differences in the actual nest-sites used. More importantly they feed in the same waters but on different sets of fish. The cormorant is a bottom feeder eating flatfish and shrimps while the shag feeds mainly on fish and eels in the upper waters (Fig. 5.7). This introduces the concept of *niche breadth* which refers to the total variety of different resources exploited by an organism.

The more heterogeneous the environment the more species it is likely to contain. With high species diversity many population interactions are possible. A high degree of regulation of the overall population will occur by the operation of these interactions rather than by the imposition of external controls, such as climatic severity. We find, as a general rule, increasing species diversity from the Arctic to the Tropics. In harsh environments populations are more subject to violent fluctuations, while in the climatically favourable environments stability is more marked. In the former, the battle for survival is often against the physical environment: in the latter, the problem is often one of surviving in relation to the many other species in the community. It must be remembered that the work of Lotka, Volterra and Gause was based on simple populations and uncomplicated laboratory experiments. While their findings appear to hold in some field situations, the complexity of natural communities is such that we should expect to find exceptions to, and differences of opinion on, this simplified competition theory.

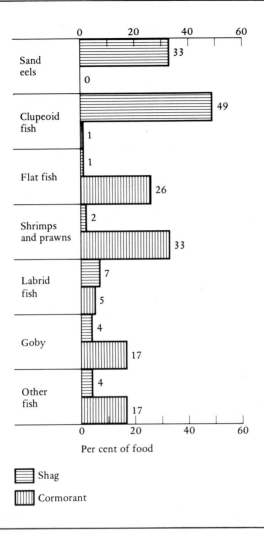

Fig. 5.7 Food sources of the shag (*Phalacrocorax aristotelis*) and the cormorant (*P. carbo*), two closely related sea-birds with similar ecologies. The shag mainly consumes eels and free-swimming fish while the cormorant feeds on bottom fish and bottom invertebrates. These food differences largely prevent direct competition between the two species. (Data from Lack, 1945.)

ECOLOGICAL EFFICIENCY

Previous sections have demonstrated the complexity possible in even simple ecosystems. We may now consider how man fits into this picture and the implications for him of the way in which these natural systems function.

However, first it is necessary to examine in outline two other related aspects of the functioning of ecosystem which have a bearing on these considerations, namely, *ecological efficiency* (of energy transfer) and *ecosystem productivity*.

Several workers have studied the efficiency of energy transfer from one trophic level to another. Unfortunately, over the years confusion has arisen because many ways of measuring and expressing ecological efficiency have been proposed (Kozlovsky lists 19 such definitions from the literature). One common approach has been to consider the energy flow along the food chain as a percentage for each trophic level based on the relationship between the amount of energy taken in by that level and the amount passed on as food for the next trophic level above. Green plants during photosynthesis fix only about one per cent of the solar radiation received on the leaf surface. But once energy has been fixed by the autotrophs the efficiency of energy transfer to other trophic levels is higher. Slobodkin, studying a simple, artificial community in a laboratory experiment, showed the ecological efficiency to be about 13 per cent when transfer occurred between a primary producer and a primary consumer. In the field, Odum found at the Silver Spring ecosystem in Florida an efficiency of 11.4 per cent in a herbivore-to-carnivore transfer and 5.5 per cent in a carnivore-to-carnivore transfer. Teal, examining a less complex and smaller springwater ecosystem in Massachusetts, derived an ecological efficiency of 9 per cent for the energy transferred from the herbivore to the carnivore trophic level. These early studies suggested that values around the 10 per cent level were to be expected in most ecosystems. However, these studies mainly concerned aquatic ecosystems and later work on terrestrial ecosystems, summarized by May in a study of patterns in multi-species communities, points to much lower percentage values, frequently not exceeding 1–2 per cent where large herbivores are concerned.

ECOSYSTEM PRODUCTIVITY

The amount of biological material in an ecosystem at any given time is known as the *standing biomass*. Thus the standing biomass of the wheat plant in a wheatfield is represented by the standing crop as we view it, including the roots, leaves, shoots and wheat kernels. This is not the same as the productivity of the wheatfield since *productivity* refers to the amount of energy fixed in a given time, i.e. a *rate* of growth. Gross primary productivity is the total rate of photosynthesis and includes organic matter used in the process of respiration during the period over which productivity is being measured. Net primary productivity is a measure that excludes the organic matter used in respiration. It therefore represents the rate of accumulation of organic matter, all of which is theoretically available for passing on to the next trophic level. Only the autotrophs can show primary productivity; the heterotrophs show secondary productivity, gross and net, since they must depend on food already in existence (consider Fig. 5.1).

There are many difficulties in measuring productivity in natural ecosystems. Measurements of primary production are best regarded as estimates rather than exact values since they are subject to many errors. Despite these difficulties a list of annual net primary productivities for a selection of ecosystems is now available from the recent work of the International Biological Programme and is well worth more than a few moments' attention (Table 5.3). The main points from this table are perhaps best considered in the context of our examination of man's place in the ecosystem.

MAN IN THE ECOSYSTEM

Mature, natural ecosystems usually exhibit complex structural and functional relationships, high species diversity, niche diversification and many homeo-static mechanisms. Man, on the other hand, in modifying an existing ecosystem for his own use or in creating an artificial agricultural ecosystem, will concentrate his attention on one species, or at most a few, to the exclusion of the other species present. With modern agricultural methods, there is a ruthless elimination of all plant and animal species which compete in any sense with the particular species population man is trying to encourage. The extreme case of this is monoculture, a common feature of modern agriculture. As a general rule, nature attempts to diversify in habitats whereas man simplifies. There are, of course, examples of natural monocultures (e.g. bracken) and these may be highly stable. However, May concludes that 'the thing which destablizes man's agricultural monocultures is not so much their simplicity per se, as their lack of an evolutionary pedigree'. Many interesting points stem from this difference and several can be made in conjunction with the data presented in Table 5.3.

In the early stages of a plant succession, the site resources (light, soil nutrients, etc.) may be only partly used because of an incomplete or simple plant cover (fractional utilization). Many ecologists believe maximum utiliz-ation occurs when the climax state or a stage close to it has been reached in a community's development. There is then conservation and efficient use of nutrients. Productivity (i.e. rate of growth) is highest in the youthful stages but maximum biomass is achieved at the mature stage in most communities. This biomass is, of course, the maximum reserve of potential energy for use by the various trophic levels in the system, other than the autotrophs. Most cultivated plants have a high productivity rate for several months (the growing season) but there may well be almost zero production for long periods at the site. This stems from the use of annuals as our main food crops (wheat, barley, maize, rice, etc.). These will only utilize light energy for part of the year. The natural vegetation on these sites would utilize resources for much longer periods because most of it is dominated by perennial species.

Table 5.3 reveals that many of the natural or non-cultivated ecosystems have higher productivity (mean NPP.) than the cultivated crops which man has put

Table 5.3 Net primary production and related characteristics of various natural ecosystems and cultivated land (Modified from Whittaker and Likens, 1975)

1	2	3	4	5
			Net primary production (dry matter)	
Ecosystem type	Area (10^6 km^2)	Normal range (g/m^2/year)	Mean (g/m^2/year)	Total (10^9 t/year)
Tropical rain forest	17·0	1,000–3,500	2,200	37·4
Tropical seasonal forest	7·5	1,000–2,500	1,600	12·0
Temperate forest:				
evergreen	5·0	600–2,500	1,300	6·5
deciduous	7·0	600–2,500	1,200	8·4
Boreal forest	12·0	400–2,000	800	9·6
Woodland and shrubland	8·5	250–1,200	700	6·0
Savanna	15·0	200–2,000	900	13·5
Temperate grassland	9·0	200–1,500	600	5·4
Tundra and alpine	8·0	10–400	140	1·1
Desert and semidesert scrub	18·0	10–250	90	1·6
Extreme desert – rock, sand, ice	24·0	0–10	3	0·07
Cultivated land	14·0	100–4,000	650	9·1
Swamp and marsh	2·0	800–6,000	3,000	6·0
Lake and stream	2·0	100–1,500	400	0·8
Total continental	149	–	782	117·5
Open ocean	332·0	2–400	125	41·5
Upwelling zones	0·4	400–1,000	500	0·2
Continental shelf	26·6	200–600	360	9·6
Algal beds and reefs	0·6	500–4,000	2,500	1·6
Estuaries (excluding marsh)	1·4	200–4,000	1,500	2·1
Total marine	361	–	155	55·0
Full total	510	–	336	172·5

in their place. Forests provide the bulk of the world's NPP, and in many cases agricultural land has been established at the expense of woodland covers. Many swamps and marshes are highly productive communities but these also have been frequently converted to farm land. Highest productivities in the ocean are associated with algal beds, reefs and estuaries, just the communities which are most vulnerable to human interference. According to Odum, 'man has not increased maximum primary productivity beyond that which occurs in the absence of man'.

6 **Biomass (dry matter)**	7	8
Normal range (kg/m^2)	Mean (kg/m^2)	Total (10^{9}t)
6–80	45	765
6–60	35	260
6–200	35	175
6–60	30	210
6–40	20	240
2–20	6	50
0·2–15	4	60
0·2–5	1·6	14
0·1–3	0·6	5
0·1–4	0·7	13
0–0·2	0·02	0·5
0·4–12	1	14
3–50	15	30
0–0·1	0·02	0·05
–	12·2	1,837
0–0·005	0·003	1·0
0·005–0·1	0·02	0·008
0·001–0·04	0·001	0·27
0·04 –4	2	1·2
0·01 –4	1	1·4
–	0·01	3·9
–	3·6	1,841

Many crops reveal wide variation in their yields. This obviously reflects the different environments under which the same crop can be grown but it also points to differing management techniques and agricultural efficiencies. Table 5.4 shows this clearly: here the main contrast is between extensive and intensive agriculture (cf. North America and Japan).

Not only do most of man's ecosystems have lower primary production than those they have replaced but they also entail losses from the system far in excess of any in natural systems. We have already referred to poor management

Table 5.4 The population density per acre of food-producing land and the food production per acre compared for different regions (Data from Brown, Bonner and Weir, 1957)

Regions	Cultivated acres per capita	Calories produced per cultivated acre per day
United States	2·4	4,500
North America	3·6	2,500
Western Europe	0·9	7,500
Latin America	1·0	4,700
Asia (all)	0·6	4,000
USSR	2·8	2,300
Japan	0·17	13,200
India	0·9	2,500

techniques but there are more important losses due to the removal of 'wastes' such as decomposition products, unused plant and animal remains, and human faeces by garbage combustion or sewage disposal to the sea. In natural ecosystems these are recycled but increasingly this does not happen with modern farming.

To get reliable, high crop productivities, man often has to add additional energy and mineral supplies to his systems. The use of fertilizers is an obvious example. By supplying chemicals and water we can greatly increase primary productivity in situations where these materials were limiting at the original site. But we should never forget the cost of these additional inputs by man when working out the efficiency of our systems. As Ridge pointed out, in the United Kingdom the additional energy subsidy necessary to maintain a wheat-field as a monoculture costs about £28 per acre (£70 per ha) at 1974 prices. This takes the form of fertilizers, seeds, tractor and fuel costs; many inputs represent the use of finite resources accumulated as fossil fuels by a previous natural ecosystem. It is interesting to compare the energy ratio (of energy output as food with energy input as fossil energy used) for various agricultural systems. Studies have shown that for many intensive industrial forms of food production the ratio is much less than one, i.e. a negative energy balance. Blaxter derives a ratio of 0·34·1 for edible food output at the UK farm gate to fossil energy used in its production. Livestock production systems are particularly inefficient users of energy subsidies. Strong positive balances are only seen with the so-called primitive forms of food production (e.g. ratios above 5·1 for some hunters and gatherers, above 10·1 for tropical shifting agriculture and approaching 60·1 for shifting cultivation of cassava, according to Leach).

We also frequently have to inject 'foreign' chemicals such as herbicides and pesticides into the system. The hope is that these will act as controlling mechanisms in situations where the original homeostatic mechanisms have been broken down by the very creation of a simplified ecosystem. Such toxic substances may then become concentrated along the food chain, causing death to harmless or beneficial species. One of the earliest demonstrations of this was at Clear Lake, California. Carson showed that when the chemical DDD was sprayed at a concentration of only 0·02 p.p.m. to control gnat swarms it was taken up by plankton (5 p.p.m.) passed on to fish (40–300 p.p.m.) and finally reached the birds (grebes) in sufficient amounts (1,600 p.p.m.) to cause whole-sale deaths.

While we cannot improve on nature's primary productivities without massive inputs of energy and materials we can sometimes improve on secondary productivity through a thorough understanding of how an ecosystem works. A classic example is provided by the introduction of the grass carp (*Ctenopharyngodo idella*) into some tropical ponds and streams. The fish is a herbivore of temperate latitudes (Siberia, North China) and, although with-out pituitary injections it does not breed south of 22° north, it has proved highly successful in some tropical ecosystems which previously lacked a suitable herbivorous fish species. Hickling has shown how the carp will directly help to keep in check the massive growth of aquatic vegetation by its feeding habits. Left unchecked this vegetation is a troublesome weed. The fish also indirectly checks this vegetation because its excretions encourage the growth of plankton which, in turn, shade out some of the weed. Further, as a food source itself for carnivorous fish in the system it makes available much of the nutrient material bound up in the aquatic weed. Other fish feed on the droppings of the carp which contain about half the protein and carbo-hydrate of the original intake of aquatic weed. The carp itself is a food fish of high quality and has a rapid growth rate. When set with other fish in this ecosystem, the total fish crop available to man from the same initial weight of plant food is doubled. (Incidentally, experiments are being conducted in East Anglia using the carp in aquatic weed control and as a possible food source for man.)

Nevertheless, it should be remembered that usually much less than 10 per cent of the energy entering one trophic level is passed on to the next level above. This low value seems to be the maximum amount available for transfer without that level suffering such a loss of population as to imperil its chances of survival. In mature, complex natural ecosystems many energy-transfer path-ways may exist in the network, perhaps up to 1,000 in some systems. But if man is to obtain sufficient chemical energy (food) he must keep the energy transfers down to a reasonably small numbers (i.e. a short food chain) and minimize respiration losses (i.e. diversion of energy for work purposes). This is precisely what he tries to do when he develops battery hen systems or raises beef by bringing the food to the animal in insulated cattle sheds. This cuts down 'unnecessary' animal movements and heat outputs. However, in stopping farm animals 'wasting' energy he has to provide an input of his own energy (labour, tractor fuel, etc.). It thus becomes a question of economics. Interest-ingly enough, economics and ecology as words have the same Greek root (*eco*, the home) and more in common as studies than superficial examination would suggest.

In all but a few ecosystems man is now the dominant species. This status has been reached either by virtue of our actual presence or because our indirect activities influence profoundly the number and type of other species that we allow to co-exist with us. Our direct superimposition on ecosystems has occurred rapidly, unlike natural dominants which grow up with and within their slowly evolving natural systems. We are not subjected to the same homeostatic mechanisms which operate for other species. This allows us to dominate ecosystems for purposes of short-term gain or exploitation. The great danger, however, is our lack of a good understanding of ecosystem dynamics, especially those involving long-term changes resulting from the impact of man. The implications of this, as the human population currently increases by about 70 million people per annum (see Fig. 5.6), should be quite obvious.

DISCUSSION SECTION

An essential characteristic of geography is its emphasis on synthesis. Synthesis is equally central to the idea of the ecosystem, so are the essential features of the ecosystem concept similar to what has always been the core of the geographical approach?

Yes, so much so that Stoddart points to four properties of ecosystems which make the concept of particular importance and suitability in geographical investigation:

1. The concept is *monistic*, that is, it brings together environment, man, plants and animals within a single framework, allowing interactions to be analysed.
2. Ecosystems are *structured*. They are orderly arrangements and once these structures have been recognized they can be studied further.
3. Ecosystems *function*. They are dynamic entities with a throughput of energy and cycling of matter. These aspects can be *quantified* and accurate measurements will lead to a more scientific approach, largely replacing mere description.
4. Ecosystems are a type of *general system* and possess the properties of general systems. The world is made up of many differing types of system (organic systems, geomorphological systems, machine systems, etc.). The ecosystem is just one of the types. This allows us to make comparisons with other general systems and permits us to bring to bear on our study of ecosystems any appropriate techniques and ideas developed in the other fields of system study. For example, increasing use is made of mathematical techniques, information theory and cybernetic techniques. Cybernetics is the theory of communication and control mechanisms and is largely a field for engineers. But ecosystems are also characterized by the possession of control mechanisms (homeostatis) and so experience and problems may be shared in the analysis.

Much of the new development in geography centres on the use of *models*. A model is a simplified structuring of reality in which the significant relationships of the real world are stressed and trivial material, as far as possible, is excluded. The ecosystem concept functions as a model, allowing us to reduce the bewildering complexity of biotic communities to a simplified structure of essential relationships.

Table 5.5 A structural classification of systems applicable to the real world (From Chorley and Kennedy, 1971)

Animate world	11. Human ecosystems
	10. Social systems
	9. Man
	8. Ecosystems
	7. Animals
	6. Plants
	5. Self-maintaining systems
Inanimate world	4. Control systems (transducers)
	3. Process–response systems
	2. Cascading systems
	1. Morphological systems

The term 'ecosystem' crops up in studies outside biogeography and I am not sure just how widely used the concept has become in geography.

Chorley and Kennedy argue for a systems approach to the whole of physical geography. They present a hierarchy of increasing complexity in a structural classification of systems applicable to the real world, as shown in Table 5.5. Some branches of physical geography (e.g. geomorphology, climatology) are mainly concerned with the first four systems covering the inanimate world. But these low-complexity systems also interact with systems of the animate world and so they too become relevant to the physical geographer in certain studies. Biogeography, which emphasizes the interactions between individuals and their environment, nicely links together the inanimate and animate worlds. The basic idea of the concept is incorporated into the higher systems where the increasing complexity is due to the overwhelming influence of man. Farming may thus be described in terms of agricultural ecosystems and likewise, cities as urban ecosystems. Because the definition has no scale limitation we can view much of regional geography also in these terms.

Ideas from the study of natural ecosystems have already been applied to urban ecosystems: see, for example, the recent book by Douglas. Holling has also adapted the predator-prey model by considering land developers and speculators as the 'predator' and the land as the 'prey', and he draws some interesting conclusions on the relative stability of large-scale speculations as compared with small-scale unplanned, piecemeal developments. According to Margalef, 'when two systems of different maturity meet along a boundary that allows an exchange, energy (production) flows towards the more mature sub-system, and the boundary or surface of equal maturity shows a trend to move in an opposite direction to such an energy flow'. He states this applies not only to ecosystems but also to human social systems (e.g. 'urban centres represent localized elements that have accumulated high amounts of information, fed on the production of neighbouring subsystems, and have exerted a directive action'). Thus energy flow is from less mature (rural) areas to more mature (urban) centres. Margalef notes that many concepts from ecosystem studies may apply equally well to human organizations.

*If ecosystems are being altered so rapidly and destroyed by man is it really worth
the effort to conserve the few remaining natural and semi-natural examples?*

We can justify the effort and expenses of conservation on several grounds:
scientific, economic, aesthetic and ethical. Aesthetic and ethical views are based
on highly personal interpretations and perhaps no more need be said. The
scientific and economic justifications can be related in a more practical, direct
way to a basic self-interest in our own preservation.

Natural ecosystems provide a yardstick against which to measure the efficiency of our food-producing systems. Often, when we do this, we realize that
our so-called 'improvements' have not been that successful. Only a handful of
animal species have been domesticated for regular food purposes. Recent
studies in East Africa show the possibilities of using native fauna (gazelles,
elands) for protein production. These species appear to be far superior to
introduced European cattle. Likewise, studies on the Isle of Rhum in Scotland
show that native red deer (*Cervus elaphus*) on moorland grazings have better
productivity than introduced sheep. There was no substantial decline in meat
production when the previous sheep and deer regime was changed to deer only.
The Nature Conservancy is now exporting a greater weight of venison from
Rhum than it previously did of mutton. A careful study of the natural faunal
components which have evolved within an ecosystem and are therefore well
adapted to it may provide several solutions for feeding increasing human
populations.

With plants, some 25,000 species are currently threatened with extinction
but only a minute fraction of the world's flora has been examined for possible
benefits to man. For example, some threatened plants could contain useful
drugs. An alternative source for a main component of the contraceptive pill
has recently been discovered in fenugreek (*Trigonella ornithopodioides*), a
common hedgerow weed. This has come at a time when the main source (yams)
for this drug may become much scarcer in the future. Plants also represent a
renewable source of energy. As supplies of fossil fuel decline and costs soar,
the possibilities are being explored of biomass energy supplies from crops of
widespread plants that are easy to grow, such as heather, bracken or willow.
Callaghan *et al.* report a cost of £61 per hectare at 1980 prices to establish
upland pasture in the UK on bracken-infested land. But this same bracken is
five times more productive than the pasture and could be used as an energy
crop.

Natural ecosystems are also vitally important for tourism. A few rare animals
in a locality can attract thousands of visitors and have a great impact on the
local economy. Kenya benefits by more than £100 million per annum from
tourists who are mainly attracted there by the wild animal populations.

What does it mean when an ecosystem is described as a 'black box model'?

Because of the complexity of natural ecosystems (see Fig. 5.5) all the variables
in such a system can seldom be studied. To make some progress we can

concentrate on an input variable and a related output variable of the system, all the internal variables being ignored for the purpose of the study so as to reduce complexity to a level which we can handle. The ecosystem is thus regarded as a 'black box', the internal variables are 'unseen' in the sense of not being included in the study.

In a small, woodland ecosystem the number of variables is enormous: all the different plant and animal species, each with its own set of properties, the soil and all its properties, the environmental factors, etc. No model could include all this variation. By concentrating our studies on one input (say solar radiation) and a related output (say seasonal growth rates of certain ground flora species) at least some detailed quantifiable knowledge begins to emerge. Even in the apparently simple situation represented by this black box approach there are still many practical problems. Enormous variations in the solar radiation pattern of a woodland occur and there is great difficulty in measuring this pattern exactly. Likewise, so many other factors could be influencing the growth rates measured for the selected ground flora species. So the black box approach treats the whole system as a unit. A 'grey box' model involves a partial view of the system, centred on a limited number of subsystems. A 'white box' approach tries to obtain as much detail as possible on the internal structure of the system and the way this is related to a given output from a given input.

REFERENCES

Anon, 1967. *Red Deer Research in Scotland*, Progress Report 1, The Nature Conservancy, London, 33–41.

Ayala, F. J., 1968. 'Genotype, environment and population numbers', *Science*, **162**, 1453–9.

Begon, M. and Mortimer, M., 1981. *Population Ecology: a unified study of animals and plants*, Blackwell Scientific Publications, Oxford and Boston.

Birch, L. C., 1957. 'The role of weather in determining the distribution and abundance of animals', *Cold Spring Harbor Symposium on Quantitative Biology*, vol. **22**, 203–15.

Blaxter, K. L., 1975. 'The energetics of British agriculture', *Biologist*, **22**, 14–18.

Boughey, A. S., 1968. *Ecology of Populations*, Macmillan, New York.

Brown, H., Bonner, J. and Weir, J., 1957. *The Next 100 Years*, Viking Press, New York.

Callaghan, T. V., Lawson, G. J., Scott, R. and Whittaker, H. A., 1980. 'Energy crops – an alternative use of land', paper delivered to Production and Decomposition Ecology Group meeting, reported in *Bulletin*, **11**, (3), 111, (British Ecological Society).

Carson, R., 1963. *Silent Spring*, Houghton-Mifflin, Boston.

Chorley, R. J. and Haggett, P., 1967. *Models in Geography*, Methuen, London.

Chorley, R. J., and Kennedy, B. A., 1971. *Physical Geography: a systems approach*, Prentice-Hall, London.

Darling, F. F., 1938. *Bird Flocks and the Breeding Cycle*, CUP, Cambridge.

Douglas, I., 1983. *The Urban Environment*, Arnold, Baltimore and London.

Dyer, M. I., 1980. 'Mammalian epidermal growth factor promotes plant growth', *Proc. Natl. Acad. Sci. USA.*, **77**, 4836–7.

Ehrenfeld, D. W., 1970. *Biological Conservation*, Holt, Rinehart and Winston Inc., New York and London.

Evans, F. C., 1956. 'Ecosystem as the basic unit in ecology', *Science*, **123**, 1127–8.

Fosberg, F. R., 1963. 'The Island Ecosystem', in *Man's Place in the Island Ecosystem, a Symposium* (ed. Fosberg, F. R.), Honolulu.

Golley, F. B., 1960. 'Energy dynamics of a food chain of an Old-Field Community', *Ecol. Monogr.*, **30**, 187–206.

Gosz, J. R., Holmes, R. T., Likens, G. E. and Bormann, F. H., 1978. 'The flow of energy in a forest ecosystem', *Sci. Amer.*, **238**(3), 92–102.

Hall, A. D. and Fagen, R. E., 1956. 'Definition of system', *General Systems*, **1**, 18–28.

Hardin, G., 1960. 'The competitive exclusion principle', *Science*, **131**, 1292–7.

Hickling, C. F., 1965. 'Herbivorous fish in a water economy', Paper presented at 2nd Symposium of the East African Society for Biological Research, Kampala (abstracted in *J. appl. Ecol.*, **2**, 413).

Holling, C. S., 1969. 'Stability in ecological and social systems', in *Diversity and Stability in Ecological Systems* (eds Woodwell, G. M., and Smith, H. H.) *Brookhaven Sym. Biol.*, No. **22**, 128–41.

Jenkins, D., Watson, A. and Miller, G. R., 1964. 'Predation and Red Grouse populations', *J. appl. Ecol.*, **1**, 183–95.

Keith, L. B., 1963. *Wildlife's Ten-year Cycle*, University of Wisconsin Press, Madison.

Kormondy, E. J., 1969. *Concepts of Ecology*, Prentice-Hall, New Jersey.

Kozlovsky, D. G., 1968. 'A critical evaluation of the trophic level concept, I. Ecological efficiencies' *Ecol.*, **49**, 48–60.

Lack, D., 1945. 'Ecology of closely related species with special reference to cormorant (*Phalacrocorax carbo*) and shag (*P. aristotelis*)', *J. Anim. Ecol.*, **14**, 12–16.

Leach, G., 1976. *Energy and food production*. IPC Press, London.

Maarel, E. van der, 1975. 'Man-made natural ecosystems in environmental management and planning', in *Unifying Concepts in Ecology* (eds van Dobben, W. H., and Lowe-McConnell, R. H.), Junk, The Hague, pp. 263–74.

MacArthur, R. H. and Wilson, E. O., 1967. *The Theory of Island Biogeography*, Princeton University Press, Princeton.

Margalef, R., 1963. 'On certain unifying principles in ecology', *Amer. Naturalist*, **97**, 357–74.

May, R. M., 1975. 'Stability in ecosystems: some comments', in *Unifying Concepts in Ecology* (eds van Dobben, W. H., and Lowe-McConnell, R. H.), Junk, The Hague, pp. 161–8.

May, R. M. (ed.), 1981. *Theoretical Ecology: principles and applications*, (2nd ed.), Blackwell Scientific Publications, Oxford and Boston.

Murdock, W. W. (ed.), 1975. *Environment*, Sinauer, Stamford, Conn.

Murdock, W. W. and Oaten, A., 1975. 'Predation and population stability', *Adv. Ecol. Res.*, **9**, 2–132.

Odum, E. P., 1971. *Fundamentals of Ecology* (3rd edn.), W. B. Saunders, Philadelphia.

Odum, E. P., 1969. 'The strategy of ecosystem development', *Science*, **164**, 262–70.

Orians, G. H., 1975. 'Diversity, stability and maturity in natural ecosystems', in *Unifying Concepts in Ecology* (eds van Dobben, W. H., and Lowe-McConnell, R. H.), Junk, The Hague, pp. 139–150.

Owen, D. F., 1980. 'How plants may benefit from the animals that eat them', *Oikos*, **35**, 230–35.

Owen, D. F. and Wiegert, R. G., 1981. 'Mutualism between grasses and grazers: an evolutionary hypothesis', *Oikos*, **36**, 376–8.

Packham, J. R. and Harding, D. J. L., 1982. *Ecology of Woodland Processes*, Arnold, London.

Paine, R. T., 1966. 'Food web complexity and species diversity', *Am. Natur.*, **100**, 65–75.

Phillipson, J., 1973. 'The biological efficiency of protein production by grazing and other land-based systems', in *The biological efficiency of protein production* (ed. Jones, J. G. W.), CUP., London, pp. 217–37.

Pianka, E. R., 1981. 'Competition and Niche Theory', in *Theoretical Ecology: principles and applications* (ed. R. M. May), (2nd edn.), Blackwell Scientific Publications, Oxford and Boston, pp. 167–196.

Pimentel, D., 1966. 'Complexity of ecological systems and problems in their management', in *Systems Analysis in Ecology* (ed. Watt, K. E. F.), Academic Press, London, 15–36.

Reichle, D. E., O'Neill, R. V. and Harris, W. F., 1975. 'Principles of energy and material exchange in ecosystems' in *Unifying Concepts in Ecology* (eds van Dobben, W. H., and Lowe-McConnell, R. H.), Junk, The Hague, pp. 27–43.

Ricklefs, R. E., 1980. *Ecology*, Nelson, London.

Ridge, I., 1975. Open University broadcast, BBC 2, London, 14 Feb. 1975.

Slobodkin, L. B., 1959. 'Energetics in *Daphnia pulex* populations', *Ecol.*, **40**, 232–43.

Southwood, T. R. E., 1977. 'Habitat, the templet for ecological studies?' *J. Anim. Ecol.*, **24**, 337–65.

Stearns, S. C., 1977. 'The evolution of life history traits: a critique of the theory and a review of the data', *Ann. Rev. Ecol. Syst.*, **8**, 145–71.

Stoddart, D. R., 1965. 'Geography and the Ecological Approach. The ecosystem as a geographic principle and method', *Geog.*, **50**, 242–51.

Tansley, A. G., 1935. 'The use and abuse of certain vegetational concepts and terms', *Ecol.*, **16**, 284–307.

Teal, J. M., 1957. 'Community metabolism in a temperate cold spring', *Ecol. Monogr.*, **27**, 283–302.

Temple, S. A., 1977. 'Plant-animal mutualism: coevolution with Dodo leads to near extinction of plants', *Science*, **197**, 885–6.

Watts, D., 1971. *Principles of Biogeography*, McGraw-Hill, London.

Went, F. W. and Stark, N., 1968. 'Mycorrhiza', *Bioscience*, **18**, 1035–9.

Whittaker, R. H., 1975. *Communities and Ecosystems*, Macmillan, New York.

Whittaker, R. H. and Likens, G. E., 1975. 'The biosphere and man' in *Primary productivity of the biosphere* (eds Lieth, H., and Whittaker, R. H.), Springer-Verlag, Berlin, Heidelberg and New York, pp. 305–28.

Yodzis, P., 1981. 'The stability of real ecosystems', *Nature*, **289**, 674–6.

6

ECOLOGICAL FACTORS AND ENVIRONMENTAL VARIATIONS

INTRODUCTION

In Chapter 3 the many influences at work which determine the properties and characteristics of soil types were outlined. In a similar manner, each biotic community represents the response of a set of species to a set of ecological factors. An ecological factor is defined as one having an influence, maybe decisive, on the presence, absence or performance of a species. This broad definition reflects the fact that there may be almost countless influences at work on a community: each varying in time and space and acting reciprocally with the other factors. Although some progress has been made in assessing the significance of certain ecological factors on some individual species or communities, we still know little about how every environmental factor operates on a single species, let alone a simple community.

As a convenient starting point for some understanding of this complexity, Major advocates a *functional, factorial approach* to plant ecology similar in many respects to Jenny's factorial approach to soils (Chapter 3). In this case the equation reads:

vegetation = f (climate, parent materials, relief, organisms, time)

For the parent material factor, the edaphic factor could be substituted as a more appropriate term. These factors may be arranged in a two-way table, so that we may consider either the influence of each factor in turn or how various factors interact to produce the observed patterns in plant communities. Whatever the approach it becomes complicated because the time factor influences all other environmental parameters. However, as Odum states, 'the aims of environmental analysis are not to make long uncritical lists of possible "factors" but rather to achieve these more significant objectives: (1) to discover, by means of observation, analysis and experiment, which factors are "operationally significant" and (2) to determine how these factors bring about their effects on the individual, population or community, as the case may be'.

To achieve this, certain concepts are particularly valuable. The first of these stems from the work of Liebig in 1840 on agricultural crops. He noticed how the growth of a plant is often controlled by the amount of nutrient which is supplied to it in minimum quantity. This gave rise to Liebig's *law of the minimum*: there is a critical minimum level for any essential plant requirement and if that material is at or below this level in the environment it will act as a limiting factor.

Shelford expanded this idea by showing how too much of something can be as limiting as too little. Each species has an upper and lower limit of tolerance (*tolerance range*) for each factor. However, because factors interact, one may compensate for deficiencies in another; some factors may be more critical at one stage in a species life cycle than at other times (vulnerability is often marked in the early reproductive stages); a species can have a wide tolerance range for one factor but a very narrow range for a second factor (see Fig. 6.1). As Toumey states, 'when the absence of a particular vegetation is accounted for by one factor becoming limiting, invariably other factors, either directly or indirectly, play an important part in determining at what point the so-called limiting factor actually becomes most significant'.

These considerations make it very difficult to determine accurately the limiting and the optimum conditions for a species. Nevertheless, an awareness of these concepts aids the search for the most critical ecological factors. We can then focus attention on these, ignoring all environmental phenomena which, in the phrase of Mason and Langenheim, are not 'operationally significant'.

Difficult though it is to measure precisely the response of plants to ecological factors, this knowledge, apart from being vital for an understanding of plant and community patterns, enables plants to be used as environmental indicators. Some species have narrow tolerance ranges and the prefix *steno* denotes this response to particular factors, e.g. stenothermal – a narrow temperature range, stenophotic – a narrow light range. Others have broad tolerance ranges and the prefix *eury* is used in these cases, e.g. euryhaline – a wide salinity tolerance, euryphotic – a wide tolerance of light conditions. Obviously, 'steno' species are better environmental indicators than 'eury' species because of their more precise requirements. Plants and animals can thus provide much useful information about environmental conditions, often more quickly and cheaply than instrumental methods. Much environmental measurement is available only in the form of average values (particularly the climate parameters) which do not relate to the immediate environment of the plant. The plant, through its growth pattern, is a good integrator of all environmental variables. Best results are obtained when the response of whole populations or communities are considered rather than a single species.

An example where such investigations are proving successful is in studies of atmospheric pollution. Here the sensitivity of certain lichen communities indicates atmospheric levels of sulphur dioxide. Hawksworth and Rose present a detailed 0–10 point scale of lichen response to pollution. A simplified version of this, applied by Rose to the area around London, gave the following zonation:

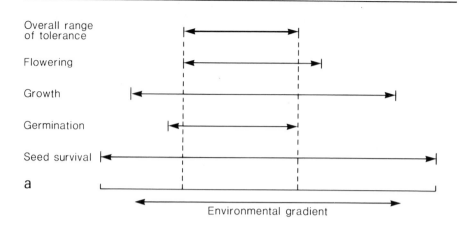

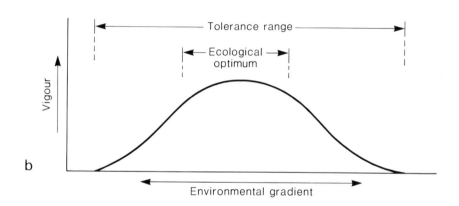

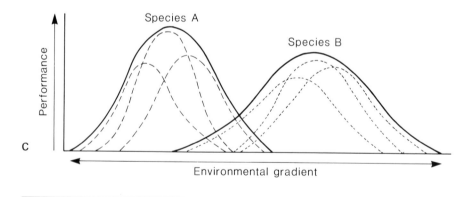

Fig. 6.1 Tolerance ranges along environmental gradients: (a) at different life cycle stages of a hypothetical plant species; (b) the ecological optimum of a plant species; (c) inter- and intra-specific differences in performance of two plant populations. (Modified from Kellman, 1980.)

1. Built-up central area of London, no lichens on trees.
2. A zone in which the grey-green crustose lichen, *Lecanora conizaeoides* appears at base of trunk first and then gradually higher up the tree.
3. A zone in which the grey, narrow-lobed foliose lichen, *Parmelia physodes* appears (corresponding to the beginning of open country in Kent and Surrey due south of London). This species is now known as *Hypogymnia physodes*.
4. A zone of grey, broad-lobed foliose lichens, *Parmelia saxatilis* and *Parmelia sulcata*.
5. A zone where the yellowish, broad-lobed foliose lichen, *Parmelia caperata* appears (at about 20 miles from the city centre on the south, 50 miles to the north).
6. A zone with two smooth, grey foliose lichens appearing, *Parmelia perlata* and *Parmelia reticulata*, together with *Usnea subfloridana*, a fruticose lichen (at about 30 miles to the south, but up to 100 miles to the north-east).
7. A fragmentary zone, confined to old forest and parklands in sheltered valleys, where the large lungwort lichen, *Lobaria pulmonaria* is found.

Rose and Hawksworth have recently demonstrated the sensitive response of lichens to cleaner air: some species are now recolonizing parts of Greater London following the falling sulphur dioxide levels of the last 15 years. However, Andre *et al.* have recently suggested that *Humerobates rostrolamellatus*, a small mite common in temperate orchards, might be used as a quicker method of monitoring air pollution. Its response time to sulphur dioxide levels is one week compared to several months for lichens.

A second example is the demonstration by Cannon of the usefulness of some deep-rooted pines and junipers in prospecting for uranium. Uranium taken up by the roots becomes incorporated in the foliage and even the presence of a minute quantity may point to a site with commercial mining possibilities.

It is beyond the scope of this book to attempt a coverage of all environmental factors in plant ecology. Indeed whole books have been devoted to studying just one factor in depth. Here only two aspects will be considered in some detail. It is hoped that these examples will guide students in the approach to adopt when pursuing other examples on their own. The two selected are both influenced at times by the actions of man; and this brings out another feature of environmental variation which is of increasing importance.

First, we shall consider a major botanical phenomenon, the altitudinal limit of tree growth (the tree-line), and examine the factors which are thought to determine the position of this important plant boundary. To quote Odum: 'oftentimes a good way to determine which factors are limiting to organisms is to study their distributions and behaviour at the edges of their ranges. In such marginal situations one or more environmental factors may undergo sudden or dramatic change, thus setting up a natural experiment which is often superior to a laboratory experiment, because factors other than one under consideration continue to vary in a normal manner instead of being "controlled" in an abnormal, constant manner'. Some tree-line studies are classic examples of this type of investigation. However, field studies do not always provide the full answer and laboratory experiments are sometimes needed to isolate the effective causes of the phenomenon under study.

Secondly, fire as an ecological factor will be examined in terms of its impact on ecosystem components. Particular stress will be placed on man's use of fire in ecosystem management. Both examples should demonstrate clearly the intricacies of factor-species interactions.

TREE-LINE STUDIES

The term tree-line has sometimes been used loosely and requires a definition. It implies a more or less clearly marked line where tree growth ceases. It is either a latitudinal limit to tree growth, such as the taiga-tundra boundary of Canada, Scandinavia, Northern Russia and Siberia, or an altitudinal limit on mountain sides, marking the point where subalpine communities give way to low-growing alpine vegetation. But *inverted* tree-lines can also occur when cold air drains into mountain basins. There is then a lower limit to the tree zone; a pattern well seen on some Australian mountains (Fig. 6.2) and reported by Billings for mountain basins in Nevada (Fig. 6.3).

It might also be legitimate to speak of tree-lines where trees give way to grasslands because of the increasing continentality of the climate as in the vast prairie tracts of interior North America and Asia and in the savannas of Africa.

Fig. 6.2 Inverted tree-line surrounding a 'frost hollow' at 1,520 m on Mount Buffalo, Victoria. The snow gums (*Eucalyptus pauciflora*) are replaced by herbfield, grassland or dwarf-shrub vegetation below the tree-line.

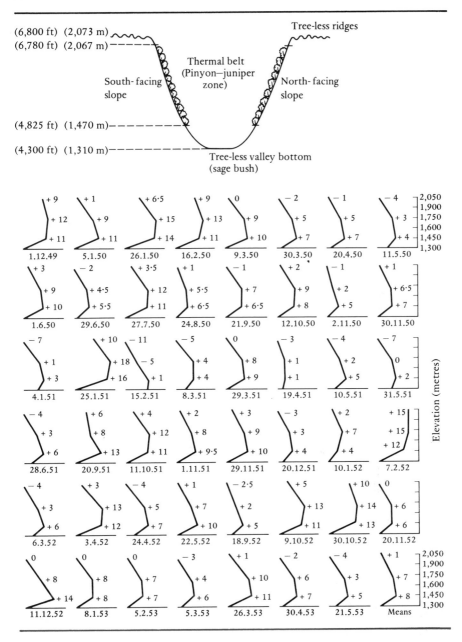

Fig. 6.3 Temperature inversions in the Pinyon-Juniper zone (*Pinus monophylla-Juniperus osteosperma*) of a Nevada mountain range. Temperatures were recorded at six stations at various elevations for the period 1 December 1949 to 21 May 1953. Only the values for the south-facing slope are shown here. For the three higher elevations on this slope the values are expressed in terms of their departure from the value recorded at the lowest site in the tree-less valley bottom. North-facing stations were usually only a degree or two colder than those south-facing at the same elevations. The thermal belt coincides closely with the Pinyon-Juniper zone. (Modified from Billings, 1954.)

Fig. 6.4 Altitudinal limit of spruce forest near Mt. Amery (3,140 m) in the Banff National Park, The Rockies, Canada. The limits of trees growing at forest densities (waldgrenze) and the upward extension of open, poorly grown trees to form the tree-line (baumgrenze) are shown.

Fig. 6.5 Close-up of deformed spruce trees showing flag-like growth-forms (krummholz condition) at the altitudinal tree-line in the Banff National Park. In the foreground, a dense alpine willow scrub.

But the complicating factors of soil influence and a long history of human interference (mainly through use of fire – see later sections of this chapter) mean that abrupt termination of tree growth in these cases is almost certainly artificial. In natural conditions there would be, and often is, a broad transition zone or interdigitation of the forest and grassland vegetations. So it is not usual to use the term tree-line in these particular cases. In this chapter study will be confined to the altitudinal tree-line.

European ecologists use three terms which help to sharpen the focus on altitudinal limits of tree growth. The upper limit of tall, erect tree growth occurring at forest densities is known as the timber-line or *waldgrenze*. The *kampfzone* refers to the zone between timber-line and tree-line (or baumgrenze). The *baumgrenze* is the line through the last few scattered trees on the mountain slopes. Many of the trees in the kampfzone and up to the baumgrenze will be deformed, dwarfed or prostrate in appearance due to the severe environmental regime (i.e. they show the *krummholz condition*, Figs. 6.4 and 6.5). These three features need not always be present. In the Southern beech (*Nothofagus*) forests of New Zealand the timber and tree-lines sometimes coincide and Wardle defines tree-line here as the upper limit of trees and shrubs more than 2 metres high.

Figs. 6.6 and 6.7 illustrate the general position of the tree-line in two areas which will be referred to subsequently when tree-line ecology is discussed. In the Cairngorms the actual tree-line runs at about 490 m, but this is an artificially low level caused by centuries of grazing and burning at the forest edge. The main tree here is *Pinus sylvestris* var. *scotica* (Scots pine), although locally *Betula pubescens* (birch) and *Sorbus aucuparia* (rowan) may also be prominent.

In the high core of the Central Alps in Austria (e.g. Mt. Patscherkofel near Innsbruck) *Picea abies* (spruce) and *Larix decidua* (larch) are well represented at the waldgrenze, which runs at 1950 m, *Pinus cembra* forms the baumgrenze at about 2,150 m and dwarf specimens of all three are found in the kampfzone between (see Figs. 6.8 and 6.9). The Central Alps demonstrate the '*Massenerhebung effect*' whereby biotic boundaries occur higher on larger mountain masses than on smaller or isolated mountain systems in the same latitude. Here the tree-line runs much higher on the siliceous core than on the less massive calcareous mountains to the south and north (e.g. South Tyrol). This area also illustrates another phenomenon teaching us to be wary of immediate interpretations. Whereas the restricted growth-forms of most tree species in the kampfzone are due to environmental impact, this is not true of *Pinus mugo* (dwarf mountain pine). In the northern Alps, it occurs as an extensive prostrate tree scrub at the tree-line but retains this growth-form even when grown at low altitudes. It is therefore a genetically determined krummholz and occurs as such, irrespective of external environment.

TREE-LINE THEORIES

It is now generally agreed that climate is the major determinant of the Alpine

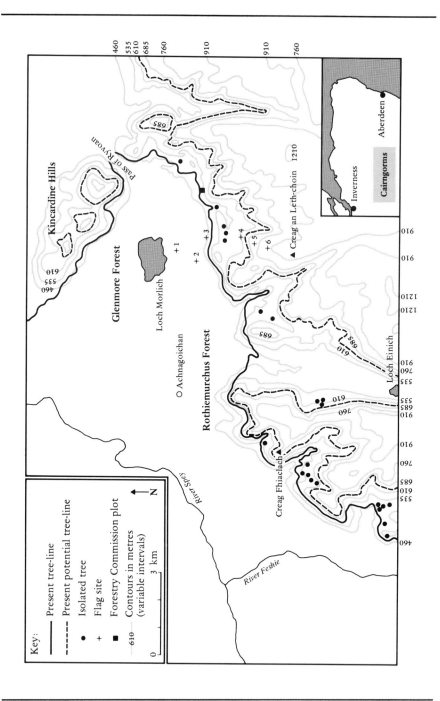

Fig. 6.6 The present tree-line and present potential tree-line in the Western Cairngorm Mountains, Scotland.

tree-line position (although locally other factors may be vital, as Veblen *et al.* demonstrate for the Chilean Andes where massive slope disturbances, mainly due to volcanism, are particularly important in determining the altitude of forest and its floristic composition and structure). Despite this, there has always been the problem of deciding whether one climatic variable acts as the limiting factor or whether a factor-complex (several interrelated climatic parameters) operates to limit tree growth. Only recently has refined research isolated a probable mechanism to explain alpine tree limits. The early theories may be briefly stated as follows:

1. *Excessive light.* The increase in light intensity with altitude was thought to impair leaf functioning and set a critical upward limit for tree taxa.
2. *Carbon dioxide deficiency.* It was argued that the decrease in partial pressure of atmospheric carbon dioxide which occurs with altitude might be the important factor limiting the upward extension of tree growth.
3. *Snow depth.* Work in North America had suggested excessive depth and persistence of snow prevented the establishment of tree seedlings.

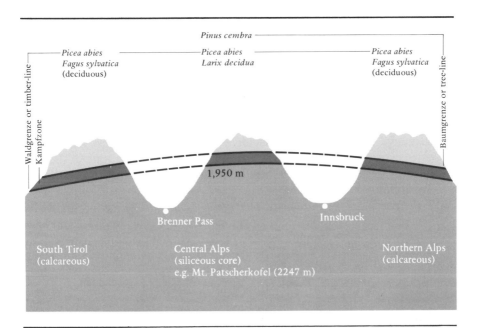

Fig. 6.7 A generalized cross-section of the Austrian Alps showing the tree species present at the altitudinal limits of forest growth and the 'Massenerhebung effect' (see text). On Mt Patscherkofel (2,247 m) the waldgrenze occurs at 1,950 m and the kampfzone extends to the baumgrenze, which is formed by *Pinus cembra* at an altitude of 2,150 m.

Fig. 6.8 Tree species at the altitudinal limits of woodland on Mt. Patscherkofel (2,247 m), near Innsbruck: *Pinus cembra* (left foreground and forming baumgrenze in background), *Larix decidua* (centre and right-foreground), *Picea abies* (between *Larix* and *Pinus*).

4. *Light deficiency*. The commonly observed increase in cloudiness in many mountain regions reduces the light intensity and perhaps this might become critical in setting forest limits.

5. *Wind*. The deformed appearance of many trees at the forest edge has long suggested wind to be a key factor in their ecology. Griggs, who worked extensively on North American tree-lines, was a firm advocate of the wind factor. Desiccating winter winds were considered to be particularly important.

6. *Desiccation during temperature inversion in winter*. The upper timber-line sometimes coincides with the upper surface of a cool, humid cloud-layer. Above this layer bright sunlight, high air temperatures, low relative humidity and scanty snow cover lead to high transpiration stress in plants. This could result in desiccation of shoot growth. These inversions were thought to be common enough in Central Europe to be a possible cause of tree limits.

7. *Heat deficiency*. Cold tree limits roughly coincide with the 10 °C isotherm for the mean temperature of the warmest month. Thus both Arctic and Alpine tree-lines represent the point beyond which heat energy is inadequate for the metabolic requirements of trees. Trees, being a life-form with a large unproductive mass (stem and roots), would need high annual requirements for growth (wood accumulation).

Fig. 6.9 Wind-shaped and stunted larches (*Larix decidua*) in the kampfzone at 2,100 m on Mt. Patscherkofel. It was on trees in this area that the research into frost-drought (see text) was carried out. (Photograph by permission of Professor W. Tranquillini.)

In 1954 Daubenmire argued that whereas each of these theories may go some way towards explaining tree-limits, or be particularly important in a local context, none by itself explained the worldwide phenomenon of the tree-line. He reasoned, 'because a great many genetically distinct trees contribute different segments of a timber-line pattern that has remarkable geographic conformity, the hypothesis is suggested that a major autecologic principle is involved that may be analogous to the wilting coefficient of the soil, in that some environmental complex abruptly exceeds the tolerance of all trees regardless of variations among them'. The factor-complex hinted at by Daubenmire affects tall and low plants differently since many species of shrubs and herbs are not prevented from extending many hundreds of metres higher on the mountain side but only the trees are so restricted. This led him to suppose that 'the critical intensities of the atmospheric complex do not extend all the way to the ground surface'. In other words, low-growing vegetation is taking advantage of the much better microclimatic conditions existing in the warmer air layer near the soil surface.

RECENT RESEARCH

More recent work helps to identify this factor-complex. Ecologists have not dismissed completely the earlier theories but rather have sought to understand fully the exact physiological workings of the factors responsible.

The key studies in establishing the most widely accepted modern theory on Alpine tree-lines were carried out at two main centres. In the Alps, the Austrian Federal Forestry Research Institute have a Subalpine Forest Research Field Station near Innsbruck. Its work is related to the realization that one of the main causes of disastrous avalanches in the Tirol (e.g. 1951, 1954) has been the retreat of the timber-line. Grazing and clearance have lowered the forest edge by an estimated 100 vertical metres in recent centuries and halved the forest area. Professor Tranquillini and his colleagues are using both field studies and sophisticated laboratory experiments with the ultimate aim of providing guidance when these upper slopes are re-afforested.

In New Zealand, Wardle has studied the growth performance of tree seedlings below, at and above the natural tree-line (about 1,300 m) in the Craigie-

Fig. 6.10 Travers Valley, St. Arnaud Range, Nelson, New Zealand, showing the level timber-line formed by the Southern Beech (*Nothofagus solandri*). Krummholz above the forest is absent. Ridge tops at about 1,900 m. (Photograph by J. H. Johns, reproduced by permission of New Zealand Forest Service and Dr. P. Wardle.)

Fig. 6.11 Experimental garden at 1,100 m in the Craigieburn Range, Canterbury, New Zealand where tree-line species from many parts of the world are grown under varying conditions of soil and shade. (Photograph by J. S. Cocks, reproduced by permission of Dr. P. Wardle and *New Zealand Journal of Botany*, **9**, 371–402.)

burn Range near Canterbury. In addition to the native timber-line species, *Nothofagus solandri* var. *cliffortiodes* (the Southern Beech), he has also grown under varying conditions of shade and soil in experimental gardens at these levels species which form the tree-line elsewhere, e.g. *Picea engelmannii* (spruce) from Colorado, *Pinus hartwegii* (pine) from Southern Mexico, *Eucalyptus niphophila* (snow gum) from New South Wales, *Sophora chrysophylla* from Hawaii and *Podocarpus compactus* from Mt. Wilhelm, New Guinea (Figs. 6.10 and 6.11). Benecke has studied upper forest levels in the Craigieburn Range using a sophisticated mobile laboratory which can measure directly the gas-exchange processes (photosynthesis, respiration and transpiration) of trees growing naturally in the field. Small branches still attached to the tree are sealed within electronically controlled gas-exchange chambers which include a highly sensitive infrared gas analyser to record carbon dioxide uptake as a measure of photosynthesis. Data from this work will be used in programmes to revegetate badly eroded mountain slopes (Figs. 6.12, 6.13 and 6.14).

Research in both areas strongly supports an idea advanced by Michaelis in 1934. He studied *Picea excelsa* (now known as *P. abies*) in the Alps and demonstrated that whereas the foliage of trees below the timber-line maintained a healthy water balance all year round, the shoots of trees in the kampfzone and at the baumgrenze often dried out to a lethal degree in winter. This winter desiccation ('*Frosttrocknis*' or frostdrought) he related to lack of

Fig. 6.12 The New Zealand Forest Research Institute's mobile physiology laboratory monitoring gas-exchange of trees at 1,100 m in the Craigieburn Range, Canterbury. The trees in the foreground are exotic pines. Several species of native and exotic trees are under trial for protection forestry on these badly eroded mountain slopes. In the background remnants of the native Southern Beech forest can still be seen on some slopes. (Photograph by J. H. Johns, reproduced by permission of New Zealand Forest Service and Dr. U. Benecke.)

'transpiration resistance'. Development of this resistance to water loss depended on conditions during the preceding summer: drought damage occurs because the short growing season at these altitudes does not permit needles to become fully mature.

Wardle defines this ripening process in the shoots morphologically 'as the completion of growth, the loss of the "soft" appearance imparted by high water content, incompletely lignified cell walls, and thin cuticles. Physiologically it is the acquisition of an ability to withstand low temperatures and desiccation, which is associated with increased osmotic concentration of cell sap, decrease of free water in the protoplasm, permeable protoplasts tolerant of considerable dehydration and tissues undamaged when ice forms between the cells'.

At different habitats on Mt. Patscherkofel near Innsbruck, the Austrian team have studied the cuticular transpiration and course of desiccation in *Picea abies* and *Pinus cembra*, two important timber-line conifers. The sites examined were at the timber-line (1,940 m), in the kampfzone (2,090 m) and at the wind-exposed and wind-protected tree-line (2,140 m). Measurements of the cuticular transpiration of excised first and second year growth of both species were made

Fig. 6.13 A gas-exchange chamber (cuvette) being used to monitor photosynthesis and transpiration in the field. Air tubes and electrical cables connect the cuvette with recording equipment housed in the mobile laboratory – Fig. 6.14 shows details of these arrangements. (Photograph by J. H. Johns, reproduced by permission of New Zealand Forest Service and Dr. U. Benecke.)

in a climatized wind tunnel under constant climatic conditions – temperature 15 °C, dew point 2·5–4·0 °C, wind speed 4·0 m/s, artificial light 8,000–10,000 lux (see Figs. 6.15, 6.16 and 6.17). These laboratory approaches permit a particular factor to be isolated and measured by controlling many of the other variables in the study. But the results must always be tested against the reality of the field situation (Figs. 6.18, 6.19 and 6.20).

Field observations by Tranquillini of shoot desiccation at various altitudes under the extreme winter conditions of 1972/3 showed a very similar development to that measured in the climatic chamber. The first year growth of *Picea abies* at the wind-exposed tree-line was already showing desiccation by February. By March, shoots in the kampfzone and at the wind-protected tree-line were similarly damaged. This first year growth of *Picea* in the field was poorly developed because the previous summer of 1972 had been short and unusually cool. Second-year branches of *Picea*, developed during the considerably better summer of 1971, were much more mature and well able to withstand the following winter severity. Samples of *Pinus cembra* from all zones below the tree-line exhibited no winter desiccation damage and this is obviously a species which can complete its ripening and control its cuticular transpiration losses better than *Picea*. It thus grows higher on the mountainside than *Picea*, forming the baumgrenze in this area at 2,140 m.

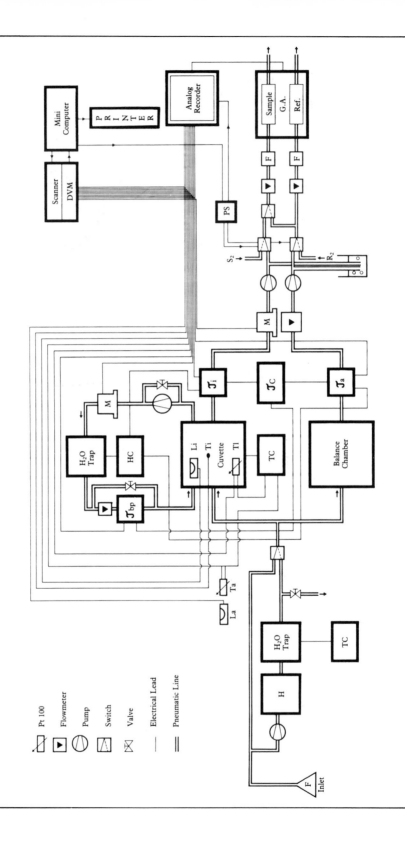

Fig. 6.14 Flow chart for thermoelectrically controlled gas-exchange cuvette (after H. Walz, Effeltrich). Differential CO_2 measurement in open system. Transpiration measurement by water removal in compensating by-pass loop.

F   Filter
G.A. CO_2 Gas Analyser UNOR2
H   Humidifier
HC Humidity Controller
La Light Sensor – ambient
Li Light Sensor – in cuvette
M   Mass Flowmeter
PS Programme Switch
R_2 Reference Gas, 2nd Gas-Exchange
    Unit
S_2 Sample Gas, 2nd Gas-Exchange Unit

Ja Dewpoint Sensor – ambient
Jbp Dewpoint Sensor – by-pass
Ji Dewpoint Sensor – in cuvette
JC Dewpoint Sensor Controller
Ta Resistance Thermometer (Pt 100)
    ambient air
Ti Resistance Thermometer (Pt 100) in
    cuvette
Tl Thermocouple, leaf temperature
TC Temperature Controller

(Reproduced by permission of Dr. U. Benecke.)

Fig. 6.15 The wind-tunnel at the Subalpine Forest Research Field Station ('Klimahaus') near Innsbruck. Spruce seedlings subjected to various wind speeds under controlled conditions for light, temperature and humidity, to assess desiccation damage and growth performance. (Photograph by permission of Professor W. Tranquillini.)

In discussing the ecological significance of 'Frosttrocknis', Tranquillini points out that the progressive decrease in the thickness of cuticular layers seen in woody species with increasing altitude is accompanied by an exponentially increasing water permeability of these layers. Trees at the upper forest margin have low cuticular transpiration resistance and thus use up their water reserves much more rapidly than mature trees of lower sites. Baig and Tranquillini have shown that for continental climates high solar radiation levels in winter result in intense heating of plant parts protruding above the snow surface. Mean

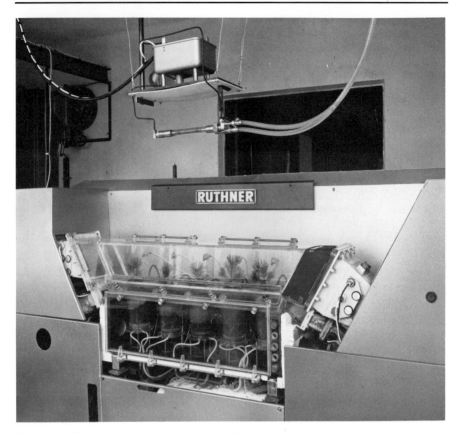

Fig. 6.16 The 'phytocyclon' at the 'Klimahaus' on Mt. Patscherkofel: *Pinus cembra* seedlings growing in a controlled environment for roots and shoots, where ecological factors for each environment can be varied separately and measured simultaneously. (Photograph by permission of Professor W. Tranquillini.)

daytime leaf temperatures in April are 14·4 °C above ambient air temperature. They conclude that high temperatures to a greater extent and wind to a lesser extent are the causative agents in the process of excessive water loss through cuticular transpiration that leads to winter desiccation at the tree-line. However, in regions where winter solar radiation is weak, such as the far north, or where persistent strong winds blow, such as in oceanic mountain climates, the wind factor may be the more important cause of increased winter transpiration of leaves.

The highest timber-lines for a given latitude are found in continental regions with severe winter regimes. As Wardle notes, this apparent anomaly is understandable when it is realized that successful shoot ripening demands, first, the completion of growth (which is dependent on the previous summer regime) and, secondly, the development of hardiness. Each species has a characteristic degree of cold hardiness and the first slight frosts of early winter induce the full development of this specific tolerance if, and only if, growth has first been completed. While many trees can withstand very low freezing temperatures without too much damage, they cannot tolerate extended loss of water from the tissues through evaporation from the leaves unless it can be adequately

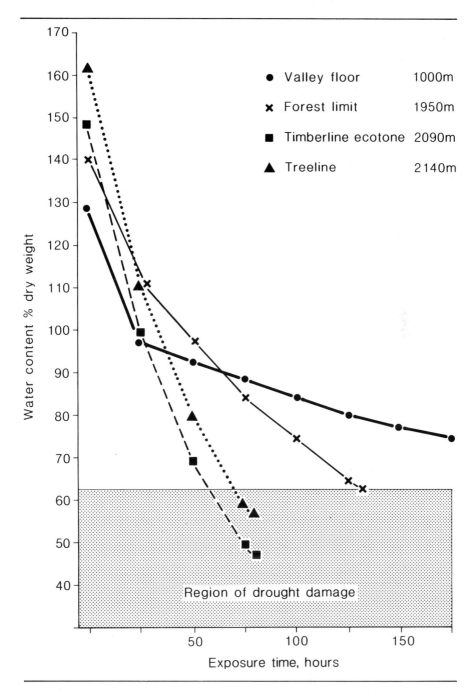

Fig. 6.17 Decrease in water content of detached 1-year old *Picea abies* shoots collected in December, during desiccation in a climate-controlled wind tunnel at 10 klx, 15 °C, 40% R. H. and 4 ms^{-1} wind velocity. The branches stood in water before the experiment, thus the initial water content was close to saturation. The shoots were from different altitudes (as indicated) on Mt. Patscherkofel. The region in which water contents are low enough to cause desiccation injury is indicated by shaded area on graph. The results show that the higher the elevation from which the shoots originate the more quickly they reach injurious levels of desiccation, under the same desiccation conditions. This is due to the sharp increase in cuticular conductance which occurs with increasing altitude. (Modified from Tranquillini, 1982.)

Fig. 6.18 Measurement in the field of girth increase in larch (*Larix decidua*) in winter at the tree-line (1950 m) on Mt. Patscherkofel using dendrographs and trunk respiration meters. (Photograph by permission of Professor W. Tranquillini.)

Fig. 6.19 Measurement of climatic conditions in the field: Mt Buller, Victoria. Rimco 'Sumner' automatic long-period recorders for temperature, humidity, wind speed and direction at the lower limit of snow gums (*Eucalyptus pauciflora*), where this species (white bark) is replaced by alpine ash (*E. delegatensis*) at about 1,350 m.

Fig. 6.20 Rimco 'Sumner' automatic long-period recorders at the upper limit (1,670 m) of snow gums, Mt. Buller, Victoria (see Fig. 6.19).

replenished by the root system. When frozen soil conditions persist this becomes a major problem. Anything which reduces the effective growing season (and thus curtails the ripening process), such as late snow pack, frost hollows, wet hollows or drought-prone slopes, will also tend to preclude the establishment of tree seedlings in the kampfzone. Tranquillini has shown desiccation damage to occur earlier in less developed shoots. Using 4-year-old *Picea* shoots, he found that those which had developed under the better growing conditions of a valley site (700 m) could withstand the test conditions of 15 °C and 43 per cent relative humidity without showing signs of drought damage for twice as long as those poorly developed shoots which had grown under conditions of a shortened vegetation period. Spruce trees at the timber-line in Austria need at least three months of full summer activity in order to survive. These investigations at continental alpine sites confirm experimentally the hypothesis of Michaelis.

To these basic ideas, supported by research in the Craigieburn Range, New Zealand, Wardle adds a note on the importance of climatic conditions near the ground surface. Leaf temperatures can vary enormously depending on whether the foliage is close to the ground or some distance above it, whether shaded or in bright sunlight and whether exposed to winds or wind protected. Tree seedlings will only become established in the kampfzone in so far as they are able to tolerate or avoid any adverse microclimatic variations found there. Some tree species, such as *Nothofagus solandri*, only become established as upright mature trees at the timber-line in shaded sites where temperatures are close to the ambient air temperature. Where temperature gradients above the ground are steepened by such factors as severe winds then seedlings will show the krummholz growth-form. So the microtopography and its associated microclimatic pattern may be all-important in determining local variations in tree form within the kampfzone.

WATER DEFICITS AND STRONG WINDS

Desiccation of the foliage implies a water deficit. This will come about when water loss through evaporation and transpiration exceeds the replenishment rate to the plant. Trees, because of their large life-form, require very large amounts of water. Thin, stony mountain soils developed on steep slopes will retain little water and in winter this may be unavailable because it is frozen. While a water deficit may arise simply through these soil conditions, it is much more likely to do so when low availability is combined with high transpiration and evaporation losses caused by exposure to bright sunlight and/or persistent drying winds.

A feature of many mountains in temperate oceanic latitudes is persistent strong wind flow, particularly in winter, and on British mountains these flows are more characteristic than bright sunshine. While the importance of wind in tree-line ecology is recognized, few specific studies exist. Odum, discussing limiting factors in ecology, stated that ecologists, being preoccupied with temperature and moisture measurements, have neglected to give due consideration to the effects of wind (studies by Sprugel (see Chapter 4) and Grace are notable exceptions). An investigation of wind as an ecological factor was undertaken by the author as part of a study of tree-line conditions in the Cairngorm Mountains of Scotland.

The present tree-line in the Cairngorms shows features of a natural forest edge along one small section only. Here, on the western flanks of Creag Fhiaclach at 640 m, exposure is severe and several pines exhibit the krummholz condition (see Fig. 6.6). Elsewhere, the present tree-line is much lower, frequently at about 490 m. It shows few signs of a natural condition because of centuries of human interference. The natural (or potential) tree-line for the area would extend to 610 m in all sections and to 690 m along sheltered slopes. These forests are the best surviving remnants of the Caledonian Forest which formerly covered much of Highland Scotland.

Wind was studied as one of the factors considered to be preventing the re-establishment of tree seedlings in the zone between the present actual tree-line and the present potential tree-line. Tree seedlings exposed to persistent strong winds suffer from: (a) severe mechanical stress which can kill windward branches, (b) abrasion near ground level from wind armed with readily available grit or frozen snow-ice particles, and (c) increased desiccation, which may be particularly marked in late winter and early spring when water is frozen in the soil.

There is a reciprocal relationship between wind and vegetation. The trees in a forest form a mutually protective system with regard to wind flow. Gloyne has shown that even a narrow belt of trees can effect a noticeable reduction of wind flow. With a medium-density tree barrier (50 per cent), wind reduction at horizontal distance beyond the barrier equal to $3h$ to $5h$ (where h = tree height) amounts to 80 per cent of the original flow. By leeward distances of

20h the speed is still only 60 per cent of the full force. The rather open type of natural pinewood typical of the upper slopes of the Eastern Highlands in the past would have functioned as a very efficient wind barrier. This is because a medium-density barrier of this type is more effective than a solid windbreak which only tends to induce vigorous turbulence which then causes wind speeds actually to increase on the leeward side.

Wind exposure in the potential zone was assessed by measuring the tattering rate of unhemmed Madapollam cotton flags. This is a simple method which Lines and Howell have compared with local anemometer records, finding high correlations. The flags, measuring on initial exposure 30·5 cm × 38·1 cm (12 × 15 in), tatter in a regular manner due to mechanical wear. Flags were exposed for eight-week periods throughout two years at six sites (see Fig. 6.6 for location). The sites were:

1. At 381 m within Glenmore Forest but centred in a large burnt tract of pinewood and therefore exposed to a reasonable wind fetch.
2. At 408 m under a mature natural stand of *Pinus sylvestris* within the main forest.
3. At 442 m along the present (artificial) tree-line of scattered and stunted pines.
4. At 634 m on the treeless lower slopes of Creag an Leth-choin.
5. At 732 m on the mid slopes.
6. At 880 m on the upper slopes of this mountain.

RESULTS AND INTERPRETATION

The seasonal and altitudinal variations in tatter rates are shown in Fig. 6.21. Comparison of sites 1, 2 and 3 shows the effectiveness of the protection system offered by mature woodland from wind exposure. Site 2 clearly enjoys the protection afforded by the surrounding pinewood which has been lost at site 1 by human activity. At site 3 – the present tree-line – tatter-rate values represent an increase out of all proportion to the moderate rise in altitude involved (only 34 m higher than site 2). These forest margins show only minimal retardation of wind flow and do not provide effective shelter because of their artificial nature.

At, and above, the present forest edge exposure is severe. Lines and Howell, who recorded tatter rates at 35 stations from the Southern Pennines to the Shetlands over a tree-year period, reported their highest values for western seaboard stations (14·3 cm^2 of flag lost per day at Carrick-Craiglure, in Ayrshire, altitude 427 m). The 'easterly' Cairngorms are generally thought to be less exposed than west coastal districts in Scotland. But the value for site 6 here (16·6 cm^2) is well in excess of Carrick-Craiglure and site 5 is only slightly below it. The winter and spring seasons are the windiest periods. Lines and Howell have argued that when conditions are otherwise favourable suitable tree species, selected from suitable sources, would succeed at sites where the tatter rate does not exceed one square inch per day (6·45 cm^2). In the Cairngorms, as Fig. 6.21 shows, this value is exceeded for most of the year even at the present tree-line. Since a short 'favourable' period would hardly compensate for longer periods of severe exposure, conditions are obviously very critical

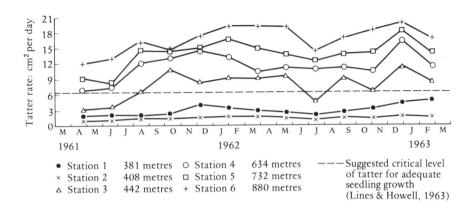

Fig. 6.21 Assessment of wind exposure by seasonal and altitudinal variations in flag tatter rates. Exposure was measured at six sites on Creag an Leth-choin, Western Cairngorms over a 2-year period (see Fig. 6.6 for site locations).

at these levels today. The tatter rate of one square inch per day is decidedly optimistic when applied to these native pinewoods. It was based by Lines and Howell on data for growth-rates of *Pinus contorta* (lodgepole pine). But this species, which is not native to the British Isles, grows significantly faster under windy conditions than the native Scots pine (as results from a Forestry Commission experimental plot at 490 m in the Cairngorms have shown: over the same period the Scots pine reached 2·7 m and the lodgepole pine grew to 4·6 m. Data from Benecke and Havranek also suggest that Scots pine has a greater heat requirement near timberline than *P. contorta*).

With the upper forest margins in Scotland in their present conditon, i.e. artificial and excessively 'open' in character, exposure to persistent strong winds would greatly increase water deficits in the tree seedlings through increased transpiration. This, in conjunction with the short summer ripening period at these altitudes and latitudes, would provide the conditions for the desiccation damage mechanism, suggested by Michaelis and substantiated by Tranquillini. It is not suggested that this is the only factor preventing the re-establishment of seedlings because animal grazing (deer, game-birds) also eliminates many individuals. Seedlings usually grow best when set among large groups of boulders, sites providing some protection from both wind exposure and biotic pressure.

ALTERNATIVE HYPOTHESES

Tranquillini notes that there has been a lively debate during the last 100 years on whether the winter damage to trees in the kampfzone is due to desiccation ('Frosttrocknis') or is induced solely by frost (injuries to the cells due purely to the freezing of contained water). He concludes that winter desiccation is the primary cause of damage on extra-tropical mountains which experience a continental climate with pronounced soil freezing and intense radiative warming of plant parts. However, this form of damage is not common to all forest limits.

At the *Eucalyptus pauciflora* timberline (*c.* 1,900 m) in South-eastern Australia (see Figs. 6.2, 6.19 and 6.20), Slatyer showed that leaves were not water-stressed in winter and he attributed the extensive damage to radiation frosts. In support of this hypothesis of frost damage, Wardle has recently pointed out that the damage seen in some conifer needles is often patchy in distribution, even among needles on the same shoot. This is difficult to explain in terms of a general water stress mechanism but would fit more readily the irregular pattern of cell freezing that might occur with frost activity.

Marchand and Chabot measured the relative water content of needles and shoots of *Picea mariana*, *Abies balsamea* and *Betula papyrifera*, three species forming the krummholz (1,280 m) on Mt. Washington, New Hampshire. They concluded that mechanical influences of wind caused the tree damage and not winter water stress; they also thought that tree-line position here was set by lack of viable seed above a certain altitude. Kincaid and Lyons, working on Mt. Monadnock, also in New Hampshire, where *Picea rubens* forms the tree-line at *c.* 800 m, likewise failed to establish water stress as a cause of shoot and leaf damage.

Most of these examples are from mountains that experience relatively oceanic climates with milder winters and heavy snowfalls. Thus mechanisms other than winter desiccation may be more important in the tree-line ecology of mountain regions which are unlike the classic continental alpine mountain systems of central Europe and the Rocky Mountains of North America.

Although uncertainties still remain in tree-line studies, the importance of research in this field has been stressed by Wardle: 'Timber-line is therefore one of the most significant boundaries in biological nature, separating two fundamentally different ecosystems. In one, living processes take place within a depth of tens of metres, and in the other they are usually restricted to a few centimetres. Since trees are at their climatic limits at timber-line they form a sensitive ecotone, and once this is destroyed the problems of restoring it are acute. The widespread disintegration of timber-lines which has occurred, and that which threatens to occur in the future, represent a serious failure by man to attain harmony with his environment.'

THE ECOLOGY OF FIRE

The examination of factors which might explain a specific biogeographical phenomenon (the tree-line) is now followed by a consideration of one particular ecological factor and how it operates to influence a whole range of biogeographical phenomena. Fires have occurred by natural means, such as lightning strikes, throughout the history of the earth, though increasingly today they are due to the activities of man.

Natural fires are quite common in certain regions: one storm crossing the State of California triggered off several hundred separate fires in its path. Many of these fires are associated with a weather pattern which includes heavy rainfall. Thus they may be short-lived and relatively unimportant in the ecosystem. However, this is not always the case. In interior Queensland, for example, the drought-prone grasslands of the Mitchell Plains are fired with sufficient frequency by 'dry' lightning strikes (unaccompanied by rain) for these to be important in their ecology. Even in areas that are cool and wet, natural fires do occur from time to time as charcoal layers found towards the base of peats in the Western Highlands of Scotland bear witness. These layers pre-date human settlement in the region and they are reasonably common. In the Western United States tree-ring studies by Weaver show average fire frequencies of approximately one every seven years or so, and the record extends back several centuries. Walker has reviewed the importance of fires over several time-scales (extending back many thousands of years) in the development of resilience by vegetation. Data from sites in the conifer-hardwood forests of eastern North America and Lake George in south-east Australia show how fire may facilitate major plant geographical changes. The fire-tolerant vegetations that have evolved can be disrupted by fires that are too frequent or too intense, but an equal threat to their resilience comes from unusual delay in the recurrence of fire. If perturbation is long delayed then competition and the emergence of a few dominants in the community lead to loss of species. This is a serious threat to the resilience of a vegetation attuned to periodic fire disturbance.

Man-induced fires may be of two types. First, the accidentally started fire, which is becoming common as the pleasure-seeking public probe further into areas of semi-natural vegetation in pursuit of recreation. Secondly, the intentional fire where there is planned use of fire to modify an ecosystem. Deliberate use like this has a long history. It is said that fire was the first 'tool' used by man for large-scale modification of his environment. Primitive people used fire to hunt game (e.g. the bison hunts of American Indians; the Australian Aborigines who burnt large areas to flush out insects, grubs and small reptiles for food). Much primaeval climax forest was cleared by the early cultivators using slash-and-burn techniques (methods still practised today in some primitive societies). Molloy et al. have established that the early Polynesian settlers of New Zealand destroyed much of the southern beech forest cover of South Island by fire. This was long before the arrival of European colonists, who carried on the practice of regular burning. This climax forest was poorly adapted for fire survival and its loss has been accompanied by heavy soil erosion.

In Britain fire was used regularly by Mesolithic hunting populations in the southern Pennine uplands over a 4,000 year period and Jacobi et al. show that this led to permanent suppression of closed tree cover at altitudes above c. 350 m and may have caused the onset of soil deterioration and the initiation of blanket peats. However, burning probably imposed a degree of predictability on the movement of deer herds. It may also have led to increased productivity of ungulates, a point explored by Mellars who argues that firing produced substantial improvements in the economic potential of Mesolithic environments. Improvements in the quality and quantity of food supplies (grass, etc.) led to increased carrying capacities and improved growth and reproduction rates for deer; the production of ungulates may have increased by a factor of $\times$ 10.

At later dates, fire was used as part of military stratagems by the Romans (to flush out the natives!) and by the Danes, Norse and Vikings during their raids around our coasts. Many so-called natural upland moors are, in fact, the product of centuries of burning and grazing. In the Scottish Highlands, 'muirburning' as a practice goes back a thousand years at least and occurred at any convenient time of the year. Only in the last century or so has the muirburning been regulated to a three-week period in the spring.

Ecologically, it is more meaningful to classify fires in terms of their impact once initiated. The dominant trees in a forest may have taken nearly a century to reach their present heights. If destroyed by fire it might take as long again for a similar forest to grow up. A wheatfield destroyed by fire may be a disaster to the farmer but it takes only twelve months or less to replace it with another one. It is not surprising therefore that much research into fire ecology is associated with forestry interests. Most of the world's forest cover is now essentially viewed as a crop. The extreme case of this is the monoculture of certain conifers, e.g. the large areas of exotic spruce plantations in Britain. With these interests in mind, the forester recognizes three types of fire: the crown (or canopy), the surface and the ground fire.

The crown fire is a severe form, involving the whole tree, and usually leads to the death of many species. The surface fire has less ecological impact and burnt material is confined to undergrowth species and organic matter on the forest floor. The ground fire, which might be thought similar to the last type, is a severe form since it penetrates below surface level and often destroys rooting systems. To penetrate in this way implies a supply of readily burnable material in the soil. Foresters are nearly always forced to plant on land unsuitable for agriculture. In Britain this often means upland moorland, much of which is underlain by peat. These peats are drained before planting and any subsequent fire may readily burn for a considerable time.

FIRE IMPACT

The ecological impact of fire is partly determined by its frequency. If fires are common and regular, the ecosystem over a long period tends to become adapted to this pattern. Certain species will be eliminated; others will adjust more readily to the fire disturbance. This can give rise to vegetation which is fire-dependent: a *pyric climax*. The Long-leaf Pine forests of south-east United States are an example. Likewise, if fire is not regularly experienced in a community, when it does occur its impact may be much more drastic. The weather pattern which precedes and follows a fire may have an important control over its ecological influence. A marked dry season will ensure plant material in a readily burnable state to 'feed' the fire. Strong, gusting winds following a fire will do much to widen its area of impact. A large fire will generate its own air turbulence. In these, and other obvious ways, the accompanying weather pattern influences the ecological role of the fire. The nature of the site where the fire begins must also be considered. The topography (slope angles and aspects) and characteristics of the soil (e.g. its structure and its subsequent susceptibility to erosion once the plant cover is destroyed) will each partly determine the impact of the fire, especially when considered in terms of fire frequency and associated weather pattern.

However, an influence often overlooked is that of the vegetation itself. The reciprocal nature of ecological factors was stressed before: fire influences vegetation but vegetation also influences fire. Plants vary in terms of their combustability and resistance to firing. There is also great variation in the amount of material they produce in a form which is easily fired. Some of the difficulties of fire-fighting in Australia are caused by the widespread dominance of *Eucalyptus* species. Many of these trees shed not only leaves but also bark. The bark may be shed as powder or flakes or as long strips, festooned about

Fig. 6.22 *Eucalyptus rubida* (candlebark) at Daylesford, Victoria: the strips of bark may act as tapers, causing fires to spread rapidly.

Fig. 6.23 Contrasting bark types in the genus *Eucalyptus*, Warby Range, northern Victoria. *E. macrorrhyncha* (red stringybark) retains its fibrous bark but *E. polyanthemos* (red box) and *E. blakelyi* (Blakely's red gum) have smooth or scaly barks which flake off to varying degrees. Most trees show fire scars around their base.

the trunk or lower branches (Figs. 6.22 and 6.23). This paper-thin material quickly ignites and an insignificant surface fire can be transformed into a severe canopy fire by the taper-like action of the bark strips. The *Eucalyptus* also contains volatile oils and these may vaporize to form gas pockets as fire temperatures build up. A spark carried on the wind can cause sudden ignition away from the main fire. The fire thus appears to 'jump', making the fire-fighting task extremely difficult. Some *Eucalyptus* forests produce vase amounts of litter so that a good supply of burnable matter is always on hand. *Eucalyptus regnans* (mountain ash, which can grow to 100 m or more) may produce 8·1 metric tonnes per ha per annum total litter. These values should be compared with those of other community types: Arctic-Alpine communities, 1·0; Cool Temperate Forest, 3·5; Warm Temperate Forest, 5·5; and Equatorial Rain Forest, 10·9. These *Eucalyptus regnans* communities, although examples of Warm Temperate Forest, have annual litter productions approaching those of the Wet Tropics. In addition, the prolonged dry summers will greatly increase the likelihood of devastating fires (e.g. the fires of February, 1983, in Victoria which caused the deaths of 69 people and $100 million damage).

Fire impact may be examined in terms of likely disadvantages and advantages to man. Expressing it this way obviously reflects a very anthropogenic view of this ecological factor. But since man often causes fire in ecosystems this approach is considered legitimate.

DISADVANTAGES

In a severe fire, apart from the obvious removal of much macrovegetation (which is, in most natural and semi-natural forest and grassland communities, either an actual or potential 'crop' for man), there may be much destruction of microflora and microfauna in the topsoil layers. These ecosystem components play a vital role in organic decomposition and soil fertility. Although potentially capable of quickly building up their numbers again, this recovery may be delayed since their development closely parallels that of the macro-vegetation, which may take many decades in the case of a forest climax community destroyed by fire.

Each plant community has an associated animal community and destruction of the former by fire has severe repercussions for the latter. The competitive balance achieved in the macrofauna (expressed in terms of food catchment areas, breeding territories, etc.) may be destroyed. Although larger animals may move away from the fired area it will not be easy for them to re-establish in adjacent communities. There will have been an overall reduction in carrying capacity for the region. Ecological niches, territories and food sources in these undisturbed areas will already have been 'claimed' by the macrofauna in possession. While these animals are not burnt many subsequently starve to death. Animals seeking new territories will also be very susceptible to predation.

Frequent fires may promote ecological instability and hold the vegetation well below the true climax level. This implies periods of only fractional utilization of ecosystem resources – e.g. fixation of solar radiation may be virtually zero for a time following a pronounced burn. Susceptible plants and animals will be eliminated and this leads to an impoverished biota. Well-adapted species come to dominate, and virtually monospecific plant communities may arise. An example is the large area of Western Australia now dominated by a *Lepto-spermum* community (a prickly shrub known locally as tea-tree). This plant is so adapted to fire that seeds are only released after scorching by flames of sufficiently high temperature. In Minnesota, *Pinus banksiana* (jack pine) shows a similar dependence on fire for the release of seed from the cones. This has led to the conversion of many mixed *Pinus resinosa* (red pine) and *Pinus strobus* (white pine) stands to jack pines. In other areas annuals may replace perennials, simply because of their quicker recovery rate after a burn. Likewise, deep-seated rhizomes may be favoured in that the topsoil offers some protection from the worst ravages of fire. In this country *Pteridium aquilinum* (bracken) is a plant with this growth-form and fires may give it a competitive edge, allowing it to spread and dominate.

In all ecosystems large amounts of chemical nutrients may be locked in the standing biomass, i.e. the leaves, branches, trunks, fruits, cones, etc. of the green plant, for considerable periods. Fire releases them, but whether they then return to the soil for re-use by the autotrophs depends on the severity of the fire and any subsequent soil erosion. Most fires are followed by some soil erosion and chemicals may be lost. For example, the leaching of phosphorus may be particularly unfavourable on coarse-textured soils. When fire temperatures exceed 400 °C, and they frequently do, important chemical nutrients may be removed in the smoke. However, the picture is complex as the case of nitrogen illustrates. Humus contains much nitrogen but only a fraction is in a form immediately available to plants. This is the mineralized nitrogen, existing as ammonia or nitrate ion. While large amounts of nitrogen literally 'go up in smoke' (well over half the original content in the vegetation can be lost this way), at the same time the burning has a favourable effect on the production of mineralized nitrogen in the top soil. So the end result might be a strong gain in available nitrogen which outweighs the loss to the atmosphere. The nitrogen exported from the community in the smoke may become an input to other ecosystems, returning to the soil dissolved in the rain water. But this may not always be advantageous: excess nutrients elsewhere may be superfluous to requirements and in this sense a waste. These are not the only losses. Hopkins, studying savanna burning in Nigeria, shows that about three-quarters of the energy in the herbaceous aerial shoots is lost to the biological system annually through fires (Fig. 6.24).

Extensive fires cause the redevelopment of vegetation over a wide area to commence from approximately the same date. This leads to an even-aged structure in forests, a phenomenon well represented in the pinewoods on Speyside, Scotland. Here, many of the trees are just over 200 years old. Their development can be traced to well-documented fires in the Spey valley during the eighteenth century. Cooper, examining the Ponderosa Pine forests of South-West United States, shows how the pattern in these forests is maintained by fire. The process involved is a self-regulating feedback mechanism governed by the density of the stand. These examples show that the temporal and spatial characteristics of a community may both reflect the influence of past fires and partly determine how fire may spread during some future burn. All the trees in an even-aged stand reach maturity about the same time, thereafter entering a declining stage together. Such communities show reduced ecological diversity when compared with the uneven-aged forest with its many more ecological niches.

We have been considering above the direct influences of fire in forest ecology, but there are several indirect consequences which may be of even greater importance. Statistics issued by the United Nations show wood consumption for various countries. Not unexpectedly those countries with a western-style economy head these tables, particularly the USA and USSR. But the tables are incorrect because the largest consumers of wood are the insects who probably devour more timber each year than both the Americans and Russians put together! Many trees have a degree of immunity to massive insect or fungal attack; damage usually being localized and the tree frequently recovering. However, extensive insect and fungal outbreaks often occur in forests which have been recently burnt. Even if the fire has been slight, damage to branches takes place and fallen, rotting wood accumulates. At these sites insect and fungal populations quickly multiply, spreading rapidly to points of damage on other trees. In New Zealand the *Sirex* wasp has done immense damage to plantations: while it will attack forests which have not been fired, a recent burn greatly assists its establishment and development. It has now

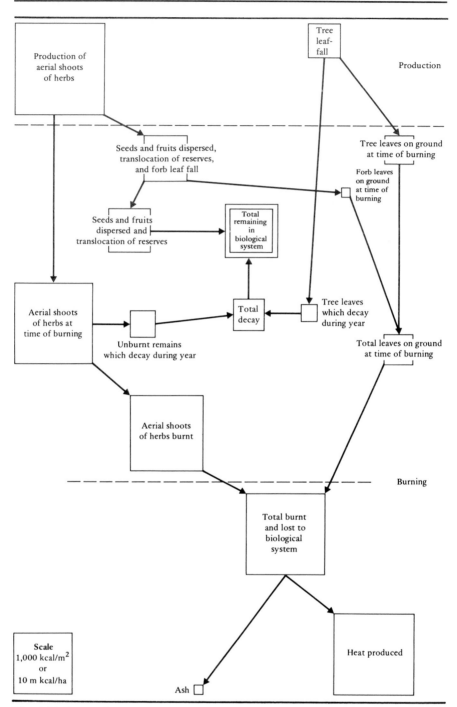

Fig. 6.24 Energy-flow diagram for Savanna burning in the Olokemeji Forest Reserve, Nigeria. Areas of squares proportional to calorific values per unit area. Forbs are associated herbaceous species other than grasses. (From Hopkins, 1965.)

become well established in Tasmania and small outbreaks are reported from the Australian mainland. In all these regions forest fires are very frequent (Figs. 6.25 and 6.26).

Another problem in some Australian *Pinus radiata* plantations is severe damage by a fungus, *Diplodia*. Australia lacks suitable native conifers for wood and pulp production and so this tree, a native of California, has been widely used. Once again, recently burnt forest is more susceptible to this fungal spread. In the Ovens Valley of North Victoria many thousands of hectares of this conifer were planted in the early 1920s, mainly to give employmet to returning ex-servicemen after the First World War. *Diplodia* causes three types of damage to these trees:

1. Dead top, where the terminal buds are destroyed and the tree ceases to grow.
2. Kinking, where the main trunk twists or kinks instead of growing straight.
3. Forking, where the main trunk branches out, perhaps repeatedly, instead of continuing to show monopodial or single stem growth.

These deformations make much of the timber unsuitable for timber merchants: up to 40 per cent may have to be rejected. This has meant a complete loss of profit margins for the original undertaking of some 50 years ago. These examples concern plantations but even in natural forests regular fires encourage the spread of disease. Where fires lead to even-aged, monospecific forests, conditions for the rapid spread of infections may be almost ideal because of the lack of diversity and because all the trees may be passing through the same vulnerable growth stage together.

While fires influence the spread of pests the reverse is also true. Stands of lodgepole pine (*Pinus contorta*) resistant to low intensity surface fires, are destroyed by high intensity fires. For a severe fire to develop it requires a considerable quantity of ground fuel. Ford outlines research which has shown that the most spectacular contribution to mortality (and hence to debris on the forest floor) follows epidemics of the mountain pine beetle, *Dendroctonus ponderosae*. The insect larvae tunnel the inner bark, feeding on the hyphae of fungi introduced by the parent beetle, but epidemics are restricted to older tree stands (trees up to 60 years old show resistance). Stand destruction by fire is therefore the end stage of a complex set of interlocking processes. The pine depends on fire for natural regeneration and these biological mechanisms ensure that the necessary severe burn will occur.

ADVANTAGES TO MAN

There must be some advantages in using fire or man would not have persisted with its use. There is a long history of fire misuse but man can control this factor better than most others in ecosystems. Whereas many of the disadvantages stem from ignorance of the consequences of firing vegetation, the advan-

Fig. 6.25 A release of 'phoscheck', a chemical fire retardant designed to assist fire control in remote areas. (Photograph by permission of Forestry Commission, Victoria.)

Fig. 6.26 Severe burn in wet sclerophyll forest, about 70 km east of Queenstown, Tasmania. Many eucalypts exhibit a typical phenomenon following fire, the sprouting of shoots from formerly dormant epicormic buds lying under the surface bark of the trunk.

Fig. 6.27 A light, control burn in a good stand of eucalypt trees. The aim is to reduce litter levels and so prevent a more serious fire at a later date. (Photograph by permission of Forestry Commission, Victoria.)

tageous use of fire requires an intelligent awareness of the structure and functioning of ecosystems.

Fire can be used to control fire. A severe fire may be stopped by burning a wide strip ahead of the main blaze, which is then devoid of combustible material when the main fire reaches it. Likewise, light burning of the under-growth in the correct season can reduce the accumulation of litter which might later present a major fire hazard. In this way the chances of a severe fire starting are minimized (Fig. 6.27).

Some mature woodlands (e.g. the pinewoods of the Scottish Highlands) produce vast quantities of seed each year but little regeneration of the trees takes place. This is because litter accumulates to such a depth on the soil surface, perhaps up to 0·5 m of conifer needles. The seeds are unable to grow a rooting system quickly enough to make contact with the underlying moisture reserves in the mineral soil. They thus die of drought even though the rainfall might exceed 250 cm per year. A fire will 'thin out' this litter layer and it is nearly always followed by a wave of extensive pine regeneration. This, of course, is the mechanism producing the previously noted even-aged woodland structure (Fig. 6.28).

Fires may greatly increase for man the productivity of an ecosystem. Chemi-cals locked up in old herbage are released and this promotes new growth. It is the basic reasoning behind the long-established practice of heather burning on upland moors to produce fresh grazing for sheep or grouse. Viro's review of the effects of prescribed forest fires on soils in Finland concludes that the

Fig. 6.28 Use of fire in Australia to clear sites rapidly and prepare ground for tree planting. A tanker is in attendance to control spread if necessary. (Photograph by permission of Forestry Commission, Victoria.)

advantages mostly outweigh the disadvantages, as long as fire is not employed more frequently than is absolutely necessary (and on the best soils with good forestry management techniques it may never be necessary). He sees improvements in nutrient and thermal conditions, marked seedling regeneration and the opportunity to select and encourage the most suitable tree species for a site as the main advantages of fire use. If properly done, it produces an entirely positive effect on the availability of nutrients, particularly with regard to nitrogen mobilization. Fires may also encourage certain soil bacteria and nitrogen-fixing legumes, thus enhancing the fertility of the site. Likewise, Heinselman's studies of fire in Minnesota forests led him to observe that protection of living forests from fire will cause a loss of species, increase the probability of occasional but very severe fires and greatly reduce nutrient cycling by the long accumulation of litter.

We can control with the aid of fire the development of the whole community, removing undesirable plant species while fostering the expansion of desired ones. The same applies to animal communities. Woodlands may be fired to reduce scrub cover and increase the grass component in the ground flora. This might enable a higher population of herbivorous mammals, such as deer, to exist. In this way fire may be used as part of conservation practice, maintaining a seral stage characterized by much faunal diversity rather than allowing one to develop which is less rich.

Fires, while indirectly assisting the spread of pests and diseases, may also be used to control such spreads. In Georgia, USA, the brown-spot fungus spends part of its life cycle on the forest undergrowth and part on the trees. Burning out the undergrowth at the correct season can prevent the fungus from completing its life cycle on the nearby trees.

In using fire we all too readily see only its obvious effects on the ecosystem. Detailed knowledge of all its influences, direct and indirect, is slow in accumulating. Because of the destructive nature of fire there is great difficulty in setting up complex and delicate instrumentation to monitor its precise operation. Equally, because of the unacceptability of wholesale devastation any experimental use of fire to elucidate its *modus operandi* must be strictly limited. In this respect, therefore, the sophisticated approaches employed in recent tree-line research cannot yet be matched in the study of fire ecology.

DISCUSSION SECTION

Why is it important to go to so much trouble to establish the level of the tree-line and its ecology, especially in this country where little natural forest remains?

But even in Austria and New Zealand little natural forest now exists. There, the financing of tree-line studies grew out of concern over the dangers and economic costs of avalanches and the problems of soil erosion. In Britain it is important to know the ecological status of our uplands before we can frame correct land-use policies for them. The Forestry Commission controls a vast upland acreage much of which could be planted with trees. The enormous cost of land will increasingly force us to use this poorer land for plantations. Because of the long-term nature of forestry operations it is vital to have as much knowledge as possible about these marginal situations for tree growth.

It has been argued that a cover of trees is the best form of land use in mountains, bearing in mind their poor soils, steep slopes and severe climatic regime. Woodland probably represents the most efficient ecosystem in terms of energy fixation, microclimatic modifications, utilization of deep-seated soil minerals, slope stabilities and soil erosion prevention. In New Zealand, destruction of the forest cover by the grazing of excessive populations of deer (introduced species) has already led to much soil erosion on the high watersheds. Consequently, the silting-up of reservoirs at lower levels is becoming a problem. We must also consider the impact of people on uplands. The presence of a few dozen people on open moorland soon gives the site a crowded appearance. Woodland, on the other hand, can absorb far greater numbers of tourists while still providing a sense of privacy for those using the area. With many uplands now experiencing a vast influx of tourists, the type of landscape we create there is of great importance.

The tree-line is a fundamental plant boundary, separating distinctive landscape types. These landscapes differ not only in their vegetation cover but may also do so in terms of soils, geomorphic processes and microclimate. So field scientists other than ecologists have an interest in the determination of the tree-line. This last point can be nicely illustrated by reference to several studies on the boundary of the Tundra-Taiga zones (the latitudinal tree-line). Bryson in 1966 demonstrated for central Canada a strong link between the position of the Tundra-Boreal forest boundary and the mean summer location of the Arctic Frontal Zone, which marks the southerly limits of Arctic air-masses. Krebs and Barry have examined Eurasia for a similar relationship and, in fact, find a near-coincidence of the Tundra-Taiga boundary and the median location of the Arctic Front in this region. This is strong support for Bryson's hypothesis that not only is the position of the latitudinal tree-line probably determined by air-mass dominance but other major biotic regions may equally bear a close correlation with different types of air-mass dominance. But Hare points out that the cause of this relationship is not certain. Perhaps the vegetation zones reflect the dominance of particular air-masses or it could be that the position of the front may itself be determined by the differing meteorological characteristics of the two biotic zones, such as albedo (reflectivity by a surface of received solar radiation), aerodynamic roughness and moisture budget. For these meterologists, this presents a nice problem of which is cause and which is effect. These examples represent just some of the reasons why a detailed knowledge of forest limits is required.

To pinpoint the exact effect of a factor seems to require very sophisticated approaches and years of work. Are climatic data, presented in the usual form as average values, therefore of much use in ecological studies?

This type of data does provide useful background information which might indicate the factor or climatic variable to look at more carefully. However, ecologists are really concerned with the *immediate* plant environment. Studies of this aspect often need very sophisticated methods, as Benecke's work shows (see Fig. 6.14). A further example is provided by Peacock's study of the relationship between temperatures and leaf growth in *Lolium perenne* (perennial ryegrass, an important agricultural grass species). Whereas other workers have used temperature data measured in the standard Stevenson's screen and related this to crop growth-rates, he has gone a stage further by considering the different effects on plant growth of soil temperature, air temperature, and the temperature of specific plant parts. Heating cables were used in the field to vary the soil temperature (and thus the temperature of the plant parts within the soil) independently of the air temperature. This work was followed by the use of 'heating collars' to examine the effect of localized temperature changes on a specific plant part, the stem apex, of single specimens (Figs. 6.29 and 6.30). There is conflicting evidence about the relative importance of soil and air temperatures in determining plant growth. The results of Peacock demonstrate that temperature was effective in the sensitive, specific region of the stem apex rather than there being a general effect of soil or air temperature on leaf growth. Such detailed studies, it is hoped, provide a clear picture of the present-day ecology of a species. In another context, this can be vitally important when the biogeographer tries to interpret the occurrence and distribution

Fig. 6.29 The control of soil temperatures in an experiment to relate leaf growth to temperature in perennial ryegrass: layout of heating cables prior to burial in the plot. (Photograph by permission of Dr. J. M. Peacock, The Grassland Research Institute, and the *Journal of Applied Ecology*.)

of fossil species. Often these species are taken as indicators of specific habitat conditions, so the fewer assumptions we make about their ecological requirements the better will be our reconstruction of past environments.

When studying the immediate plant environment, we should bear in mind the axioms for environmental research suggested by Platt and Griffiths:

1. The environment of any experimental unit is the result of all external conditions and influences directly affecting that unit. (For 'experimental unit' we may also read community, species or habitat under study as appropriate.)
2. The many factors of the environment exist as an interacting complex and react as a whole.
3. Certain of these environmental factors or combinations of factors may be limiting, compensating, or triggering.
4. The magnitudes of these factors continuously change in space and time.
5. The experimental units continuously modify those factors peculiar to their own environment.
6. The focal point of research is the investigative unit – not the environment.
7. Environmental data obtained in one area cannot be used directly for the understanding of experimental units in another area, 40 ft or 40 miles away.

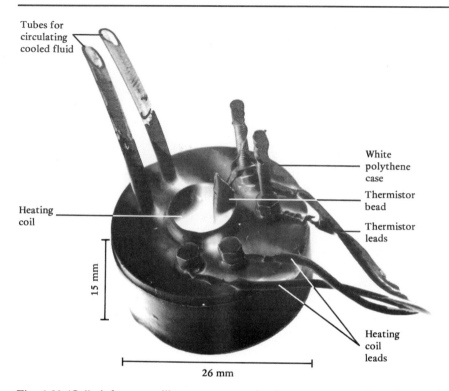

Fig. 6.30 'Collar' for controlling temperature in the stem apex region of perennial ryegrass. See Fig. 6.29. (Photograph by permission of Dr. J. M. Peacock, The Grassland Research Institute, and the *Journal of Applied Ecology*.)

8. The environment possesses the dimensions of space and time.
9. Data obtained at one level above the ground may not be valid for other levels.
10. Environmental effects observed on a given unit cannot be applied without question to the population.
11. Environmental data obtained at one particular time cannot be applied without discrimination to an experimental unit, if it is observed at other times.
12. Data obtained in controlled environments cannot be applied directly to experimental units in their natural environment.
13. Conclusions cannot be made on the behaviour of experimental units within a particular environment without knowing something of the reactions of these units to the factors involved.
14. All environmental factors must be taken into consideration even though all cannot be measured or controlled.
15. Experimental units must be precisely identified.
16. Conclusions and correlationns cannot be stretched beyond that permitted by the limitations of the instrumentation used.

On balance, should the use of fire in ecosystems be discouraged?

In some ecosystems fire is an essential ingredient. The giant sequoia (*Sequoia-dendron giganteum*) forests of California only survive because competing conifers, which would shade out sequoia seedlings, are regularly removed by fires against which the sequoia is insulated by a thick, asbestos-like bark. The Pocosin wetlands of North Carolina, evergreen shrub bogs which once covered millions of hectares on the south-eastern coastal plain from Virginia to north Florida, only achieve some environmental heterogeneity through the impact of fire. Christensen *et al.* state that 'fire adds heterogeneity to an otherwise monotonous landscape and its potential role in maintaining species diversity cannot be overemphasized'.

When used correctly, fire is a quick and effective method of bringing about a desired change in an ecosystem: it might only cost the price of a box of matches compared with the use of costly labour and equipment to clear a site or hold back a seral development. But the more we know about ecosystems the more we realise the complexities and dangers involved. One example will bring this out. A practice on Scottish moors is the burning of old heather to produce patches of fresh heather growth. These are much more suitable as feeding grounds for grouse. Research by Miller and by Picozzi has shown how the creation of very large burnt strips (i.e. greatly increasing the feeding area) does not necessarily lead to higher grouse populations. If the strips are too wide, grouse venturing out onto them to feed become much more vulnerable to predation. They need the nearby presence of taller, older heather as a refuge. Thus best results are achieved when narrow burnt strips and differing ages of heather are maintained in close proximity.

It would seem that for every good use of fire there is a corresponding bad use. For example, its use to control fungal disease or its role in assisting fungal spread. Whether the regular occurrence of fire is acceptable in an area will depend on a full appreciation of its short-term and long-term effects on that ecosystem.

You mentioned three ways in which wind influences plant establishment. Are there any other aspects to wind as an ecological factor?

Yes – in addition to the direct physical impact of wind (mechanical stress) or the temperature of the air-flow and its influence on the water budget of the plant, there is the question of materials carried by the wind. We previously mentioned how on steep upland slopes fine mineral particles may act as a natural sand blast. But air can also be heavily polluted or charged with salt particles. In the Southern Pennines, Tallis has pointed to polluted winds carrying sulphur dioxide from adjacent industrial areas as responsible for the marked decline in *Sphagnum* mosses on these moorlands. These upland peat surfaces constitute the most important natural water reservoir in the British Isles. Once erosion sets in they are quickly destroyed. *Sphagnum* plays a vital role in growing over and healing erosion scars so its reduction in the flora could have serious implications.

Around Aberystwyth, Edwards and Claxton studied the distribution of salt brought inland by winds. Quite appreciable differences in the amounts of salt on windward and leeward leaf surfaces were detected up to nearly 2 km from the coast. In south-east Australia soil scientists attribute a significant part of the salt content of saline soils found in interior locations to oceanic sources. In some cases, soils up to 240 km from the coast are thought to have had this input of atmospheric salt, brought by regional wind flows.

Returning to an earlier point, data presented as average values can hide the main ecological impact of the wind factor. The infrequent extreme values may have a far greater influence on the ecology. For example, in 1953 one storm in north-east Scotland damaged $39\frac{1}{2}$ million cubic feet of timber. The regeneration phase which followed this wind-throw destruction will produce in some areas a new, even-aged forest. This characteristic will persist for the life of these trees and will have a great bearing on the structure of the forest, possibly for some 300 years or more.

REFERENCES

Ahlgren, I. F. and Ahlgren, C. E., 1960. 'Ecological effects of forest fires', *Bot. Rev.*, 26, 483–533.

Andre, H. M., Bolly, C. and Lebrun, Ph., 1982. 'Monitoring and mapping air pollution through an animal indicator: a new and quick method', *J. appl. Ecol.*, 19, 107–11.

Baig, M. N., Tranquillini, W. and Havranek, W. M., 1974. 'Cuticuläre transpiration von *Picea abies* und *Pinus cembra* Zweigen aus vershiedener Seehöhe und ihre Bedeutung für die winterliche Austrocknung der Bäume an der alpinen Waldgrenze', *Centralblatt für das gesamte Forstwesen*, 91 (4), 195–211.

Baig, M. N. and Tranquillini, W., 1980. 'The effects of·wind and temperature on cuticular transpiration of *Picea abies* and *Pinus cembra* and their significance in desiccation damage at the Alpine treeline', *Oecologia (Berl.)*, 47, 252–6.

Benecke, U. and Davis, M. R. (eds), 1980. *Mountain Environments and Subalpine Tree Growth*. New Zealand Forest Services Technical Paper No. 70, Wellington, New Zealand.

Benecke, U. and Havranek, W. M., 1980. 'Gas-exchange of trees at altitudes up to timberline, Craigieburn Range, New Zealand', in *Mountain Environments and Subalpine Tree Growth* (eds Benecke, U., and Davis, M. R.), New Zealand Forest Services Technical Paper No. 70, Wellington, New Zealand.

Billings, W. D., 1954. 'Temperature inversions in the Pinyon-Juniper zone of a Nevada Mountain Range', *Butler Univ. Bot. Studies*, 11, 112–18.

Bryson, R. A., 1966. 'Air-masses, streamlines and the Boreal Forest', *Geogr. Bull.*, 8(3), 228–69.

Caborn, J. M., 1957. 'Shelter-belts and microclimate', *Forestry Comm. Bull.*, Vol. 29, HMSO, London.

Cannon, H. L., 1954. 'Botanical methods of prospecting for uranium', *Mining Engineering*, Feb., 217–20.

Christensen, N. L., Burchell, R. B., Liggett, A. and Simms, E. L., 1981. 'The structure and development of Pocosin vegetation', in *Pocosin Wetlands* (ed. Richardson, C. J.), Hutchinson Ross, Stroudsburg, Pennsylvania, 43–61.

Cooper, C. F., 1961. 'The ecology of fire', *Sci. Amer.*, **204**, 150–60.

Daubenmire, R., 1954. 'Alpine timber-lines in the Americas and their interpretation', *Butler Univ. Bot. Studies*, **11**, 119–36.

Daubenmire, R., 1968. 'Ecology of fire in grasslands', *Advan. Ecol. Res.*, **5**, 209–66.

Edwards, R. S. and Claxton, S. M., 1964. 'The distribution of air-borne salt of marine origin in the Aberystwyth area', *J. appl. Ecol.*, **1**, 253–63.

Ford, E. D., 1982. 'Catastrophe and disruption in forest ecosystems and their implications for plantation forestry', *Scottish Forestry*, **36**, 9–24.

Geiger, R., 1965. *The Climate Near the Ground* (3rd edn.), Harvard University Press, Cambridge, Mass.

Gloyne, R. W., 1955. 'Some effects of shelter-belts and windbreaks', *Met. Mag., Lond.*, **84**, 272–81.

Grace, J., 1977. *Plant Response to Wind*, Academic Press, London.

Griggs, R. F., 1938. 'Timber-lines in the Northern Rocky Mountains', *Ecology*, **19**, 548–64.

Griggs, R. F., 1946. 'The timber-lines of Northern America and their interpretation', *Ecology*, **27**, 275–89.

Hare, F. K., 1968. 'The Arctic', *Quart. Journ. Royal Meteorol. Soc.*, **94**, 439–59.

Hawksworth, D. L. and Rose, F., 1970. 'Qualitative scale for estimating sulphur dioxide air pollution in England and Wales using epiphytic lichens', *Nature, London*, **227**, 145–8.

Heinselman, M. L., 1973. 'Fire in the virgin forests of the Boundary Waters Canoe Area, Minnesota', *Quaternary Research*, **3**, 329–82.

Hopkins, B., 1965. 'Observations on Savanna burning in the Olokemeji Forest Reserve, Nigeria', *J. appl. Ecol.*, **2**, 367–81.

Jacobi, R. M., Tallis, J. H. and Mellars, P. A., 1976. 'The Southern Pennine Mesolithic and the Ecological Record', *J. Archaeological Sci.*, **3**, 307–20.

Kellman, M. C. 1980. *Plant Geography* (2nd edn.), Methuen, London and New York.

Kincaid, D. T. and Lyons, E. E., 1981. 'Winter water relations of red spruce on Mount Monadnock, New Hampshire', *Ecology*, **62**,1155–61.

Kozlowski, T. T. and Ahlgren, C. E. (eds), 1974. *Fire and Ecosystems*, Academic Press, New York.

Krebs, J. S. and Barry, R. G., 1970. 'The Arctic Front and the Tundra-Taiga boundary in Eurasia', *Geogr. Rev.*, **60**(4), 548–54.

Liebig, J., 1840. *Chemistry and its application to agriculture and physiology*, Taylor and Walton, London.

Lines, R. and Howell, R. S., 1963. 'The use of flags to estimate the relative exposure of trial plantations', *Forestry Comm. Forest Record*, **51**, 1–31.

Major, J., 1951. 'A functional factorial approach to plant ecology', *Ecology*, **32**, 392–412.

Marchand, P. J. and Chabot, B. F., 1978. 'Winter water relations of treeline plant species on Mt. Washington, New Hampshire', *Arctic and Alpine Research*, **10**, 105–16.

Mason, H. L. and Langenheim, J. H., 1957. 'Language analysis and the concept environment', *Ecology*, **38**, 325–40.

Mellars, P. A., 1976. 'Fire ecology, animal populations and man: a study of some ecological relationships in prehistory', *Proc. Prehist. Soc.*, **42**, 15–54.

Michaelis, P., 1934a. 'Okologische Studien an der alpinen Baumgrenze, IV. Zur Kenntnis des winterlichen Wasserhaushaltes', *Jahrbuch für wissenschaftliche Botanik*, **80**, 169–247.

Michaelis, P., 1934b. 'Okologische Studien an der alpinen Baumgrenze, V. Osmotischer Wert und Wassergehalt Während des Winters in den verschiedenen Höhenlagen', *Jahrbuch für wissenschaftliche Botanik*, **80**, 337–62.

Miller, G. R., 1964. 'The management of heather moors', in *Land use in the Scottish Highlands*, (ed. L. D. Stamp), *Advmt. Sci.*, Lond. 21, 163–9.

Molloy, B. P. J., Burrows, C. J., Cox, J. E., Johnston, J. A. and Wardle, P., 1963. 'Distribution of sub fossil forest remains, eastern South Island, New Zealand', *New Zealand J. Bot.*, 1, 68–77.

Mutch, R. W., 1970. 'Wildland fires and ecosystems – an hypothesis', *Ecology*, 51, 1046–51.

Odum, E. P., 1971. *Fundamentals of Ecology*, (3rd edn)., W. B. Saunders, Philadelphia.

Peacock, J. M., 1975. 'Temperature and leaf growth in *Lolium perenne*, I. The thermal microclimate: its measurement and relation to crop growth', *J. appl. Ecol.*, 12, 99–144.

Peacock, J. M., 1975. 'Temperature and leaf growth in *Lolium perenne*, II. The site of temperature perception', *J. appl. Ecol.*, 12, 115–24.

Pears, N. V., 1967. 'Present tree-lines of the Cairngorm Mountains, Scotland', *J. Ecol.*, 55, 815–29.

Pears, N. V., 1968. 'The natural altitudinal limit of forest in the Scottish Grampians', *Oikos*, 19, 71–80.

Picozzi, N., 1968. 'Grouse bags in relation to the management and geology of heather moors', *J. appl. Ecol.*, 5, 483–8.

Platt, R. B. and Griffiths, J., 1964. *Environmental Measurement and Interpretation*, Reinhold Pub. Co., New York.

Rose, F., 1970. 'Lichens as pollution indicators', *Your environment*, no. 5, 7.

Rose, C. I. and Hawksworth, D. L., 1981. 'Lichen recolonization in London's cleaner air', *Nature*, London, 289, 289–92.

Shelford, V. E., 1913. *Animal Communities in Temperate America*, University of Chicago Press, Chicago.

Slatyer, R. O., 1976. 'Water deficits in timberline trees in the Snowy Mountains of South-eastern Australia', *Oecologia*, 24, 357–366.

Tallis, J. H., 1964. 'Studies on southern Pennine peats, III. The behaviour of *Sphagnum*', *J. Ecol.*, 52, 345–53.

Tikhomirov, B. A., 1962. 'The treelessness of the Tundra', *Polar Rec.*, 11, 24–30.

Toumey, J. W., 1947. *Foundations of Silviculture upon an Ecological Basis* (revised by C. F. Korstain), Wiley, New York.

Tranquillini, W., 1967. 'Uber die physiologischen Ursachen der Wald–und Baumgrenze', *Mitteilungen der Forstlichen Bundesyersuchsastalt*, Wien, 75, 457–87.

Tranquillini, W., 1974. 'Der Einfluss von Seehöhe und Lange der Vegetationszeit auf das cuticuläre Transpirationsvermögen von Fichtensämlingen im Winter', *Ber. Deutsch. Bot. Ges. Bd.*, 87, 175–84.

Tranquillini, W., 1979. *Physiological ecology of the alpine timberline*, Ecol. Stud. vol. 31, Springer-Verlag, Berlin, Heidelberg, New York.

Tranquillini, W., 1982. 'Frost-drought and its ecological significance', in *Encyclopaedia of Plant Physiology*, New Series, 12B, *Physiological Plant Ecology II*, (eds Lange, O. L., Nobel, P. S., Osmond, C. B. and Ziegler, H.), Springer-Verlag, Berlin, Heidelberg, New York, 379–400.

Veblen, T. T., Ashton, D. H., Schlegel, F. M. and Veblen, A. T., 1977. 'Plant succession in a timberline depressed by vulcanism in south-central Chile', *J. Biogeography*, 4, 275–94.

Viro, P. J., 1974. 'Effects of forest fire on soil', in *Fire and Ecosystems* (eds Kozlowski, T. T. and Ahlgren, C. E.), Academic Press, New York, 7–45.

Walker, D., 1982. 'The development of resilience in burned vegetation', in *The Plant Community as a Working Mechanism* (ed. E. I. Newman), Blackwell Scientific Publications, Oxford and Boston, 27–43.

Wardle, P., 1965. 'A comparison of Alpine timber-lines in New Zealand and North America', *New Zealand J. Bot.*, 3, 113–35.

Wardle, P., 1971. 'An explanation for Alpine Timber-line', *New Zealand J. Bot.*, **9**, 371–402.

Wardle, P., 1974. 'Alpine timber-lines', in *Arctic and Alpine Environments* (eds Ives, J. D. and Barry, R. G.), Methuen; London, 371–402.

Wardle, P., 1981. 'Winter desiccation of conifer needles simulated by artificial freezing', *Arctic and Alpine Research*, **13**, 419–23.

Weaver, H., 1974. 'Effect of fire on Temperate Forests: Western United States', in *Fire and Ecosystems* (eds Kozlowski, T. T. and Ahlgren, C. E.), Academic Press, New York, 279–319.

7

DISTURBED ECOSYSTEMS

INTRODUCTION

So far statements in the text have been mainly illustrated with examples from the plant world. In this chapter the emphasis is largely on examples from animal ecology. The intricate nature of biotic communities was stressed in previous chapters. Plant succession, usually leading to an increase in plant diversity, is broadly paralleled by animal succession. The complex mature ecosystems that evolve are characterized by control mechanisms. These help to stabilize the natural community and numbers of any one species tend to remain reasonably constant over long periods. Nevertheless, some species do show fairly regular, dramatic population fluctuations, e.g. Arctic lemmings. However, recent research, such as the work of Krebs, shows that these fluctuations are probably just as common in tropical and temperate ecosystems.

Most natural ecosystems are balanced in the sense that changes within them are slow and orderly, giving the species-mix time to adjust to new conditions (though it must be stressed that the 'balance of nature' concept stems from studies which, for the most part, have been restricted to only a few of the many species that make up most communities). Since the ecology of animal communities is closely allied to that of plant communities each exerts a powerful influence on the other. Many animals are capable of moving considerable distances in pursuit of their necessary requirements. Despite this mobility there are numerous examples of animals restricted to particular plants or plant communities. Some are restricted because of a highly specific food requirement while many larger animals have a strong territorial sense. For example, the northern limit of the North American beaver generally coincides with the northern limit of the aspen tree, which is its main food source. The elf owl of California and Arizona is only found where a particular species of giant cactus grows; it nests exclusively in this plant. A plant community in balance with its environment usually has an associated animal community in a similar state of balance. Disturbance of one will cause disturbance to the other.

INTERRELATIONS IN ANIMAL COMMUNITIES

A full understanding of the nesting ecology of the elf owl can only be obtained by considering other important animals. Two species of woodpecker use the cactus, probing its surface for grubs and water. These activities create the nesting holes for the owl. In addition, several species of scavenger beetle and body parasite live exclusively in these nests and on these birds. Thus, in an apparently simple ecosystem, intricacies are introduced through the development of species interrelationships. We have already noted other examples of two-species interactions (see Table 5.1) and part of the web of animal interactions associated with just one plant (the food web of the wild cabbage, Fig. 5.5). Elton stated four principles which aid the understanding of these interrelationships in most animal communities. Some of these ideas have already been outlined in Chapter 5.

First, the principle of the *food chain* goes a long way towards explaining many species interactions. All animals have a driving force in the form of a necessity to find food of the correct type and in sufficient quantity. But, at the same time, each animal is being looked upon by other animals as a possible source of food. Tracing food webs in terms of energy flows reveals much about the animal components of an ecosystem and how they are organized.

The second principle concerns *food size*. For any particular animal there is a well-defined range for the size of food that it can obtain. As a general rule, large animals cannot feed entirely or mainly on small organisms because of the huge numbers needed for satisfaction (a lower limit). Likewise, small animals have an upper limit to their food size set by inability to procure larger animals. The whale and man are obvious exceptions to this rule, as are small scavenger carnivores which feed on large dead animals. Moreover, animals can effectively increase their size to obtain food or prevent themselves becoming food. These mechanisms help to explain many of the social and behavioural patterns of animal groups. For example, Elton mentions wolves attacking in packs, deer grouping themselves for defence, and forest toads around Lake Victoria inflating themselves to prevent them from being swallowed by snakes.

The concept of the *ecological niche*, already considered in Chapter 5, is the third principle. Just as human beings are placed in society (we speak of their role or social level), so the ecologist notes an animal's place within its community in terms of relationships with food sources, enemies and occupants of other niches.

The fourth principle is called the *pyramid of numbers* and refers to animal population structures. Each animal has a minimum area it must occupy for existence. Usually, the smaller the animal the smaller the area and hence the commoner is the animal. Animals at the lower end of a food chain are numerous and breed rapidly. Those at the upper end are scarce and breed slowly. So in terms of their numbers, the animals in any community are arranged in a pyramidal form.

The operation of these four principles produces a balanced, integrated community in any given area but not a static position. It is one in which controls maintain a broad equilibrium and where minor environmental changes can be accommodated. The biogeographer has a special interest in what happens when balanced communities are grossly disturbed, for the results are often expressed on the ground as great changes in the spatial patterns of the component species.

FORMS OF DISTURBANCE

Ecosystems can be disturbed by natural processes. However, most natural changes are usually spread over many years, giving the community time to adjust. There are, however, natural catastrophes such as volcanic eruptions, droughts, floods or very severe winters which can destroy a community. However, we are not concerned here with these types of rapid change for on a world view they are usually minor or highly localized incidents. There is another form of rapid disturbance which has a far greater impact on the world's flora and fauna. Many of the changes brought about by man are rapid in operation and highly destructive of natural communities.

Upset may be caused by man: (a) destroying or replacing either the native plant or animal groups; (b) introducing new (or alien or exotic) species; (c) altering one or more environmental factors; or (d) adding 'foreign' substances to the ecosystem as pollutants or pesticides. Often actions under one of these headings also produce changes under another. For example, the clearing of forest for crop production not only means the removal of natural vegetation but it also profoundly alters the native animal populations, the local micro-climate and the soil types in the area. Further, it may necessitate the use of herbicides to check the growth of 'weeds' and scrub. This regrowth is often an attempt by the natural vegetation to recolonize the site by seral developments.

In this chapter all these various sides of the problem cannot be covered. Rather, we will concentrate on one aspect, population imbalance (others will be mentioned in the Discussion Section). This will be illustrated by a series of examples, each pointing to the need for better appreciation of ecological relationships. Some main conclusions from these examples will then be drawn, followed by an examination of the methods used to correct such disturbances. Finally, the ecological problems associated with rapid land-use changes in developing countries are reviewed; an area of much interest to geographers.

POPULATION IMBALANCE

THE GIANT AFRICAN SNAIL

Our first example concerns a species of giant African snail (*Achatina fulica*) native to East Africa. These snails may be up to 30 cm long and are voracious herbivores that greatly damage crops. Mead has traced their impact as they have rapidly spread eastwards into the tropical islands of the Pacific. With the help of man, the snail had reached Mauritius and India by the nineteenth century and beyond into Sri Lanka, Malaya, Borneo and Thailand during the early twentieth century. It quickly became a major pest on many Pacific islands (e.g. Guam, Hawaii, the Marianas) during the Second World War, being accidentally transported on Japanese war equipment. It also reached California on returning US Army equipment, but never established itself there. In 1968 it entered Florida when, apparently, a boy brought in three snails as pets from Hawaii! These were allowed to go free and their numerous descendants have now become locally well established as a pest. The problem of the giant snail in the Pacific is compounded by some of the other exotic animals released there, many of them in an effort to control the snail itself. The story is perplexingly entangled but the following main points are pertinent.

In 1937 the Central American toad was introduced to these Pacific islands, in the hope of controlling a large black slug which had become a pest. Later, it was distributed to attack the giant snail. This did not work because it showed more interest in the insects infesting already dead snails than it ever did in live snails. Then the Japanese introduced the big monitor lizard to control the rat, another important pest on these islands. This was a failure because the rat was essentially nocturnal and the lizard was diurnal. Hence they seldom came into contact! Instead, the lizards developed a preference for the islanders' chickens and eggs.

The toads were also nocturnal but did not hide effectively. They were discovered and devoured by the local pigs, cats and dogs. But the toads were poisonous and many domesticated animals died as a result. This was unfortunate because the cats and dogs represented the only effective rat-catchers. The lizards also ate the poisonous toads and died. However, neither was this helpful to the situation because, about this time, the lizards were also devouring the greatest of all the islands' pests, the rhinoceros beetles. These beetles attack the coconut palms, which are vital to the economy of the islanders (food, drink, building materials, etc.). The population of beetles had greatly multiplied in the dead wood of many palms destroyed during the war.

The large land-dwelling coconut crab, another inhabitant of the islands, also damages the palms but to a lesser extent. It too, was being destroyed by the monitor lizard. This was not necessarily a good development because the crab was the only native animal to attack the giant snail successfully!

This is a clear case of the biological control of a pest going sadly astray. It points to the need for far greater knowledge of the ecology of these animals before they are introduced and to the fact that the giant snail is certainly no more harmful, and probably less so, than the creatures used to control it. Ironically, a promising control now appears to lie in the use of a much smaller snail, native to Africa, which preys on the giant snail and bores into its shell.

THE COYPU

The coypu (*Myocastor coypus*) is a beaver-like aquatic rodent about 60–90 cm long. It is a native of South American reed swamps and has edible flesh and a hide that is the source of nutria fur. About 40 fur farms were established in Britain in the 1930s but these seldom prospered. Many coypus escaped and became established in the East Anglian fens and broads. Other infected areas were mainly colonized by coypus spreading out from centres in Norfolk and Suffolk. Initially, little concern was shown as the animals did good by eating aquatic plants which clogged the waterways. At their peak in the late 1950s and early 1960s their numbers have been estimated to have reached about 70,000. The coypu has two litters per annum with 5 to 11 young per litter, and the young can breed within the first year. Thus their breeding potential is enormous. At these levels, the coypu seriously undermines dyke banks by burrowing and Boorman and Fuller show that they have probably been the main factor in the dramatic loss of reedswamp in the Norfolk Broads since 1946. They also cause extensive damage to adjacent crops: sugar beet, mangolds, various brassicas, young corn and grass. The animal readily ventures onto land during hard winters (and the severe 1962–63 winter probably did more than anything else to cut back their numbers). It is illegal to poison them but a measure of control has been achieved by a systematic trapping campaign. In 1960 the 50 per cent grant given to Rabbit Clearance Societies for rabbit control was extended to coypu control. Some 97,000 coypus were reported killed in the next year or so but a more coordinated campaign was then organized. This covered the period 1962–65 and involved some 2,645 square miles. It cost just over £70,000 to trap 11,000 coypus. A further 27,000 were trapped by separate organizations taking advantage of the 50 per cent grant paid to Rabbit Clearance Societies. It was thus an expensive campaign to operate, and had to be run with almost military precision to confine the coypus at acceptable population levels. By 1980 control costs had risen to £200,000 p.a. and policy has now changed to one of complete eradication. This may not be easy since the coypu has now moved into north-east Essex. It has also been sighted in Derbyshire – probably a case of holidaymakers taking home the young from traps in the Broads.

THE NORTH AMERICAN MINK

The North American mink (*Mustela vison*) was also introduced into Britain for fur farming. Unlike the coypu, it is an aggresive carnivore, equally at home in water as on land. Hence the feral population established by escaped mink is a threat to poultry, game-birds and fishing interests (salmon and trout). These wild mink did not initially arouse much concern, even among ecologists. The female is in season for only a few days each year and ovulates only after sexual stimulation by the male. Since both sexes appeared to be few in number and widely scattered, the chances of feral mink breeding were thought remote. The two sexes, however, are able to locate each other over great distances and, much to everyone's surprise, breeding was highly successful. Populations have now spread into many river valleys in the West Country, Wales, Cheshire, the Lake District and Scotland. Although hardier than the coypu, its natural curiosity makes it relatively easy to trap and farmers in the Lake District have recently called for a coordinated programme to combat the mink. But it is now so well established in many remote areas that the costs of a successful trapping scheme would be prohibitive. Recently it cost £2,000 to trap 100 mink in Wales. Like the coypu, in this country the adult mink has no natural enemies apart from man.

THE WEAVER BIRD AND THE SUDAN DIOCH

The weaver bird is a native of thinly wooded savanna grasslands in South Africa. It mainly lives on grass seeds and roosts at night in the scattered, isolated trees. In recent decades large tracts of this land have been developed for cereal growing. The birds now enjoy a richer and more assured store of grass seeds (or grain) to draw upon. The result has been a huge population increase. Control by poisoning and sprays was unsuccessful and the latest methods include the wholesale dynamiting of the night roosts (the trees!). In the early 1960s some 264 million birds died in a two-year campaign but the population is still on the increase.

In the same way, the Sudan dioch is proving to be an even greater pest. This small bird can eat seven times its own weight of cereal grain in a week. It has reached pest status throughout most of East Africa, from the Sudan to South Africa. Attempts at control include explosives and fuel placed under night roosts and nesting sites. Half a million birds may perish in a single blast and the following fire. Some 500 million died in these early campaigns. Despite this, the dioch population is now greater than it was at the start of extermination measures.

Before we leave these examples of bird pests, reference to the experience of the Chinese is of interest. In 1960 the Chinese government ordered the slaughter of all wild birds in the country. This followed a report showing each bird yearly consumed an average of 2·7 kg of food suitable for human consumption. A wave of indiscriminate killing of seed-eaters, insect-eaters and birds of prey occurred. The result was an enormous population increase in rodents, insects and other pests. Disastrous losses to standing crops and stored food followed. The World Wildlife Organization has now stated that the wholesale killing of wild birds is today a punishable offence in China.

THE DISEASE OF THE NORTH AMERICAN SWEET CHESTNUT

The next example concerns a plant, the North American sweet chestnut (*Castanea dentata*). Earlier this century the tree developed a fungal disease (*Endothia parasitica*) and by 1950 most chestnuts throughout eastern North America had died, except those along the extreme south of their range. The disease was accidentally introduced on nursery plants from Asia. On its natural host in Asia, the Chinese chestnut (*C. mollissima*), the fungus does little harm. Introduction of the resistant Chinese chestnut to North America has been quite successful but costly. Losses due to this fungal attack run into many hundred million dollars. In Toronto, where the tree was used extensively to line streets, it costs about one hundred dollars to remove each diseased tree. Decades will elapse before replacement species grow sufficiently to fill the gaps left.

In 1938 *Endothia parasitica* reached southern Italy and began to spread northwards. Here chestnuts are grown in concentrations for commercial nut production. By 1948 it had reached Switzerland and has since done extensive damage, altering the structure and successional pattern in many woodlands. The windborne spores crossed high mountain barriers quite easily, spreading from Sicily to Switzerland in 10 years. (The current wave of Dutch elm disease in Britain has similarities with this case. In the last ten years or so, 20 million of the 29 million elms in southern England have died from a fungus disease spread by the elm bark beetle, and the disease is still spreading northwards. Jones reports that fungicide injections are costly (£10 per tree) and often fail. The cost of removing dead trees varies from £15 to £600 per tree, depending on its location. The total bill by 1980 for treatments and removals had reached £100 million. Most treatments are ineffective but Heybroek *et al.* describe a promising new method developed in the USA involving injections of a bacterium which stops the growth of the fungus *in vitro*. Tree strains resistant to the fungus are also being bred.)

THE COMMON BLACKBERRY

The common blackberry (*Rubus fruticosus*) is not a serious pest in Britain but it now is in many parts of Australia. There are some 360,000 acres of infestation in Victoria alone. Early settlers introduced the plant and sometimes deliberately spread it to the interior. Indeed, Ferdinand Mueller, the Chief Government Botanist of Victoria in the nineteenth century, is said to have always carried seeds for planting on his journeys inland and he advocated naturalization of the plant for many districts. In Australian climatic conditions, blackberry patches may grow to many metres across and up to 6 m in height. They are a serious problem on pastures and in the *Pinus radiata* plantation forests of northern Victoria. Blackberry quickly invades forest clearings and smothers large areas making the regeneration of tree seedlings virtually impossible. The costs are too high for hand clearing and so hormonal sprays have been tried. Unfortunately, although generally effective, these chemicals sometimes cause damage by drifting on to farm crops or by becoming incorporated in irrigation waters. A few types of blackberry are already showing resistance to hormonal sprays. In New Zealand the plant has long been a major pest despite the offer in 1924 of a £10,000 reward for a proven method of dealing with the weed.

Nearly 30 years ago Elton summarized the ecology of invasions by species, quoting many cases where exotics were introduced with disastrous results. The remaining examples are from that seminal work which should be consulted for added detail. Research since Elton wrote has been recently summarized by Jarvis.

THE AFRICAN MOSQUITOES

In 1929 a few African mosquitoes (*Anopheles gambiae*) accidentally reached the Brazilian coast, almost certainly from a ship. There was already present in East Brazil a local mosquito and these new arrivals did not arouse much immediate interest. A severe local outbreak of malaria followed but no systematic control of the fly was initiated. Meanwhile, the flies rapidly spread up the river valleys and became well established. In the 1936–39 period a vast outbreak of malaria occurred; hundreds of thousands were ill and an estimated 20,000 died. A vigorous but costly eradication programme finally controlled the outbreak. An important point in this example was the failure to appreciate the differing ecologies of the two mosquito species. The local species seldom went into houses, being shade-tolerant and breeding in the forest. The exotic one was a house fly, breeding in the open, stagnant pools found around poorer settlements. It also had a very rapid rate of spread – seven years between the introduction and the serious outbreak.

THE STARLING

The next example concerns a common bird in Britain, the starling (*Sturnus vulgaris*). In Europe it has a very wide range, covering a variety of habitats. It was deliberately introduced to North America in the late nineteenth century when some 80 birds were released in Central Park, New York. By 1891 successful breeding was reported. Since then the starling has spread throughout the United States (see Fig. 7.1, which shows the spread only up to 1926). Its migration range now includes regions as far apart as Alaska and Mexico. We are already aware of the problem posed by starling roosts in centres such as London and Birmingham. An increasing number of American towns are now experiencing similar difficulties.

THE SEA LAMPREY

The sea lamprey (*Petromyzon marinus*) lives mainly in the Atlantic but enters North American rivers for spawning. As an ectoparasite it secures itself to fish, sucking their tissues until they die. The lamprey infested some lakes but was unable to reach beyond Lake Ontario because of the Niagara Falls. But in 1829 the Welland Canal bypassed the Falls, opening the way for the lamprey. Initially, they made slow progress inland but after the canal was deepened in the 1930s an explosive invasion took place in the Great Lakes. The lake trout were the main but not the only victims. In a 10-year period the trout catch in Lake Huron and Lake Michigan dropped from 8,600,000 lb to a mere 26,000 lb. This had disastrous consequences for the many fishing settlements of the Great Lakes.

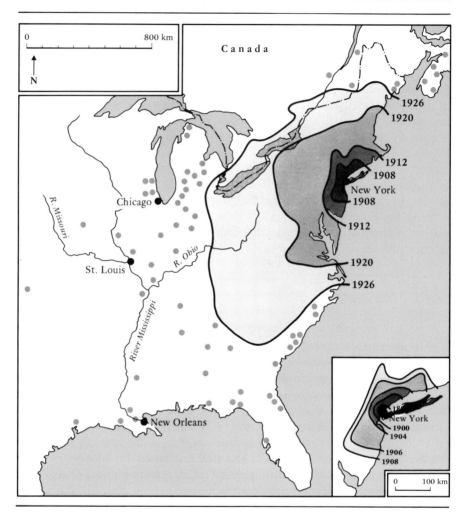

Fig. 7.1. Early spread of the breeding range of *Sturnus vulgaris*, the European starling, in North America following its release in 1891 in Central Park, New York. Dots outside the 1926 line are chiefly winter records of pioneer spread. Since 1926, the starling has become well established in most States. (Modified from Elton, 1958, after Cooke.)

THE JAPANESE BEETLES

In 1916 a dozen strange beetles were found in a New Jersey plant nursery. Because they had arrived accidentally on some Japanese plants, they became known as Japanese beetles (*Popillia japonica*). The beetles spread and by 1941 more than 20,000 square miles were infected (Fig. 7.2). Since then the beetle has extended to North Carolina, West Virginia, Ohio and northwards through New York State. They may defoliate up to 250 plant species, some of which are important crops: e.g. soya beans, clover, apples and peaches. In Japan the beetles do little damage, but their food sources have changed since arriving in the USA. (Since Elton wrote there are signs that the Japanese beetle is no

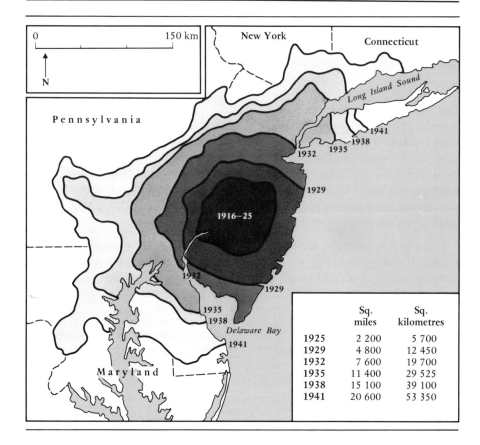

Fig. 7.2 Diffusion of *Popillia japonica*, the Japanese beetle, following its arrival in New Jersey in 1916. Since 1941 it has continued to expand its territory and only recently has it shown signs that its maximum spread may have been reached. (Modified from Elton, 1958, after United States Bureau of Entomology and Plant Quarantine.)

longer actively expanding its area. Nevertheless, extensive damage can still be a major problem within its range. Some control has now been effected by an inadvertently introduced microbe, the milky spore bacterium.)

THE FLUTED OR COTTONY CUSHION SCALE INSECT

Australia provides the next example, not only of the pest but also the eventual control mechanism. In 1868 the fluted or cottony cushion scale insect (*Icerya purchasi*) was accidentally introduced to California from Australia. Within a few years it threatened the entire citrus fruit industry. Monoculture greatly favoured the rapid spread of the pest, assuring a steady food supply to trigger off the population explosion. Fortunately, the insect was quickly brought under control, at a total cost of only $2,000, by the introduction of 139 specimens of an Australian ladybird (*Rodolia cardinalis*), the natural predator of the fluted scale insect. An Australian parasitic wasp was also successfully introduced as a control. Elton notes that this spectacular case unjustifiably made many think the use of natural predators would always be as successful.

The last example is the vine aphid (*Phylloxera vitifolii*). This native of North America does little damage to the wild vines there. But it has been deadly in France where it arrived last century, probably brought by shipping to west-coast ports. Some three million acres of French vineyards were destroyed and the situation was not rectified until European vines were grafted on to resistant American rootstocks. Many areas in northern France never re-established commercial wine production; costly remedial measures tipped the economic balance against them.

These twelve examples are but a small selection from the great number now known to ecologists. They represent a situation where a species has rapidly increased its numbers and range, either being introduced by man (accidentally or intentionally) or suddenly benefiting from the intervention of man. At this stage several main conclusions from these particular studies may be stressed:

1. Species which are harmless in one country may become deadly elsewhere.
2. Introduced animals may change their food sources and other aspects of their ecology, cutting across the established patterns of the native fauna.
3. Once firmly established, these pests are difficult to eradicate, especially as they often have no natural enemies in their new countries. With extensive new food supplies available their spread can be exceedingly rapid.
4. Disturbances can have devastating consequences, often in the most unexpected ways. Indirect influences, resulting from the introduction, may pose as many problems as the direct or obvious changes.
5. We know little about these complex inter-relationships between species. We know even less about man's part in influencing these relationships.

The dramatic impact that is possible when exotic species are introduced into a relatively small but highly vulnerable area is well demonstrated by the classic case of New Zealand (Fig. 7.3). As a remote and long isolated island it developed a unique flora and fauna which was largely defenceless against most introduced species. Many exotic plants and animals were deliberately brought to New Zealand by the Acclimatisation Societies. In the nineteenth century 23 such societies were established throughout New Zealand with the expressed aim of introducing 'suitable' species, mainly from Britain. Table 7.1 gives an indication of part of the activities of just one of these societies for the period 1865–6. Such work continued until the New Zealand Wildlife Act was passed in 1953. The societies still exist and now play a part in the overall management of New Zealand's biogeography, as Fig. 7.4 shows. The very complexity of this diagram is itself largely a reflection of the past activities of the Acclimatisation Societies through their establishment of freshwater fishing, their introduction of most of the noxious animals and agricultural pests and the resultant threats to the native flora and fauna (see also Figs. 7.5 and 7.6).

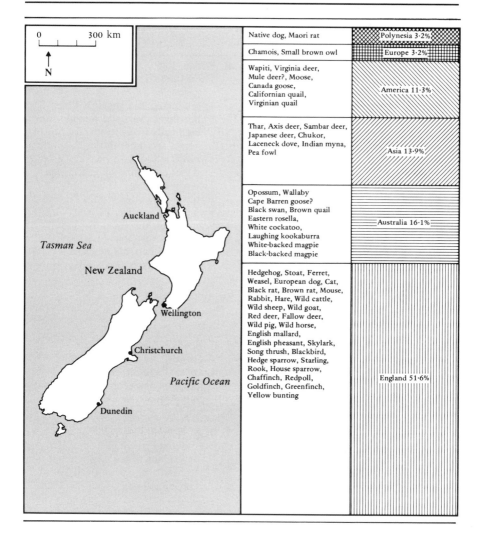

Fig. 7.3 The contributions of various countries to the introduced bird and mammal fauna of New Zealand. (Modified from Elton, 1958, after Wodzicki.)

CONTROL METHODS

Conway notes that the traditional classification of pests has been according to taxonomic order or the species or materials that pests attack. He argues for a potentially more illuminating classification in terms of the spectrum of ecological strategies that pests exhibit: we can thus speak of r-pests and K-pests (see Chapter 5 for definitions). Classic pests which form ravaging swarms are r-pests. K-pests are less obvious but their damage may nevertheless be significant. The codling moth, for example, damages apples only superficially but

Table 7.1 Contributions of plants and animals received by the North Canterbury Acclimatisation Society, 1865–66.

W. Hislop, Esq.
Forsythia veridissima
Two Hibiscus, red and white variety
Two Silver-edged Hollies
Golden-edged Hollies
Cape Mulberry
Two Photinia Seratifolia
Pomegranate
White Bean, or Service Tree
Mountain Ash
Rhododendron
A collection of Native Veronicas
Two Geulder Roses
Seed of the Scotch Heath
California Prairie Grass
Two seeds Riddles
A large selection of Plants, Cuttings,
    and Seeds, at various times.

Captain Rose, of the Ship *Mermaid*:
A pair of Blackbirds

Edward Flood, Esq., of Sydney:
An Emu

Dr Mueller, of Melbourne:
Two large packets of Australian Seeds
One pair of Black Swans

Governor Gore Brown:
A packet of Tasmanian Seeds

His Honor Samuel Bealey:
A quantity of Wire Netting
A pair of Wekas, and Young One

D. Innes, Esq.:
A pair of English Pheasants

Mr H. Smith:
Seeds of the Tea Tree

R. Wilkin, Esq.:
Three Carob Trees, or Locust Beans
    (*Ceratonia Siligua*). A most
    productive and valuable Tree,
    bearing a sweet and nutritious pod,
    used extensively for the feeding of
    horses and other cattle.
Two Black Swans
Two pair of Bronze-winged Pigeons
Two pair of Doves
Two pair of Wonga Wonga Pigeons
Two pair of Squatter Pigeons
Four pair of Australian Magpies
One dozen Australian Sparrows
Two pair of Laughing Jackasses
A packet of Australian Seeds, Plants
    and Bulbs

W. Wilson, Esq.:
A packet of Tasmanian Seeds
Silkworms' Eggs

J. C. Brooke, Esq.:
A quantity of large Gum Trees
Leeches (these increased largely in a
    pond set apart for their reception,
    but owing to some heavy floods,
    they made their escape)

J. B. A. Acland, Esq.:
Seed of the Parsnip Cheval. A new
    description of vegetable, about the
    size and shape of an Early Horn
    Carrot, and of a flavour between
    the Potato and Chestnut; it prefers
    a humid rich soil.

J. H. Potts, Esq.:
Two Paradise Drakes

Mrs I. Stemson:
A Native Owl

Consul General of Turkey:
Seeds, including some of the (*Sesamum Orientalis*). A plant growing like Rape, the
    seed of which is much valued in the manufacture of Confectionery, and believed to
    be well suited to the New Zealand climate.

(This list represents but two pages of a six-page list appended to the *Second Annual
Report of the Acclimatisation Society*, dated 7 May, 1866. It is based on the original
format.)

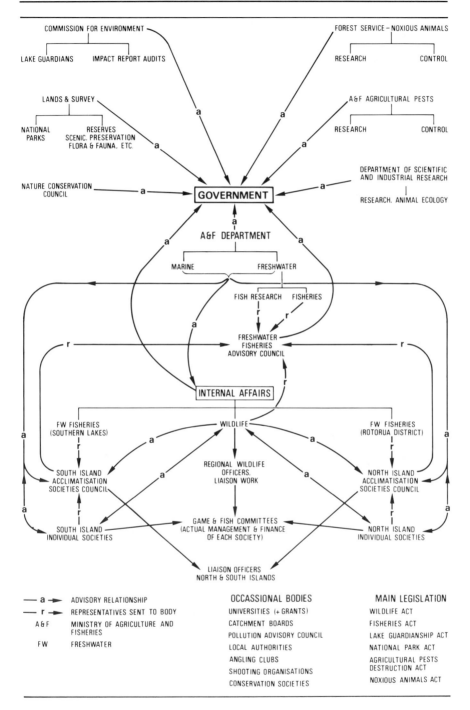

Fig. 7.4 The position of the Acclimatisation Societies within the Environmental Wildlife and Fisheries Administration and Management structure of New Zealand, as in
1976. (Modified from diagram supplied by Internal Affairs Department.)

Fig. 7.5 Canada geese (*Branta canadensis*) nesting in the High Country of the Southern Alps, New Zealand. This exotic species, introduced by the Acclimatisation Societies, annually migrates to the agricultural lowlands, causing widespread crop damage. See Fig. 7.6 for control measures. (Photograph supplied by J. Adams, Internal Affairs Department.)

Fig. 7.6 Control of the Canada goose: eggs at nesting sites in the mountains are injected with formaldehyde. Only one or two chicks hatch – see Fig. 7.5. (Photograph supplied by J. Adams, Internal Affairs Department.)

the blemished surface is enough to reduce considerably the commercial value of the crop.

Once the dangers of a pest outbreak are appreciated, precautions are normally taken to prevent it from spreading to new areas. These *quarantine* measures attempt to control the spread by checking the movement of potential carriers of the new species (people, goods or other plants and animals) or the species itself. They have to be vigorously applied and are, at best, holding operations for, in many cases, the pest does eventually succeed in becoming established.

A good example is provided by the Colorado beetle (*Leptinotarsa decemlineata*), a major pest of potatoes. Discovered in the 1820s as a rather insignificant insect in the dryish Eastern Rockies, the beetle lived on members of the potato family but not the potato itself, which had yet to be introduced to this area. When potato crops arrived from the East a massive population explosion occurred, eventually spreading to the Atlantic coast. It reached Europe towards the end of the nineteenth century, but only became firmly established around Bordeaux. Eradication measures during the First World War were then handicapped by a labour shortage caused by the war effort. A rapid diffusion occurred throughout France, Spain, Germany and into Italy and Eastern Europe. Fortunately, the English Channel protected Britain where vigorous quarantine measures were enacted. These included wide publicity and incentives to the public (£5 reward for each beetle handed in). Despite this we still get an average of 40 small outbreaks each year. In 1982 beetles arrived in imported Italian spinach: 40 were found at Bradford and, 24 hours later, 27 were discovered in a similar consignment at Spalding. At the moment these outbreaks are easily dealt with but this may not always be the case.

Once the pest (or weed) is firmly established, the only approach then is a campaign of containment or eradication. This usually takes the form of either *biological control* or *non-biological control*. The former method utilizes some aspects of the natural ecology or biology of the pest. The latter mainly employs chemical or physical treatments which have no exact counterpart in the natural world. This distinction is not always clear-cut because there are also *cultural methods* of control based on man's land management techniques and these approaches may combine aspects of biological and non-biological control. Examples of each are now examined, briefly noting their effectiveness and some of the problems arising from their use.

BIOLOGICAL CONTROL

One method of effecting control of a pest is by introducing or encouraging either its natural enemies or some other species which will readily adopt the pest as a new food source. This is usually thought of in terms of using the classic predator-prey relationship. However, the 'attacking' organism may be minute or microscopic, e.g. a nematode (round worm) or a pathogenic fungus or bacterium. The classic case where the predator-prey approach worked well was the infestation of California by the Australian cottony cushion scale insect. As outlined above, the introduction of its natural predator, an Australian ladybird, quickly controlled the pest. In New Zealand the introduced ragwort (*Senecio jacobaea*) was a serious pasture weed. Each plant produced more than 100,000 seeds. It greatly reduced grass growth and was poisonous to cattle if eaten in large quantities. In this instance the introduced cinnabar moth has brought some control; its grubs are voracious feeders on the weed. When used in Australia, however, it was far less successful. Ragwort control here is now largely by means of aerial hormonal sprays. Examples of classical biological control have been reviewed by Caltagirone. However, these approaches have often backfired. The mongoose was introduced into Jamaica to control the black rat but showed more interest in the existing enemies of the rat. It also severely reduced some bird populations and caused the extinction of several reptiles. In 1959 105 egrits were introduced into Hawaii to control insect pests of cattle (themselves introduced). These birds, now numbering thousands, are now major pests at the inland freshwater shrimp breeding ponds developed for the tourist industry. The involved case of the giant snail provides another example.

A second approach is to breed resistant stock or hybrids which will greatly reduce the incidence of attack, as in the case of *Phylloxera* in France. Conway notes that it takes usually 15–20 years to perfect a new resistant variety. In the USA new races of wheat rust appear every three to four years, necessitating a continuous effort to produce new resistant wheats. Needless to say, these methods are costly.

Much recent research has sought to develop control methods based on interfering with the enormous breeding potential of the pest. It is the removal of natural checks which causes a fuller realization of this potential, turning the insect into a pest. The United States has pioneered chemical *birth-control methods*. Birth-control pills are administered in food supplies and act as gametocides, killing the reproductive cells, Some success is reported with fox populations and a major pest of fruit-growing regions, the red-wing blackbird. They are also used in New York for rat control where, in poorer districts, the ratio of rats to people is thought to be 1:1.

A variation of this approach is the '*flooding*' technique. This initially proved very effective in the control of screw worm fly, a serious pest of cattle in Texas. Cattle infection fell from 50,000 in 1962 to 153 in 1970. However, by 1972 the figure had risen to 92,197 due to reinvasions from Mexico and evolutionary adaptations by the fly. The same technique is also being used for control of fruit fly in Queensland. Sterilized males are released into the natural population of fruit flies, competing with the fertile males for food and mates. Many of the females will have unsuccessful matings and if the releases are repeated several times the population drops towards an acceptable level. Sterlization is a highly selective method, being species-specific. It may be brought about by irradiation of the males (the principal technique) or through the use of chemosterilants. In some ecosystems, the latter pose a hazard to non-target species because they may cause mutations and cancerous growths. Another variation of this genetical approach, still mainly experimental, is to introduce a lethal hereditary factor into the population. The mutant conditions will be passed on and in time autocidal eradication should take place.

Regulation of body processes, such as growth, moulting and reproduction, by *hormonal treatments* is another new approach. The insect is exposed to an outside source of synthetic juvenile hormone which prevents the onset of sexual maturity. Until recently, it was thought that these techniques were highly specific but McNeil produces evidence for damage also to the natural enemies (e.g. parasites) which prey on insect pests. The population of predators may take longer to build up again than the pest. This could cause even worse pest outbreaks in the future. Huffaker similarly concludes that these hormonal actions may not be as specific as was first thought. Cases are now known where insects have developed resistance to juvenile hormones and growth regulators. Some geneticists claim that an insect population could make such a change in about 10 generations, i.e. in only a few years, if the right mutations occur. The use of these promising hormonal techniques might thus have to be limited and carefully monitored.

Under test at the moment are various *viruses* which are specific to certain insect pests. These pathogens offer exciting prospects for insect control. As far as we know they are safe to use (though once released they would be difficult to contain if shown to infect other animals or man). They are quickly effective and, unlike many chemicals, biodegradable. Viruses sprayed on food supplies (e.g. vegetation) will kill the insect. Sprayed at varying concentrations they will allow some grubs to mature to moths which then disperse the virus further. As Huffaker points out, already some 1,100 species of microbes associated with arthropods (spiders, mites, etc.) have been identified and there is now much interest in microbial insecticides. The rhinoceros beetle (which featured in the giant snail saga!) is now controlled by a virus which attacks its larvae. It should not be forgotten that in Britain the control of the rabbit (introduced in Norman times) was effected in the mid–1950s by the virus myxomatosis transmitted by the rabbit flea. At its peak the rabbit cost British agriculture about £50 million a year. Unfortunately, the rabbit is now making a dramatic return to the British countryside. Resistance to myxomatosis has grown and crop damage is again running into millions of pounds.

Another promising biological method is the use of *pheromones*. These compounds are secreted by animal species and have a powerful influence on the behaviour of other individuals of the same species. Pheromones constitute a form of 'chemical language' between individuals and there is an obligatory reaction to such signal odours, e.g. causing aggregation for sexual purposes. Each pheromone was thought to be specific but there is less certainty about this now. In addition to the pest they could perhaps attract useful species or the natural enemies of the pest. Pheromones have not yet had widespread use although many can now be synthesized. Only minute amounts are necessary, e.g. the female emperor gum moth can attract the male from over a mile away by releasing only a few molecules. One suggestion is to use pheromones to attract pests to bait which has been infected with a virus disease.

Other new biological methods may well be developed since this work is still in the initial stages. However, for many new techniques there is uncertainty about long-term effects and possible disturbance to non-target species.

NONBIOLOGICAL CONTROL

CHEMICAL METHODS

The most commonly used control method for pests and weeds in recent decades has been, and still is, chemical treatment (chemical warfare!). Pest problems often have to be solved quickly and this explains part of the appeal of the chemical approach. However, the dangers of relying too heavily on these chemical approaches is now being appreciated. Chemical pesticides and herbicides are costly to produce and dosages may need to be repeated frequently. *Resistance* soon builds up: about half the known pests in the world are now resistant to at least one chemical formerly used against them and the number is increasing each year. Frequent spraying builds up resistance quickly because it is a genetical phenomenon. The few individuals possessing the right genetical structure to confer resistance against a given pesticide are just those which survive to breed the next generation. These genetically resistant individuals must come to form an ever-increasing proportion of the subsequent populations.

Whereas some chemicals are quickly biodegradable, others are persistent in the environment. Use of this latter group is now banned or severely restricted in some countries (particularly the chlorinated hydrocarbon insecticides like DDT, endrin and dieldrin). Many persistent chemicals become concentrated along food chains. This concentration is more marked in predators than in prey. Further, predators may be more susceptible to the chemicals than their prey which are the targets of the treatment. For example, cyclamen mite populations on strawberries may increase by a factor of about 20 after parathion spraying because a mite which preys on cyclamen mites is much more easily killed by the spray then they are. Similarly, in 1962 weedkiller used as a corn dressing in eastern England resulted in many birds dying while populations of other birds in the same region increased relatively. It was then discovered that birds of prey, such as the sparrow hawk, could only tolerate 20 p.p.m. of the chemical in their body tissues while sparrows could tolerate up to 200 p.p.m. The result was an increase in the sparrow population through disruption of the predator-prey chain. Many similar cases have now come to light. In the classic case at Clear Lake, California, the original concentration of the insecticide DDD (related to DDT) spread in the water to exterminate midge larvae was 0·02 p.p.m. But at the end of the food chain the grebes had as much as 1,600 p.p.m. in their fat tissues. Mellanby was provided a detailed and balanced review of these chemical treatments, noting the many successful uses as well as the hazards involved. They still represent the best approach to particular pest problems in the absence of a proven alternative control method. Likewise, Samways concludes that chemical control is still the only practical and economic method for certain insect pests, e.g. some social insects and pests of stored products.

No doubt many more chemicals will be developed in the fight against pests but it is not simply a matter of finding out what will kill a species. This was the approach in the early days. We now recognize the need to understand the *ecology*, and even the *psychology*, of pests if treatments are to be fully effective. This is the situation with modern rat control. No one knows the rat population of Britain or the cost of their damage: it has been put as high as £50 million per year, with possibly one rat for every person. Rats are cautious animals, suspicious of new objects in their environment. So the introduction of traps and bait which are moved about frequently is ineffective because it arouses their suspicion. If they eat a sub-lethal dose of poison, rats for a considerable time will link the resulting pain with the bait causing the discomfort. Such baits will be avoided thereafter. This has led to the development of anti-coagulants, which cause delayed-action brain haemorrhages. The time-lag between ingestion and death prevents the rat associating any discomfort with the bait. Warfarin is the main commercial form of this poison. However, this sophisticated technique has now foundered somewhat with the appearance in Scotland and Wales of mutant rats resistant to Warfarin. Further research has yielded norbormide, a toxic chemical highly effective and, apparently, highly specific as a rat poison. Norbormide might solve the problem by replacing the anti-coagulants wherever necessary.

PHYSICAL METHODS

Research is now active on new methods for the physical control of pests. A promising technique, pioneered by Canadian ecologists, is the use of *sound*. In one experiment sound waves beyond the human range, at 50,000 cycles per second, were broadcast in 1962 over cornfields in the USA. This produced a dramatic fall in the number of cornborer moths. (the exact reason for this was not clear but it could be that the moths mistook the sound for a bat invasion and quickly moved out of the area). This technique could represent an important breakthrough in controlling this introduced European moth, which consumes about 10 per cent of the maize crop each year. Sound has also been used under experimental conditions against the mosquito. When broadcast at 200,000 cycles per second into infested water, the high vibrations damage the windpipes of the mosquito larvae, leading to suffocation.

The use of *fluorescent light* is another new experimental development. While not likely to be employed on a large scale it could find a use in commercial greenhouses and small orchards. The light forces the early maturity of larvae ahead of their food supplies. They then starve to death.

Static *electrical charges* of the type naturally found around trees and bushes, have also been studied as the basis of a possible new control method. Apparently mosquitoes congregate where a positive electrical charge builds up. A pattern of such charges could be used to attract mosquitoes along a predetermined path towards a trap, where they could then be electrocuted.

CULTURAL METHODS

Huffaker reminds us that cultural approaches and cropping schemes have provided control of many pests for centuries. The timing of sowing and harvesting, the rotation of crops, the management of water levels and the use of controlled burning (see Chapter 6) are traditional attempts at pest control. In other words, we manipulate the environment rather than mounting a direct attack on the pest. Control is achieved because the site's carrying capacity for pests is reduced. All too often these techniques have been abandoned in favour of some quick and apparently successful chemical treatment. There is now a growing realization that all forms of pest control have a part to play. Whenever possible the main approach should be biological, i.e. taking advantage of naturally occurring phenomena. These methods should be supplemented by cultural, chemical and physical techniques whenever necessary, leading to an *integrated system* of pest management.

Evidence suggests benefits would arise from reintroducing diversity to agroecosystems. Doutt and Nakata show how the leafhopper pest of Californian vineyards is controlled at some sites by a parasitic insect which first develops on nearby blackberry patches. On these 'weeds' the parasite overwinters on another species of leafhopper. By early spring the parasite is abundant and invades the vines, effectively suppressing the leafhopper pest. The regular distribution of blackberry patches throughout the vineyards and an acceptance of integrated control techniques has brought about reasonable control at one-sixth of previous expenditure.

Not only has vegetation become less diversified but far fewer cultivated strains of crops are now used in agriculture. The dangers of reducing the genetic base by concentrating cultivation on a narrow spectrum of high-yielding varieties is beginning to be recognized. This leads to a reduction in the genetic factors available for resisting pest attack. These potential dangers are seen in the so-called Green Revolution where, for example, a few 'miracle' high-yielding strains of rice are replacing many native varities (up to 600 in Indonesia, according to Smith). About 90 per cent of the United States corn crop in 1970 was based on the same genetically constituted varieties. This is what made outbreaks of the southern corn blight so potentially disastrous. To paraphrase Huffaker: natural enemies, evolved plant resistance and traditional cultural practices are *the great triumvirate* in insect containment. By concentrating on the use of chemicals these three prime controls have been neglected.

RAPID LAND-USE CHANGES

Uvarov draws attention to the particular problems in pest ecology for developing countries; those being 'opened up' extensively by man for the first time, using modern agricultural techniques. They include parts of the USSR, Canada and Australia and we may also add those where traditional native agriculture is quickly giving way to 'European' farming methods, as in parts of Africa. The rapid changes in land use accompanying these modern practices are a feature of such areas. Uvarov summarizes Russian experiences in this context.

The Russians, particularly under Khrushchev, had a vigorous policy for the conversion of their natural grasslands (the virgin steppes) into wheatlands. These projects were troubled from the outset by massive outbreaks of insect pests. Ecologists, comparing the insect population structures of the original native grasslands with that of the new cereal crops, noted the former had twice as many insect species as the latter. However, in the wheatlands the insect population density was nearly twice that of the steppes. Further, in the wheat- lands a few insect types dominated the whole population. In other words, the steppe population was balanced, with no single species dominating. In the wheatfields the dominant species nearly all became important pests within a matter of a few years. Insect eradication was very difficult, as a reserve of them always existed in the natural vegetation. After repeated application of costly pesticides the population soon built up again from this reserve stock.

Other examples where rapid changes in land use have led to sudden pest outbreaks, include:

1. The clearance of forest land for crop cultivation, resulting in a vast invasion of the cleared areas by insects adapted to open, drier terrain. This occurred with locusts in the Philippines and grasshoppers in Siberia.
2. The irrigation of dry land which favoured the rapid spread of riverine pests. The most serious pests of the Gezira cotton irrigation scheme in the Sudan are now the thrips. In Swaziland, where 8,000 ha were brought under irrigation in the 1950s, two new pests, not regarded as such in the area before, are the tadpole shrimp and the water snail. Several rare or previ- ously unknown insects are now major pests in the Israeli irrigated areas created during the last 30 years.
3. The planting of tree shelter-belts to improve crop microclimates has resulted in vast numbers of insects moving into these wooded tracts, as in the southern Russian steppes. Here, ecologists demonstrated that the increase in yields due to the direct effect of the shelter-belts on the crop was often reduced or nullified by the losses due to these new pests.

A recent review of integrated pest control approaches in the developing world is provided by Brader: several of the cases are associated with problems arising from rapid land-use changes.

Uvarov presents several important conclusions from the Russian studies referred to earlier in this section:

1. The insect pests of a given crop may arise from the segregation of a reduced number of species already present in the wild. Pest status is reached because improved conditions are produced for the existence of this reduced number by the actions of man. (Of course, pests may also be exotic species, but again man is frequently responsible for their introduction.)

2. These pests retain reserve populations in the wild. Plenty of breeding stock remains even after the crop pest is destroyed.
3. In the underdeveloped areas of a country any native insect has the capacity to become a pest if subsequent changes by man favour this development. The pest potential of a native insect population is thus virtually unknown.

What has been said in general terms for insect pests probably applies equally well to other pest species and to weeds. Even in well-developed countries like Britain rapid change in agricultural practices are still possible, e.g. the wholesale removal of hedges in eastern England. These changes may subsequently cause significant alterations in pest ecology, leading to a grossly disturbed ecosystem.

DISCUSSION SECTION

Are pest problems new? Did primitive man suffer to the same extent?

Pest problems are not new but they have changed their character with the development of modern agricultural practices. In the early days of agriculture, some 10,000 years ago, Huffaker's triumvirate in insect containment operated. The first cultivated crops were similar to types existing in the wild, which had evolved over many thousands of years in an approximate balance with the insects living on them. Murdock makes just this point: a major cause of instability in the form of pest outbreaks in modern crops is the fact that the interacting species, unlike those in natural communities, do not have a long-shared *coevolutionary history*. In primitive agriculture many weeds existed, crop patches were small and often scattered and plant diversity was reasonably high. This allowed the natural enemies of potential pests to exist and effect some control. Today we aim for huge blocks of weed-free, monospecific crops under controlled environments. The agricultural techniques of the early cultivators also introduced further control. An important concomitant of these techniques was the exposure of pests to the rigours of the climate, the encouragement of natural enemies and of plant resistance.

As Mellanby notes, the pests of early man were more likely to be the larger carnivores while he remained a hunter and the larger herbivores once he became a cultivator (and he either exterminated both these groups in the well-settled regions or domesticated some of the herbivores for his own benefit). Insects were probably not a major problem in most areas, though plagues of locusts are an obvious exception. But these were triggered off by climatic mechanisms, not human activity. Of course, in some areas man has grossly disturbed the ecosystem from very early times. There is increasing evidence of early soil erosion and salinization problems. Overgrazing of semi-arid lands in many regions has greatly extended the desert area. Indeed, several early civilizations (e.g. Ur, Mohenjadoro) were buried by shifting sands whose mobility was probably increased by land mis-use. Even in this country, Darling sums up the position reached in the Scottish Highlands thus, '. . . the underlying rock is so poor, the slopes so steep, the climate so adverse, and the glacial history so devastating that the soil could persist only when the climax vegetation was preserved. The soil was then in circulation, implying the trapping of solar energy with a minimum of loss, and the slow decomposition of the bed rock meant that there was actual accretion under climax vegetation of oak, birch and pine.' He argues that because of wholesale forest clearance and subsequent overgrazing and burning over the last 1,000 years many parts have virtually been turned into a Temperate Latitude version of desert.

Why do we always seem to be caught unawares with these ecological disasters?

This century is characterized by rapid changes in land use and rapid transport of people and goods around the world. This greatly increases the chances of disturbance to the native biota, or of introduction of exotic species. In both cases, through the operation of population dynamics, a species may quickly reach pest status. These changes are over such very *short timescales* when compared with the thousands of years over which natural ecological changes usually operate. As a measure of the extent to which we have changed just one small part of the earth's surface consider the situation in New Zealand as shown in Fig. 7.3. This diagram refers only to faunal changes and takes no account of some equally great flora changes: Healey lists 1,083 species in the adventive flora of just the Canterbury Province of New Zealand. In addition, some 3,000 million sheep are said to have trod the pastures of New Zealand since white settlement while, at their peak, the introduced rabbit outnumbered the introduced ewe by about 10 to one.

There is also the problem of disturbances which work slowly through the environment, often hidden from direct observation and not appreciated. A case in point is cadmium. Hutton reports that this heavy metal pollutant is increasing in Europe and can cause several diseases in man. Most studies have concentrated on atmospheric sources of cadmium. However, the most significant source now appears to be from agricultural phosphate fertilizers and this has only recently been realized. Chemical pesticides provide a second example. Only years after their first application do we realize the hazards. With DDT (now banned by some authorities) it may take another twenty years before nearly all of this chemical that has already been used becomes incorporated in world food chains. Some countries, however, are still increasing their use of DDT. These insidious developments may be just as dramatic in the long run and vastly more difficult and costly to control.

Are the economic aspects of pest control as important as the ecological ones?

Without doubt the economic aspects are vitally important. After all, most pests by definition are only such when they compete with us for food supplies. This competition partly determines the cost of food production and the profit

Table 7.2 Examples of gains in terms of yield and net dollar returns from pesticide use on various crops in the United States (From Conway, 1975 after Rudd)

Crop	Pest	Yield per acre		Cost of treatment ($ per acre)	Net gain ($ per acre)
		Untreated	Treated		
Seed cotton	Bollworm	7,203 lb	7,860 lb	16	126·50
Pea seed	Fungi	456 lb	610 lb	0·70	21·25
Tomatoes	Diseases	5·4 tons	11·8 tons	40	100
Sugar beets	Root maggot	11·8 tons	14·5 tons	2·25	32·20

margins. Table 7.2 shows one side of the economics, the benefits of pesticide use. A more dramatic case is provided by Mohyuddin and Shah who show that the introduction of a parasitic wasp into New Zealand to control the army-worm, a pest of several crops, cost about US $21,700 in research and initial programme expenditure. The benefits for the 1974–75 period exceeded US $10,000,000. Each year about a third of human food is lost to pests, either from the standing crop or while in storage. Without doubt insects should be regarded as man's chief competitors. But in attempting control by pesticides more attention must be paid to the economic thresholds (see Fig. 7.7). In too many cases the level at which cost of control exceeds the cost of yield losses has never been determined.

However, there is a debit side: increasing development costs of effective pesticides and the eventual cost to the community as a whole of worldwide pollution. Conway neatly summarized these arguments in a recent paper. Rudd

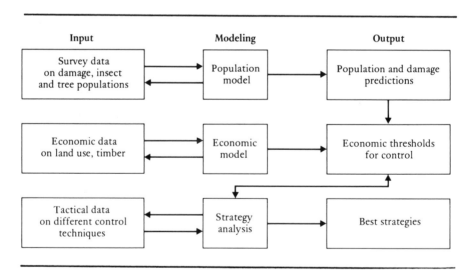

Fig. 7.7 A scheme for analysing the stages in the development of a pest-control system for the pine bark beetle in the United States, illustrating the need for careful checking at each point in the process before a best strategy is determined. (From Conway, 1975, after Stark.)

puts total sales of pesticides produced in the USA at \$3 billion annually and demand worldwide is increasing at about 10 per cent per year. ICI are said to test some 150 compounds per fortnight for possible use as insecticides. Most are rejected for various reasons while a successful product may take 5 years to develop fully at an average cost of well over £1 million (for initial research, field trials, etc.). This applies even to highly specialized products with small sales. Inflation has caused the prices of many chemicals to soar: many involve energy-intensive processes or are derived from oil. This increase has stimulated more research into biological methods of control. However, many of these are highly specialized fields, where scientific manpower is scarce and costly. Now that the complexity of pest ecology is realized, attempts at correcting disturbed ecosystems will require careful monitoring (see Fig. 7.7). Simple, direct chemical control may no longer be feasible in many situations or may lead to far greater costs when all the consequences are appreciated.

Are we winning the battle against ecological disturbance?

Most ecologists are pessimistic. Ehrlich only gives us a one or two per cent chance of surviving to AD 2000! Much of this pessimism is related, of course, to our own population growth-rate (similar to that of an insect pest in that traditional controls have been largely removed). A main difficulty is that the answer is not entirely an ecological one. Either at the local or the global scale, ecological disturbances also have vitally important social, political and economic consequences. The various sections of society make differing demands on the same landscape, e.g. when recreational interests run counter to those of wildlife conservation.

There is, of course, a good side to the use of chemicals in the environment which Mellanby demonstrates quite clearly. The case is also stated by Gunn and Stevens. Some scientists believe that technological advances will be adequate to deal with our environmental problems. Maddox produces a reasoned argument against too much pessimism which should be read as a counter to the recent flood of books on the impending ecological catastrophe. Nevertheless, it should be emphasized that this chapter has concentrated on only one form of ecosystem disturbance.

Soil erosion (see Figs. 7.8–7.11), chemical enrichment of lakes (eutrophication) by run-off water from agricultural land carrying excessive fertilizers (particularly nitrogen and phosphates), possible man-induced climatic changes, etc. are other forms. A particularly serious pollutant, only recently recognized in many regions, is *acid rain* ('pure' rain has a pH value usually well above 5·6; 'acid' rain values of 4·3 are common). Sulphur and other oxides released into the atmosphere by industrial processes and fossil fuel burning can be carried well into the Arctic, some 5,000 miles from their source. They descend as acid rain, accumulating in lakes, rivers and ground water. Acid rains dissolve copper and iron (from pipes) and certain heavy metals (such as cadmium): these toxics then enter public water supplies. More importantly, they leach out calcium and magnesium ions from the soil, leading to lower soil fertility and

Fig. 7.8 Intense soil erosion on steep slopes near Queenstown, Tasmania. This followed the destruction of dense cool temperate rain forest by copper toxicity from nearby smelters.

Fig. 7.9 Deep soil erosion on a terrace of the River Murray, about 25 km east of Mildura, northern Victoria. Originally the site was covered in continuous eucalypt woodland but was cleared for pasture development without much appreciation of the underlying soil structure and its susceptibility to heavy downpours once the protective tree cover was removed.

Fig. 7.10 Same area as Fig. 7.9. Poor, drought-prone pasture in the foreground. The erosion channels, many up to 10 m deep, are rapidly working towards the Sturt Highway in the background.

Fig. 7.11 A network of erosion channels eating back into native farmland and home-steads, north-west of Ladysmith, Union of South Africa.

reduced crop productivity. Eventually aluminium ions are removed and accumulation of these in water bodies are particularly toxic, especially to fish. Because of wind trajectories the Scandinavian countries and Canada receive 'imports' of acid rain from their heavily industrialized neighbours to the south. More than 9,000 lakes in Sweden are now sterile and widespread tree destruction is reported from Scandinavia, Germany and Canada. It has only recently been appreciated that Britain also suffers from acid rain and has some of the highest concentrations in the world. Research shows that pollutant oxides in combination (e.g. sulphur dioxide plus a little nitrous oxide) can cause much greater problems than either alone. A Dutch study by van Breemen *et al.* even demonstrates how cow manure spread on fields leads to higher acid rain values locally: the ammonia released combines with sulphur dioxide to create ammonium sulphate which is then washed into the soil downwind to give some of the highest acidity values yet measured. To complicate matters, Charlson and Rodhe have argued that the natural sulphur cycle can produce large geographical and temporal variations in the amount of natural sulphur dioxide in the atmosphere. Thus the use of global average values of natural sulphur concentrations may be misleading in the evaluation of local acid rain causes and therefore all high values may not be entirely due to pollutant sources. However, man is clearly the cause of most acid rain and the several recent international symposia (e.g. see Falkenmark) bear witness to the growing concern.

Taken together, these various forms of ecosystem disturbance point to the intricate nature of the biosphere and our, as yet, rudimentary knowledge of how it operates. A very gloomy scenario is painted by 'The Global 2000 Report to the President' (Barney), a US Government study commissioned by President Carter on the probable changes in world population, natural resources and environment during the rest of this century. Deforestation, severe soil erosion, desert encroachment, wildlife extinctions, dependence on fewer genetic crop strains, increased vulnerability to pests, water shortages, etc., are the main problems set in a context of a world population nearing the planet's carrying capacity. This is a far from cheerful report to read but when pollution can be detected high in the atmosphere and deep in the ocean it is time to question seriously what we are doing to the ecosystems of the world (Figs. 7.12–7.15).

It is a difficult situation to appraise because, as in other fields, scientists may disagree over the interpretation of the data. A classic case is the influence of DDT residues on eggshell thinning found in some species of wild bird. The decline in eggshell thickness in recent decades appears to correlate well with the spread in use of DDT. Many workers favour a cause-and-effect relationship. Hazeltine challenged this and went further by suggesting that the raids on nests by scientists to collect eggs for analysis were just as likely to upset the birds and cause reproduction failure. While later detailed work has now established the link with DDT quite clearly, this case does illustrate the uncertainty that can arise even in areas where many studies have been made. Occasionally, some of the battles we fight today may cease to exist tomorrow just simply because our perception of an ecological problem may change. Fitz cites the case of the tall larkspurs (*Delphinium barbeyi* and *D. occidentale*) in the foothills and intermontane basins of the Rockies. These serious rangeland

Fig. 7.12 Heavy infestation by a tumbleweed in wheat fallow near Dimboola, Victoria. The plant spreads readily and can cause serious damage to agricultural machinery.

Fig. 7.13 Foothills of the Southern Alps, New Zealand. Overgrazing of pastures derived from land formerly forested has encouraged the spread of unpalatable tussock grass.

weeds poison cattle more readily than any other range plants. Many of these areas are now National Forests. Should they become more important for recreation than for agriculture then the larkspur may cease to be a problem. Indeed, as a widely distributed flower, attractive in appearance, it would become an asset, a valued component of the tourist landscape.

Fig. 7.14 Overgrazed sheep paddock near Desert Camp, South Australia. Invasion by the introduced Scotch thistle (*Onopordum acanthium*), which often reaches heights of 1.5 m or more under Australian climatic conditions.

Fig. 7.15 The same area as Fig. 7.14. Here invasion of the pasture land is by a native species, the Austral grass-tree (*Xanthorrhoea australis*), a spiky, unpalatable member of the lily family whose spread is encouraged by fire.

REFERENCES

Anon., 1972. 'A blueprint for survival', *The Ecologist*, **2**, 1–43.

Barney, O. G. (Study Director), 1982. *The Global 2000 Report to the President. A Report Prepared by the Council on Environmental Quality and the Department of State*, Penguin Books, Harmondsworth, Middlesex.

Bedford, G. O., 1980. 'Biology, ecology and control of palm rhinoceros beetle', *Ann. Rev. Entomol.*, **25**, 309–39.

Boorman, L. A. and Fuller, R. M., 1981. 'The changing status of reedswamp in the Norfolk Broads', *J. appl. Ecol.*, **18**, 241–69.

Brader, L., 1979. 'Integrated pest control in the developing world', *Ann. Rev. Entomol.*, **24**, 225–54.

van Breemen, N., Burrough, P. A., Velthorst, E. J., van Dobben, H. F., Toke de Wit, Ridder, T. B. and Reijnders, H. F. R., 1982. 'Soil acidification from atmospheric ammonium sulphate in forest canopy throughfall', *Nature.*, London, **299**, 548–50.

Caltagirone, L. E., 1981. 'Landmark examples in classical biological control', *Ann. Rev. Entomol.*, **26**, 213–32.

Charlson, R. J. and Rodhe, H., 1982. 'Factors controlling the acidity of natural rainwater', *Nature*, London, **295**, 683–5.

Conway, G. R., 1975. 'Better methods of pest control', in *Environment: resources, pollution and society* (2nd edn.), (ed. Murdock, W. W.), Sinauer, Sunderland, Mass., pp. 355–78.

Conway, G., 1981. 'Man versus Pests', in *Theoretical Ecology* (2nd edn.), (ed. May, R. M.), Blackwell Scientific Publications, Oxford and Boston.

Crossland, N. O., 1965. 'The pest status and control of the tadpole shrimp, *Triops granarius*, and of the snail, *Lanistes ovum*, in Swaziland rice fields', *J. appl. Ecol.*, **2**, 115–20.

Darling, F. F., 1955. *West Highland Survey: an essay in human ecology*, OUP, Oxford.

Doutt, R. L. and Nakata, J., 1965. 'Overwintering refuge of *Anagrus epos* (Hymenoptera: Mymaridae)', *J. Econ. Entomol.*, **58**, 586.

Ehrlich, P. R. and Ehrlich, A. H., 1970. *Population, Resources, Environment: issues in human ecology*, W. H. Freeman, San Francisco.

Elton, C., 1958. *The Ecology of Invasions by Animals and Plants*, Methuen, London.

Elton, C., 1966. *Animal Ecology*, Methuen, London.

Falkenmark, M., 1980. 'International conference on the effects of acid rain urges more research in critical areas', *Ambio*, **9**, 198–9.

Fitz, F. N., 1972. 'Plant-insect interactions, tall larkspur and biological control', Chapter 24 in *Challenging Biological Problems: directions towards their solution* (ed. Behnke, J. A.), OUP, London.

Gunn, D. L. and Stevens, J. G. R., 1976. *Pesticides and Human Welfare*, OUP, Oxford.

Hazeltine, W., 1972. 'Disagreements on why Brown Pelican eggs are thin', *Nature*, London, **239**, 410.

Healey, A. J., 1969. 'The adventive flora of Canterbury', in *The Natural History of Canterbury* (ed. Knox, G. A.), Reed, New Zealand.

Heybroek, H. M., Elgersma, D. M. and Scheffer, R. J., 1982. 'Dutch elm disease: an ecological accident', *Outlook on Agriculture*, **11**, 1–9.

Huffaker, C. B., 1972. 'Ecological management of pest systems', Chapter 17 in *Challenging Biological Problems: directions towards their solution* (ed. Behnke, J. A.), OUP, London.

Hutton, M., 1982. *Cadmium in the European Community*, MARC. Report No. 26 (Monitoring and Assessment Research Centre, Chelsea College, University of London).

Jarvis, P. J., 1979. 'The ecology of plant and animal introductions', *Progress in Physical Geogr.*, **3**, 187–214.

Jones, P., 1981. 'The geography of Dutch elm disease in Britain', *Trans. Inst. Br. Geogrs.*, **6** (new series), 324–36.

Krebs, C. J., 1972. *Ecology: the experimental analysis of distribution and abundance*, Harper and Row, New York.

Maddox, J., 1972. *The doomsday syndrome*, Macmillan, London.

McNeil, J., 1975. 'Juvenile hormone analogs: detrimental effects on the development of an endoparasitoid', *Science*, **189**, No. 4203, 640–2.

Mead, A. R., 1961. *The Giant African Snail: a problem in economic malacology*, University of Chicago Press, Chicago.

Mellanby, K., 1969. *Pesticides and Pollution*, Fontana/Collins, London.

Ministry of Agriculture, Fisheries and Food, 1961. *The Coypu*, Advisory Leaflet No. 479, HMSO, London.

Mohyuddin, A. I. and Shah, S., 1977. 'Biological control of *Mythimna separata* (Lep. Noctuidae) in New Zealand and its bearing on biological control strategy', *Entomophaga*, **22**, 331–3.

Murdock, W. W. (ed.), 1975. *Environment: resources, pollution and society* (2nd edn.), Sinauer, Sunderland, Mass.

Norris, J. D., 1967. 'The control of coypus (*Myocastor coypus* Molina) by cage trapping', *J. appl. Ecol.*, **4**, 167–90.

Norris, J. D., 1967. 'A campaign against feral coypus (*Myocastor coypus* Molina) in Great Britain', *J. appl. Ecol.*, **4**, 191–200.

Parsons, W. T., 1963. *Blackberry: a widely distributed noxious weed*, Pamphlet No. 7, Vermin and Noxious Weeds Destruction Board, Melbourne.

Pears, N. V., 1982. 'Familiar aliens: the Acclimatisation Societies' role in New Zealand's Biogeography', *Scott. Geogr. Mag.*, **98**, 23–34.

Rudd, R. L., 1975. 'Pesticides', in *Environment: resources, pollution and society* (2nd edn.), (ed., Murdock, W. W.), Sinauer, Sunderland, Mass., pp. 325–53.

Samways, M. J., 1981. *Biological control of pests and weeds*, Studies in Biology No. 132, Arnold, London.

Skinner, B. J. (ed.), 1981. *Use and misuse of Earth's surface*, Readings from *American Scientist*, Kaufman, Los Altos, California.

Smith, R. F., 1972. 'The impact of the Green Revolution on plant protection in tropical and subtropical areas', *Bull. Entomol. Soc. Amer.*, **18**, 7–14.

Stark, R. W., 1973. 'The systems approach to insect pest management – a developing programme in the United States of America: the pine bark beetles', in *Insects: studies in population management* (eds Geier, P. W., Clark, L. R., Anderson, D. J. and Nix, H. A.), Ecological Society of Australia, Canberra, pp. 265–73.

Ward, B. and Dubos, R., 1972. *Only One Earth: the care and maintenance of a small planet*, Penguin, London.

Watt, K. E. F., 1974. *The Titanic Effect*, Sinauer, Stamford, Conn.

Uvarov, B. P., 1964. 'Problems of insect ecology in developing countries', *J. appl. Ecol.*, **1**, 159–68.

Part 2

SELECTED EXAMPLES FROM THE BRITISH ISLES

8

THE VEGETATION

INTRODUCTION

To attempt to cover in one chapter all the variation in vegetation displayed in the British Isles would lead to a very shallow approach. By selecting a few types for fuller treatment, superficiality can be largely avoided. First, a review of the historical development of natural and semi-natural vegetation in this country is presented. Secondly, two broad types will be examined in more detail, namely, *woodlands* and *uplands moors*. Even within these two types, further selectivity is necessary if we are to achieve any depth in description and analysis. These two types have been chosen because examples of them readily occur in Britain and they provide several links with previous chapters and with Chapters 9 and 10.

THE CLIMAX VEGETATION

Very little truly natural vegetation is left in this country today: several thousand years of various forms of disturbance by man, such as agriculture and the introduction of exotic species, have sadly reduced a continuous cover of natural vegetation to mere fragments, many of which at best only enjoy a semi-natural status now. Without such gross interference much of Britain would be covered in Temperate Forest (in Clementsian terms, the *Summer Deciduous Forest* formation), a climatic climax vegetation mainly occurring as mixed oakwood. These woodlands would occupy most lowland tracts and extend into hill regions to 300 m or more. In the more northerly latitudes of Highland Scotland (and at greater altitudes up to 600 m or more) this forest gives way to the *Northern Coniferous Forest* formation. The latter in this country is represented by the Scottish pine and birch woods (although birch is actually a deciduous species it is a common constituent of these high-latitude forests).

Of course, not all sites would carry one or other of these woodlands. Other forms of natural vegetation would occur in the seral situations already described in Chapter 4, and do so today – e.g. the psammoseres of coastal dune systems, the salt-marsh haloseres, the hydroseres of the Fens. Likewise, mountains continuing above the tree-line (about 610 m in eastern Highland Scotland) would carry open moorland vegetations, alpine grasslands and snow-bed communities. But all these naturally-occurring, non-woody vegetations, although locally important and providing the landscape with much of its character, would not account for more than about 30 per cent of our native plant cover (see Figs. 8.1–8.4). The other 70 per cent would have been various forms of climax woodland. And, of course, many non-woody communities would eventually progress towards a woodland end-form by seral developments. Peterken stresses that there are now no natural woodlands left that can be directly proved as such. The total cover of semi-natural ancient woodland in this country is only 2 per cent of our land area (cf. 20 per cent in France). Indeed, our total woodland cover (including plantations) is only 8 per cent (cf. Spain 28 per cent, West Germany 29 per cent, Ireland 4 per cent, Netherlands 7 per cent and Sweden 64 per cent).

Since so little natural vegetation now remains for study, how can we be certain about its former composition, and how can its former areal distribution be determined? *Historical records* help in this respect but they only provide a somewhat sketchy picture of former plant cover for certain periods or certain areas during the last 2,000 years or so. For example, on a map prepared by Ptolemy in the second century AD the first reference appears to the Caledonian Forest (*Caledonia Silva*). This was the descriptive term given to the whole of the Scottish Highlands, indicating the heavily wooded nature of this terrain. But as early as AD 1329 the Exchequer Rolls of Scotland (Vol. 1, p. 215) recorded the importation of Baltic timber for use in Edinburgh, thus suggesting that local supplies of timber were becoming scarce. Forest laws (*Leges Forestarum*), designed to curtail woodland destruction, appeared in the Acts of the Parliament of Scotland in the early fourteenth century, according to Steven and Carlisle. Aeneas Sylvius, who later became Pope Pius II, visited Scotland in 1405 and remarked on the lack of woodland. Likewise, Sir Anthony Weldon, touring Scotland in 1617, stated, 'there's scarce a tree to hang a Judas on'. Such records indicate rapid depletion of formerly well-wooded areas.

These timber shortages occurred first in the Lowlands, where most settlements were concentrated. The remote Highland forests of Scotland were largely inaccessible except for local needs. An Act of Parliament of Scotland dated 1609 expressed delight at finding them still largely intact. However, they soon underwent wholesale destruction and showed signs of devastation as early as the middle of the eighteenth century. Similarly, in England we have the diaries and reports of eminent citizens who toured the countryside commenting on the changes taking place. For earlier times, a partial view of the changing landscape comes from the Domesday Survey of the Normans (AD 1086). Here, the extent of woodland is often not reported directly but rather as the number of pigs it could support. From such statements, a rough estimate can be made

Fig. 8.1 Short, wind-clipped heather moorland on a stony, free-draining plateau site at 820 m near Glas Maol, Scottish Grampians. Main species are: *Erica cinerea* (bell heather), *Calluna vulgaris* (common heather) and *Trichophorum caespitosum* (deer grass).

Fig. 8.2 Salt-marsh dominated by *Spartina maritima* (cord grass) with some *Halimione portulacoides* (see purslane) fringing the drainage channel. Gilbralter Point, Lincolnshire.

Fig. 8.3 Alder woodland (*Alnus glutinosa*) forming as a scrub stage ('carr') in the hydrosere succession around an enclosed lake in the sand dunes at Pendine, South Wales. Species of sedge, rush, reed and iris in the foreground on water-logged soils; the common reed (*Phragmites communis*) dominates the centre of the lake. The floral diversity and lush growth partly reflects the exclusion of people from this area by the Ministry of Defence.

Fig. 8.4 The succession towards beechwood on the scarp slope of the Chilterns near Chinnor. On steep slopes the scrub stage is characterized by the presence of juniper (*Juniperus communis*). Also present are dog rose (*Rosa canina*) and elder (*Sambucus nigra*). The main grasses are creeping fescue (*Festuca rubra*) and false brome (*Brachypodium pinnatum*).

of former woodland areas (but several workers have pointed out just how misleading this data can be). However, many important changes were taking place in our plant cover long before the period covered by historical evidence. We thus have to look to other methods for the reconstruction of these past biogeographies. The technique of peat and pollen analysis (palynology) has provided a valuable insight into past environments and how they changed, giving a detailed picture of former plant cover.

PALYNOLOGY

Pollen analysis depends on the distinctive nature of the pollen grains of each genus. Microscopic pollen grains are variously shaped, with differing numbers of pores and furrows on their surface, and exhibit a great range of surface textural patterns. It is by these features that they can be recognized and assigned to the appropriate plant genus and, in many cases, to the plant species. The outer cell wall (exine) of a pollen grain is highly resistant to decay. Providing the pollen grain becomes incorporated in sediments which are forming under acidic, anaerobic (oxygen-free) conditions, it will persist in a clearly identifiable form for many thousands of years. Such waterlogged sites are numerous in the landscape, e.g. acid peat-bogs, lakes. There were many more of them immediately following the last glacial phase when much meltwater existed in the hummock-and-hollow terrain associated with glaci- ation. A plant community will produce countless millions of pollen grains which form an annual input to these anaerobic sediments. Because the sedi- ments accumulate in chronological order it follows that they provide a relative age sequence of embedded pollen grains.

The first stage in the technique of pollen analysis is field investigation of the preservation site, usually a peat deposit. Sites showing disturbance or contam- ination of the sediments are rejected for analysis. The peat layers themselves are carefully described and examined for any *macroscopic remains* (Fig. 8.5). These may take the form of (a) plant tissues: tree stumps, fruiting bodies, leaves; (b) animal tissues: bones, beetle cuticles and, very rarely, human beings, e.g. the Tollund man of Denmark recovered from peat about 2,000 years old; (c) archaeological material: axe-heads, pottery, charcoal layers. The visible features of the peat profile greatly assist at a later stage in the inter- pretation of the pattern shown by the included *microfossils* (the pollen grains). Small samples of peat (or lake sediment) are collected at vertical intervals of a few centimetres (say 5 cm) and taken to the laboratory. The laboratory treat- ment involves three basic steps, namely, breakdown of the organic matrix in which the grains are embedded, deflocculation (separation) of the grains and their eventual concentration.

From each sample, slides are prepared for microscopic examination. A magnification of × 400 is usually adequate, though oil-immersion techniques help with the identification of problematical grains. When counting and

Fig. 8.5 Macrofossils (pine tree stumps) exposed in a peat bank by stream dissection at a site above the present tree-line in the Western Cairngorm Mountains. The lower stump rests almost directly on the mineral soil (the pebbles of the stream course) and is separated from the upper stump by a thick band of *Sphagnum* peat. The upper stump is itself covered by peat and the present surface vegetation is mainly heather and deer grass. From a sequence like this, much can be inferred about past environmental changes in the vicinity.

recording pollen grains for each sample, the species are usually grouped into similar types, as shown in Table 8.1. Traditionally, the count proceeded until 150 arboreal pollen grains (AP) had been recorded, by which time maybe many hundred non-arboreal pollen grains (NAP) could have been noted. Increasingly, however, some workers now use a different basis for the 'pollen sum', preferring 250 NMP (non-mire pollens). Whatever 'sum' is used, the argument is that by the time this 'cut-off point' is reached enough other types (NAP, shrubs, Cryptogam spores, etc.) will have been recorded to give a good cross-

Table 8.1 Type groupings used in pollen grain counts with some British examples of species

Tree or arboreal species	Shrubs	Non-arboreal species	Non-flowering species*
Quercus (oak)	*Corylus* (hazel)	*Carex* (sedge)	*Pteridium*
Pinus (pine)	*Juniperus* (juniper)	*Juncus* (rush)	*Hypnum*
Betula (birch)	*Crataegus* (hawthorn)	*Gramineae* (the grasses)	*Sphagnum*
Fagus (beech)	*Salix* (willow)	*Erica* (heaths)	etc.
Fraxinus (ash)	etc.	*Calluna* (heather)	
Tilia (lime)		*Plantago* (plantain)	
etc.		etc.	

* These types (ferns and mosses) produce spores which are preserved and can be identified in much the same way as the pollen grains of flowering plants.

section of the total population of grains and spores in the sample. When all samples in the profile have been analysed, the data are presented in graphical form. There are various ways of doing this but again, conventionally, the usual method has been to express the counts for all individual species as a percentage of the total AP or NMP count. Figure 8.6 shows a simple pollen analysis presented in this manner and based on an AP count for a peat-bog site at about 600 m in the Scottish Grampian Mountains.

The final stage is essentially one of *interpretation*. The graphs are examined to see how the representation of each species varies with time. Frequently a range of plants will show the same relative decline (or expansion) at the same level in the profile. These strong variations are taken to represent a plant response to environmental change. Some species are excellent indicators of environmental conditions. For example, *Alnus* (alder) tolerates wet, water-logged soils and its sudden expansion in the graphs could indicate the onset of a wetter climatic phase. *Plantago major* (great plantain) is a common weed of disturbed or cultivated soils and, along with supporting evidence such as the occurrence of cereal pollen or archaeological finds, may indicate the presence of man in the area. In this way a *zonation* of the diagram is achieved and it is here that the peat macrostatigraphy aids the interpretation. The diagram is then described in terms of the relative dominance over time of various types of vegetation community. Such changes in plant covers provide environmental information (in terms of soils, climate, associated faunas, etc.). Our relative dating of these events can be greatly improved if organic layers are radiocarbon dated. In Fig. 8.6 the two tree-stump layers in the peat profile have been so dated and this gives the diagram added precision. For the latest developments in palynology and related techniques, the detailed review by Birks and Birks should be consulted. Jacobson and Bradshaw also provide a useful account of all the variables to consider when selecting sites for palaeovegetational studies.

Many hundreds of pollen diagrams from the British Isles are available. There is now general agreement on the main pattern of changing vegetation since the last glacial phase (during the Post-glacial, Post-Devensian or Flandrian, see Table 8.2). Controversy is largely centred on the finer details within this broad pattern (see further comments in the Discussion Section). The findings have been of immense value for all the environmental sciences such as climatology, geomorphology, ecology, archaeology and pedology and hence merit careful attention by the biogeographer. Figures 8.7a and b display the sequence of changing plant communities at Hockham Mere, Norfolk, and is reasonably typical for much of Lowland Britain during the last 10,000 years or so. Figure 8.6 shows a typical pattern where pine and birch form the climax vegetation on the higher slopes reaching to just over 600 m in the Scottish Highlands.

Pollen-analytical studies of interglacial organic sediments (e.g. the Hoxnian, see Table 8.2) and fossil studies of pre-Pleistocene floras in the Tertiary Period, clearly demonstrate that our Post-glacial or Flandrian flora is an impoverished one. Before the devastation of the Pleistocene ice the British Isles, and Europe generally, had a much richer complement of plants (and animals). Impover-ishment came about because the east-west mountain configurations in Europe acted to trap many species as the ice advanced from the north and down from

Carn Mor (603 m)

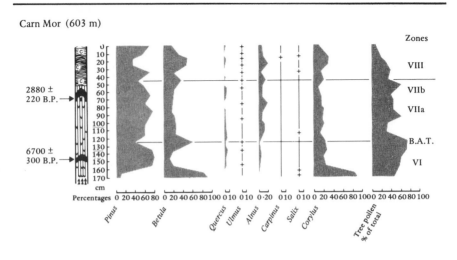

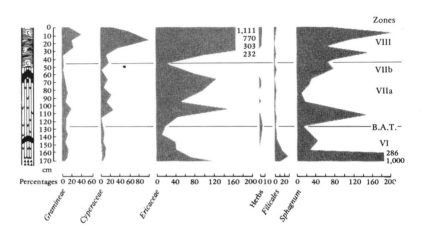

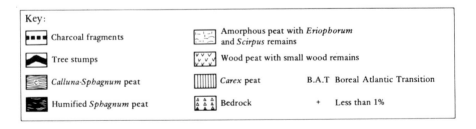

Key:

▄▄▄▄ Charcoal fragments		⌐⌐⌐ Amorphous peat with *Eriophorum* and *Scirpus* remains	
◣◢ Tree stumps		ⱽⱽⱽ Wood peat with small wood remains	
▒▒ *Calluna-Sphagnum* peat		⫼⫼⫼ *Carex* peat	B.A.T Boreal Atlantic Transition
▬▬ Humified *Sphagnum* peat		▲▲▲ Bedrock	+ Less than 1%

Fig. 8.6 Pollen diagram for Carn Mor (600 m), an upland peat bog site in the eastern Cairngorm Mountains. The two horizons of buried pine stumps have been radiocarbon dated, giving more precision to the zonation based on the microfossil pattern.

Table 8.2 Subdivisions of the Pleistocene Epoch (Quaternary Period), showing the current terminology for the British Isles and the equivalent names used on the European mainland. For subdivisions of the Holocene see Table 8.3. (Mainly based on West, 1968).

British Isles			European mainland	
Glacials	Interglacials		Glacials	Integlacials
		(Holocene)		
Devensian			Weichselian	
	Ipswichian			Eemian
Wolstonian			Saalian	
	Hoxnian			Holsteinian
Anglian			Elsterian	
	Cromerian	(0·69 million years ago)		Cromerian
Beestonian			Menapian	
	Pastonian	Although apparently same number of stages, the correlation is still very uncertain		Waalian
Baventian			Eburonian	
	Antian			Tiglian
Thurnian			Pretiglian	
	Ludhamian			
Waltonian(?)				
		(2 million years ago) Pliocene (Tertiary)		

the mountain chains. In contrast, in North America, the north-south run of the main mountain systems allowed relatively easy migration southwards as climate worsened. Hence many more species survived and were able to recolonize northwards when the ice declined. In Britain the recolonization was made even more difficult by recently created *sea barriers*: the Irish Sea, the English Channel and the North Sea. As a result, the post-Devensian woodlands of Britain were species-poor when compared with their counterparts in France and Germany, and especially so with those of North America.

LATE-GLACIAL AND FLANDRIAN VEGETATION

Table 8.3 sets out the broad divisions of this time-span. For many years palynologists attempted to zone their own individual pollen diagrams in terms of these divisions but it is now realized that forcing new diagrams to fit the established pattern is an unsuitable, initial approach. Instead, use of the *Local Assemblage Zone* concept is now urged. The Local Assemblage Zone is defined

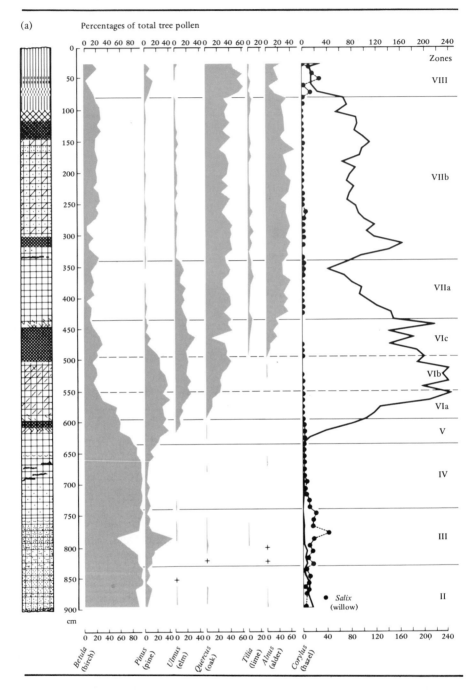

Fig. 8.7 Pollen diagram for Hockham Mere, Norfolk. (a) The tree pollen curves, showing early dominance by birch and pine woodlands followed by the arrival of deciduous forest elements at about the 600 cm level. The sharp increase in alder pollen at about the 500 cm level is taken to indicate the onset of the wetter Atlantic Period. (b) The non-tree pollen curves, showing an early phase when grasses and herbs were

(b)

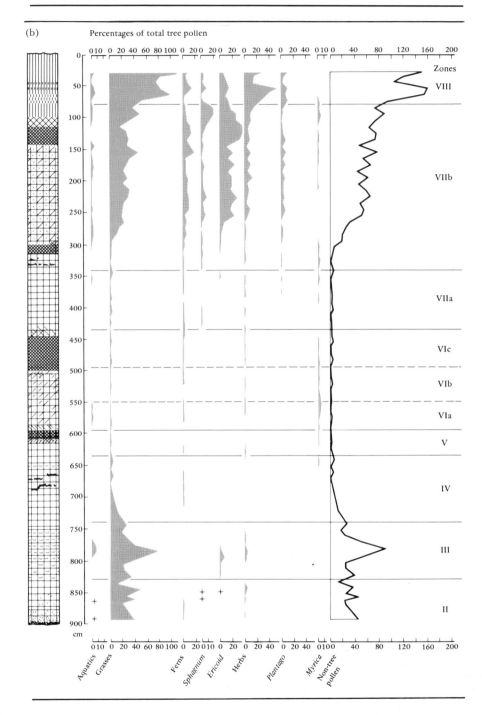

Percentages of total tree pollen

Zones: VIII, VIIb, VIIa, VIc, VIb, VIa, V, IV, III, II

Aquatics, Grasses, Ferns, Sphagnum, Ericoid, Herbs, Plantago, Myrica, Non-tree pollen

prominent in zones II and III before the establishment of denser forest growth. The non-tree pollens came back strongly towards the top of the diagram as man increasingly disturbs the climax forest. This is well illustrated by the ribwort plantain (*Plantago lanceolata*), a common weed. (From Godwin, 1975.)

on the presence and relative proportions of the contained fossils (i.e. the pollen grains and spores in this case) without any ecological, chronological, or climatic implications. It is a natural assemblage or association, distinguishable in its biostatigraphical character from adjacent strata. Cushing first suggested its use for the zonation of pollen diagrams. Such usage avoids pre-conceived ideas; only when several pollen diagrams in a region have been compared should attempts be made to fit the new data to the established general zonation for North West Europe. These conventional subdivisions of the Late Devensian and the Flandrian Interglacial (Table 8.3) are now widely used in many field sciences and the main environmental trends plus their associated plant covers for the British Isles are as follows:

Table 8.3 Conventional subdivisions of the Late Devensian and the Flandrian Interglacial (Holocene). Continental palynologists recognize two climatic ameliorations within the Late Devensian: the Bölling and the Alleröd. Ages given in years BP (before present, i.e. years before 1950)

Some approximate dates	Zones		Periods	Continental subdivisions of the Late Devensian
Present-2550 BP	VIII	Sub-atlantic		
2550–4450 BP	VIIb	Sub-boreal	Flandrian	
4450–7450 BP	VIIa	Atlantic	Interglacial	
	VI	Boreal	(or Post-glacial)	
	V			
	IV	Pre-boreal		
10000 BP				
	III	Upper Dryas		Younger Dryas
			Late Devensian	Alleröd
	II	Alleröd	(or Late-glacial)	Older Dryas
				Bölling
13000 BP	I	Lower Dryas		Oldest Dryas

THE LATE-GLACIAL

LOWER DRYAS PERIOD.　Ice persisted in the mountains with periglacial conditions elsewhere. The Arctic-type vegetation was an open, stunted growth of mosses, lichens, sedges, grasses and dwarf birch. This division is named from the occurrence in deposits of remains of *Dryas octopetala* (mountain avens). The period began about 13,000 BP.

ALLERÖD PERIOD.　Named after a Danish site, this period (also now known in Britain as the *Windermere Interstadial*) was a climatic improvement from about 10000 BC to 8000 BC. More plants were able to colonize, including willow, juniper and tree birch scrubs. The vegetation was still open in character and simple in form. Pine may have been present in the south-east of Britain.

UPPER DRYAS PERIOD.   The climate deteriorated and vegetation similar to that of the Lower Dryas returned. Ice accumulated and re-advanced in the main mountain areas.

THE FLANDRIAN

PRE-BOREAL PERIOD.   The climate improved steadily and the ice phase ended. Pioneer open vegetation of non-arboreal form was replaced by pine and birch forest.

BOREAL PERIOD.   Pine and birch woodland persisted throughout the early Boreal but soon hazel, elm, oak and other deciduous trees arrived in the British Isles and rapidly spread northwards as temperatures increased. By the end of the Boreal Period the Summer Deciduous Forest formation was largely taking over from the Northern Coniferous Forest formation in Lowland Britain; pine and birch woods now persisted at higher altitudes and higher latitudes. The appearance of *Tilia* spp (lime trees) in southern England by the close of the Boreal is a clear indication that temperatures had risen significantly.

ATLANTIC PERIOD.   Whereas the Boreal had been relatively dry, the Atlantic was wetter, although still warm. Lime reached its maximum range in Britain, indicating average summer temperatures during this *Climatic Optimum* of 2–3 °C above those of today. The rapid expansion of *Alnus* (alder) and the extensive development of peats are signs of a wetter regime – an oceanic phase replacing a continental one. Some forest was overwhelmed and is seen today as buried stumps exposed in eroding peat, e.g. in the Pennines, the Grampians, and Donegal. Our climax woodland types had now become fully established: the *Quercetum mixtum* (i.e. Summer Deciduous Forest as mixed deciduous woodland with oak predominating) and the more northerly *Coniferous Forest* (mainly pine but with admixtures of birch and rowan).

SUB-BOREAL PERIOD.   The climate may have become somewhat drier and perhaps cooler during this period (although the palynological evidence for climatic change in this period often points in contrary directions). Some bog surfaces dried out and were invaded by trees. The lime decreased as did the elm but exact causes have not yet been fully established (see Discussion Section). Undoubtedly, man was now having a pronounced influence on the climax woodlands which has continued until the present day at an accelerating rate. Before the Sub-boreal, man in Britain was essentially a hunter and food gatherer (Palaeolithic and Mesolithic cultures). We have evidence for some woodland clearance by Mesolithic populations (mainly through their use of fire – see Chapter 6) but it was not until Neolithic farming cultures became established that widespread significant changes began.

SUB-ATLANTIC PERIOD.   A climatic deterioration is how Pennington describes the onset of this phase. Rapid peat growth in many areas indicates the wetter and cooler climatic conditions. Again, some forest was overwhelmed by peat and the stumps have been preserved to this day (Fig. 8.5). The pollen diagrams show massive evidence for the increasing influence of man on the environment, e.g. a strong return of non-arboreal species adapted to open, disturbed habitats, the presence of weed and crop pollens, the occurrence of charcoal layers. Archaeological finds and radiocarbon dating can pinpoint some of these changes quite accurately.

We thus have a clear picture of our climax woodlands prior to marked human interference: oakwoods, extending northwards to cover much of Lowland Scotland and penetrating beyond into the main glens, and pine and birch woods on the upper slopes and throughout the Highlands generally. The late arrival of *Fagus sylvatica* (beech), not appearing with any abundance in the pollen diagrams until *c.* 500 BC, probably explains the restriction of its natural range to southern England and South Wales. Planted specimens grow well as far north as Aberdeen so the restriction cannot be climatic. In the south beech is frequently confined to thin soils developed on softer Oolitic and Chalk limestones (Fig. 8.8). It was argued that it could compete successfully on these sites with established oaks. But beech can tolerate a wide range of soil types, from chalky soils to acidic gravels, and Godwin suggested that man assisted its spread in the south by first opening up the oakwoods on the thinner, well-drained soils of the limestone scarplands. In the Iron Age man was able to clear the denser lowland forests and many scarpland sites were then abandoned, allowing beech to invade. Thus although pollen grains of beech are first recorded about 5000 BC they only become abundant in pollen diagrams in the Iron Age, *c.* 500 BC.

Fig. 8.8 Mature beech trees with some undergrowth of holly. Chiltern plateau, near Naphill. The profusion of low side-branches and the presence of multiple stems indicates a history of disturbance at the site. The soil is a shallow, stony, clay-with-flints overlying chalk.

Turner and West have proposed an alternative scheme based on vegetational development for the subdivision of interglacial temperate stages. This scheme applies to other Pleistocene interglacials (e.g. the Hoxnian) as well as the current Flandrian. Their fourfold division is as follows:

I   = Pre-temperate substage: development and closing-in of forest veg-
        etation, usually of boreal type (*Pinus, Betula*, etc.).
II   = Early-temperate substage: establishment and expansion of mixed oak
        forest of maximum denseness and luxuriance under rich soil conditions.
III   = Late-temperate substage: decline of mixed oak forest dominants and
        expansion of some conifers (reflecting soil degeneration rather than
        climatic changes).
IV   = Post-temperate substage: return to dominance by boreal trees (*Pinus,
        Betula*, etc.) and a thinning of the forest cover.

Of course, the Flandrian has not yet run its full term and is further complicated by the growing influence of man. However, it is not too difficult to see
where these substages fit in with the conventional subdivisions of Table 8.3.

The various forms of climax woodland have now been largely cleared,
although some interesting semi-natural fragments still persist (Peterken lists
12 types of ancient semi-natural woodland in Britain, each with subtypes or
variants). Agricultural land now predominates in lowland districts and in our
uplands extensive moorlands have come into existence, some naturally but
most derived from former woodland. The present-day ecology of selected
aspects of these two types of plant cover, the woodlands and the upland moors,
will now be examined.

WOODLANDS

OAKWOODS

In 1974 the papers of a conference held by the Botanical Society of the British
Isles on the history and natural history of the British oak were published. This
material has been freely drawn on in preparing the following section. Other
sources of valuable material on British woodlands are the earlier detailed
account of British vegetation by Tansley; Rackham's seminal work on ancient
woodland in England; Peterken's general account of woodland conservation
and management; and the 13 papers included in Holmes *et al*. Mention must
also be made of the two volumes of Ratcliffe's *A Nature Conservation Review*:
'The document as a whole is intended to serve as a compendium of information
and reference work on nature conservation interest in Britain, in which the
sites of greatest importance are presented in relation to the wider background
of wildlife and habitat over the country as a whole'. This work not only

contains much on woodlands but also on upland moors: it lists 234 woodland sites covering 67,000 ha and 207 upland and peatland sites covering 500,000 ha of national importance for conservation.

Most species of the genus *Quercus* are found in warm-temperature or even tropical montane climates. Thus, almost half of the 450 known species of oak occur in Mexico; the Iberian Peninsula has 12 species but there are only two native species of oak in the British Isles. *Quercus robur* (the pedunculate oak) is today the most common of the two and is typically associated with damp oakwood developed on heavier lowland soils. *Quercus petraea* (sessile or durmast oak) is found on shallower soils, frequently developed from siliceous rocks, and is often the dominant tree of oakwoods in northern and western districts of Great Britain. However, it must be emphasized that both oaks can be found on a wide range of soil types and topographical sites. Further, their ranges frequently overlap and hybrids may occur.

Well-developed *Quercus robur* woodland casts a moderate shade and permits a patchy under-storey of shrubs or small trees such as *Corylus avellana* (hazel), *Crataegus monogyna* (hawthorn) and *Salix* spp (willow). The main ground flora species are *Mercurialis perennis* (dog's mercury), *Ranunculus ficaria* (lesser celandine), *Anemone nemorosa* (anemone), *Endymion nonscriptus* (bluebell), *Oxalis acetosella* (wood sorrel). This list is by no means complete since the field-layer may contain several dozen species or more. The species present and their relative abundance may vary considerably as differences in soil, light, nutrient supply, grazing pressure, past and present management practices, etc. come into play. Some are more typical of the pedunculate oakwoods of wet soils while others feature more strongly on drier lowland soils. The soils associated with pedunculate oakwoods can be quite variable but are usually clays (often with gley horizons), heavy loams and silts, or fine-grained sands (see Chapter 9 for further details). Tree regeneration is reasonably widespread in these woodlands provided that there is not too much biotic interference such as grazing and defoliation. *Quercus robur* has a tap-root and on some lowland sites rooting is limited by a high water table, although it can tolerate some seasonal flooding. *Quercus robur* woodlands do not usually extend into districts above 300 m though, as we shall see, there is a notable exception on Dartmoor. In their extreme western and northern locations there is some doubt as to whether they are entirely native and many certain cases of planting are known.

Quercus petraea woodlands can tolerate thin hill-slope soils developed from coarse sandstones, grits, shales or acidic igneous rocks. They often have a more open structure with good light penetration to the field-layer. Several smaller trees or shrubs may be present. *Betula pubescens* (birch) is often a main shrub but *Ilex aquifolium* (holly), *Alnus glutinosa* (alder) and others may also occur. The ground flora is frequently continuous and commonly includes *Pteridium aquilinum* (bracken), *Holcus mollis* (creeping soft grass) and *Rubus* spp (brambles), together with other herbaceous plants that come in as habitat conditions vary. Quite a few ground flora species may be common to both types of oakwood but it is at the higher elevations, such as in the Pennines, that a more distinctive 'heathy' ground flora appears and the grasses may also change, e.g. *Vaccinium myrtillus* (bilberry), *Calluna vulgaris* (heather), *Deschampsia flexuosa* (wavy-hair grass) or *Molinia caerulea* (purple moor grass). Acidic soils are often associated with sessile oak. The higher rainfalls of these westerly locations, combined with underlying siliceous rocks low in lime, will sometimes produce marked podzols. However, large accumulations of poorly decomposed humus sometimes associated with podzols are not normally seen in oakwoods. Regeneration in sessile oakwoods is reasonably good, provided grazing pressure is not heavy, which it frequently is. The rooting system has greater lateral spread than does that of the pedunculate oak and saturated soil horizons are much less commonly encountered. *Quercus petraea* extends to the far north (Sutherlandshire) and west (Ireland) and to sites over 450 m in Cumberland. It probably formed a zone below pine on some of the higher Scottish mountains in the past, but its occurrence there now is very fragmentary.

Much present oakwood has suffered a long history of disturbance by man (Fig. 8.9). Many woods were traditional grazing areas for pigs and cattle. Even at remote upland sites sheep frequently sheltered and browsed in them. This accounts for their open nature and the lack of young tree growth at many localities. Natural oakwoods often contain other deciduous tree species, e.g. *Fraxinus excelsior* (ash), *Acer campestre* (common maple), *Ulmus glabra* (wych elm), *Prunus avium* (wild cherry). But their representation in the woodlands is often greatly increased when the oak canopy has been opened up by felling. Ash and birch, for example, grow readily in such gaps, being intolerant of heavy shading. *Acer pseudoplantanus* (sycamore) and *Rhododendron ponticum*, an introduced tree and an introduced shrub, will quickly spread through

Fig. 8.9 Sessile oak (*Quercus petraea*) in Swithland Wood, Leicestershire, developed on light acid soils associated with outcrops of pre-Cambrian rocks. A neglected woodland where formerly shrubs, especially hazel, were coppiced. Many birch trees have grown up through gaps in the oak canopy. This woodland is one of the sites included in Ratcliffe's review (see text).

Fig. 8.10 Profuse growth of willow as a weed species along a 'ride' used by the Cottesmore Hunt through degenerate oakwood at Owston, Leicestershire. Oak and ash (leafless) in the background.

disturbed woodland as will *Salix* spp (willows). All three may reach weed status in managed woodlands, requiring supression measures (Fig. 8.10). Much of our ancient oakwood contained lime trees (mainly the species *Tilia cordata*) although these trees are no longer common in most areas today (the Lincolnshire Limewoods being an exception). Indeed, Rackham has argued convincingly that lime was often more abundant than oak in southern Britain by the end of the Atlantic Period, but subsequently suffered from selective clearance by early man. Also indicative of ancient woodland are certain species of ground flora. These include *Convallaria majalis* (lily of the valley), *Milium effusum* (wood millet) and *Melampyrum pratense* (common cow-wheat).

In addition to casual or piecemeal disturbances, there has been a history of deliberate cropping of oakwoods stretching back more than a thousand years. This involved pollarding (taking a crop of the main branches 2–4 m above ground) and, more importantly, *coppicing*. The former practice was usually confined to oaks growing in the open or along woodland boundaries. With the latter, either the oak itself (usually the sessile species) was part of the coppice rotation (the stems being cut near ground level so that multiple shoots arose from the stump only to be cut themselves some years later) or the practice was 'coppice with standards'. Here the coppiced species was usually hazel, or often hornbeam in the south, and the 'standards' were pedunculate oak. The opening up of the oak canopy would allow rapid hazel growth which could then be cropped every 10 to 15 years or so. The remaining oaks and any oak saplings

Fig. 8.11 Oak woodland with strong growth of coppice shrubs. These are now 'chopped' to promote a tangled undergrowth suitable for fox cover. Botany Bay Fox Covert, East Leicestershire.

would have more room to grow into fine specimens. These, in turn, could be cropped on a much longer rotation of 100 years or so. Much oakwood was managed on this basis, providing timber for building, tanning, fencing, charcoal production and fuel. Many woods were enclosed as deer parks within which a variety of activities occurred – hunting, coppicing, arable patches, grazing areas, etc. Coppicing was a highly flexible system designed to meet local needs for a constant supply of various wood products. It gave rise to richer ground floras and some examples are worth preserving for this reason alone. Coppicing is no longer practised extensively but it is still seen in woods used for game-birds and as fox coverts (Fig. 8.11): according to Peterken, coppicing for traditional market uses is only found on some 600 ha today, most of them confined to southeast England. As Rackham noted, the twentieth-century collapse of woodland management was due not to the felling itself but to the abandonment of coppicing as a practice and the failure to replace the trees felled. This collapse of the system could lead to certain species disappearing from our woodlands. Brown has shown that species of deep shade are very vulnerable since their reserves of buried seed in the soil are low and most also have poor dispersive powers.

Tittensor described a coppice with standard system used in the Loch Lomond oakwoods of the Montrose Estate where both coppice and standards were oak; other tree species present, known as barren timber, were removed so as to increase oak dominance. The native species here was *Quercus petraea* but much later *Quercus robur* was planted, using seed from England. Initially, charcoal was produced for use in iron ore 'bloomeries' established in the wood-lands during the seventeenth and early eighteenth centuries. Later, these woods produced oak bark for tanning works in Glasgow, then charcoal for gunpowder and finally acetic acid and liquid dyes for the printing industry. By 1735 the woods were coppiced on a regular 24-year rotation and the stand-ards were allowed to grow for two to four rotation periods. These practices continued until about 1920. They have clearly left their mark on the present-day composition and structure of these woods. If this is the case in a relatively remote area then it is likely to be even more so in many southern woods where interference has a much longer history. Management by coppicing lasted just over 200 years in these particular Scottish oakwoods while Rackham has shown that 98 per cent of West Cambridgeshire was cleared of woodland by AD 1300. Then, for at least 650 years, there was intensive management along coppice with standard lines of the remaining woods and those subsequently planted.

The effects of such management are many and varied. Regular grazing suppresses a herb-rich ground flora and encourages the spread of grasses. Coppicing alters light intensities and the ground floras reflect the stage reached in coppice growth, e.g. the large spreads of pre-vernal and vernal flowering species, such as bluebell or primrose, which follow the cutting down of shrubs will be suppressed when a new shrub cover returns. The felling of large trees or shrub clearance alters microclimates and may cause the loss of bryophytes and lichens sensitive to shade and humidity changes. Steele has summarized these variations in oakwoods and the lesson should be drawn that their present-day appearance is an unreliable guide to what these woodlands would look like under truly natural conditions. However, there are a few oakwood remnants still extant which are thought to be as near to natural growth in at least some respects as we are likely to find in the British Isles. Two of the best known are Wistman's Wood and the oakwoods of Killarney, both described in detail by Tansley.

WISTMAN'S WOOD

This small wood (total area about 4 acres or 1·6 ha) lies on a steep, boulder-strewn, west-facing slope at 365–430 m in the headwater valley of the West Dart, which drains southwards from Dartmoor (Fig. 8.12). The former oakwood cover on Dartmoor, which Simmons thinks previously existed in the area to about 460 m, has been almost entirely removed by man. The Neolithic, Iron and Bronze age settlements on Dartmoor played an important part in this but the final wave of woodland destruction over most of Dartmoor came with the activities of medieval tin-mining. Wistman's Wood is thought to have escaped any serious disturbance by man or grazing animals. Some disturbance does occur, as in the severe winter of 1963 when sheep debarked some of the trees, but it is considered to be infrequent. The remoteness of the wood and the rocky nature of the site probably act as sufficient deterrent so that man or

Fig. 8.12 The beginning of Wistman's Wood, a small 4-acre patch of near-natural *Quercus robur* woodland at about 400 m along the upper reaches of the West Dart. The woodland on the boulder-strewn slope is surrounded by a fringe of *Pteridium aquilinium* (bracken).

grazing animals rarely entered the wood in the past. However, because it is now known that some high-level woods in the British Isles were planted we must allow the possibility that even Wistman's Wood could fall into this category. But its naturalness is generally accepted.

The main tree is *Quercus robur* and this is surprising since the climatic severity of the area (high rainfalls of about 180 cm (70 in) p.a., high cloud cover and strong winds) and the thin acidic soils would lead us to expect it to be *Quercus petraea*. Simmons' palynological studies show *Quercus robur* was an early immigrant to the region, arriving by the end of Zone IV. *Quercus petraea* arrived later and competed successfully with established forest on soils derived from metamorphic rocks but not on the granite. The reasons for this are unknown but he demonstrates this edaphic separation in other Dartmoor woodlands such as the West Okement Valley where hybrids of the two oaks occur on soils along the geological contact of the granite and metamorphic rocks. The oaks in Wistman's Wood show low, contorted growth-forms, seldom exceeding 7·6 m in height and frequently much less (Fig. 8.13). There is no distinct shrub stratum but other trees present include *Ilex aquifolium* (holly), *Sorbus aucuparia* (rowan) and *Salix atrocinerea* (willow). *Hedera helix* (ivy) and *Lonicera periclymenum* (honeysuckle) are common woody climbers. The oaks support much epiphytic growth on the surfaces of their horizontal branches where organic debris has accumulated. These epiphytes include ferns, mosses, rushes, herbs and even a few small specimens of rowan. Tree

Fig. 8.13 The interior of Wistman's Wood. The contorted specimens of *Quercus robur* are heavily encrusted with mosses and lichens, as are the boulders beneath. *Hedera helix* (ivy) is present on many trees. The ground flora at this point is dominated by *Pteridium aquilinum* (bracken).

trunks and boulders are often covered by a thick bryophyte carpet. The field-layer is largely of *Vaccinium myrtillus* (bilberry), *Luzula maxima* (wood rush), *Pteridium aquilinum* (bracken) and various species of *Dryopteris* (fern). Field-layer species reach their best development in pockets of deeper soil between boulders.

Longman and Coutts point out that a remarkable feature of oaks is the brevity of the shoot-elongation period. This can often be confined to a period of two to three weeks. The shoots may then lie dormant despite good growing conditions. The first flush of shoot growth is often in April/May. It may be followed by a second about midsummer, or a third in early August and, occasionally, a fourth in October. These sappy elongations, often appearing above the general canopy level, are known as 'lammas shoots'. They are particularly well developed at Wistman's Wood, being much larger and more vigorous than those normally found in other British oakwoods. An early explanation for this was that when the buds first opened in May many were killed by late frosts common at this altitude. By summer, most of the growth then became concentrated into the fewer remaining buds which thus elongated rapidly to form 'lammas shoots'. Many of these shoots in turn will be killed by the frosts of the next winter; those surviving produce more rapid height growth later. This explanation has now been challenged by Varley and Grad-well. They reported large numbers of *Operophtera brumata* (winter moth) larvae feeding on young oak leaves in Wistman's Wood and in July 1964 it was nearly defoliated when damage in woods elsewhere was negligible. Evidence suggests that this often occurs, which may explain the growth-form of the oaks and the profusion of 'lammas shoots' (as growth is concentrated into the surviving shoots). They also thought that the exceptional growth of epiphytes might result from a combination of nutrient enrichment of humus pockets on flat branches, provided by caterpillar droppings, and the high light intensities which follow defoliation.

Despite its small size, regeneration of oak at Wistman's Wood is moderately good. Seedlings establish well around the edge of the wood but these are browsed by sheep, cattle and ponies. Steele, however, noted that the luxuriant epiphytic growth in this wood appears to be declining. This could be partly because the increased growth of the oaks has raised the canopy level and perhaps caused a slight reduction in humidity beneath the trees. This increased canopy height has been clearly demonstrated by Proctor *et al.* using photographic and other evidence. In a comprehensive survey they show that there have been changes in tree growth-form (to more upright forms), a marginal expansion of the wood to nearly twice its former area, and the interior of the wood has become more open, since the first written mention of the wood in about AD 1620. These alterations they attribute to natural development changes with time, improvement in climate since the 'Little Ice-age' of the seventeenth and eighteenth centuries AD, and variations in grazing pressure. They do not detect any substantial diminution of the bryophytes. Barkham has studied a similar pedunculate oak wood, Black Tor Copse of 5·67 ha, growing in an equally severe environment on the north-western edge of Dartmoor. The trees have many morphological features in common with those in Wistman's Wood but Barkham attributes most of these differences between growth forms to the grazing history of the site.

THE KILLARNEY OAKWOODS

These woodlands were first described in detail by Turner and Watt and are included in Tansley's account. More recently Kelly has presented a detailed account. Killarney has a pronounced maritime climate with rainfalls in the range 130–150 cm (50–60 in) p.a., high humidities and no frosts or snow lay. The remaining woodland fragments (for much has been cleared) are usually confined to steep, boulder-strewn slopes developed mainly on the Devonian Old Red Sandstone (most flat surfaces in the region carry blanket bog). Turner and Watt thought that the rocky nature of the sites may have prevented persistent animal grazing but Kelly has now shown that grazing has been and still is a major factor in most parts of these woodlands. The oaks reach altitudes of 245 m but above, where the ground is flatter, an open moorland of blanket-bog vegetation occurs. Given these site conditions, the 245 m limit seems a fairly natural one. The soil is confined to pockets between boulders as alluvial and colluvial deposits. Although originally base-rich, due to parent material influence, it quickly becomes podzolised under the high rainfalls.

The species present is *Quercus petraea*. Turner and Watt recognized three types of Killarney oakwood based on the growth characteristics of the oak.

Type 1 has an oak density of 35 per acre, the trees averaging nearly 18 m in height and 2 m in girth. This type occurs under the best soil conditions: the litter layer is loose, the humus content is high and the clay particles are well distributed in the profile. The under-storey is mainly holly (*Ilex aquifolium*) to about 4 m high and the field-layer is dominated by bilberry (10–20 cm high) and heather (10–20 cm high). Most boulders and some trunks are moss-covered and epiphytic growth on trees is common. A dense shade is cast by the oaks, and few tree seedlings, apart from those of holly, were found.

Type 2 has 85 oaks to the acre, the average height being 13·5 m and average girth about 1 m. The soils are true podzols and have a thick humus layer and higher sand content. Although there are more trees they are smaller, and this allows more light to reach the ground. The bilberry grows to 60–70 cm and the heather to 100 cm or more. Bracken and several ferns are also well represented. Holly is again common and birch and rowan are more frequent. Type 2 is the main form of oakwood found in the district.

In Type 3 the oaks do not dominate although they reach average densities of 280 per acre. Their average height is only 5·2 m and girths seldom exceed 0·5 m. The woodland is really a mixture of oak, holly, rowan and birch. Also present is *Arbutus unedo* (the strawberry tree), reaching heights of 9·0 m, though usually less. This tree is a member of the *Lusitanian floral element*. The centre of distribution of this element is the western Mediterranean: south-west Ireland contains several other species belonging to it. In Killarney, *Arbutus* appears to be a pioneer species, maintaining itself well only along the margins of the oakwood. In Type 3 the soils are well podzolized and have a 5–10 cm layer of raw humus, mainly derived from the heather and bilberry which, again, grow to greater heights than in Type 1. Bracken is common and a somewhat richer herb component is present. Tree regeneration appears adequate and better than in the other two types of woodland. Although epiphytic and bryophytic communities occur in all three types, their composition and luxuriance vary with the changing structure of these woodland types.

In Kelly's study of Killarney three types of forest vegetation are distinguished: acidophilous *Quercus petraea – Ilex aquifolium* woods on the Devonian sandstone; a unique *Taxus baccata* (yew) wood on Carboniferous limestone; and a 'carr' (swamp) forest mainly of *Alnus glutinosa* on low-lying marshy ground. (A fourth type exists only as much-altered fragments on the deeper, free-draining, base-rich soils.) The woodland on the sandstone is essentially that described by Turner and Watt, but Kelly differs in his description by relating the primary 'axis of variation' in this woodland to a rainfall gradient across the area: from a relatively species-poor oakwood in the driest areas to a luxuriant oakwood in the wettest areas which has a thick bryophyte mantle and is rich in epiphytes and saxicolous higher plants (Fig. 8.14).

An interesting feature of the Killarney oakwoods is the high number of tree species with laurel-type leaves, e.g. holly, ivy, strawberry tree. These sometimes give the vegetation an evergreen appearance. This feature has become even more noticeable in recent decades by the invasion of introduced *Rhododendron ponticum*. Turner and Watt recorded this exotic shrub as flourishing in Killarney and slowly invading most woodland types, changing ground flora

Fig. 8.14 *Quercus petraea* oakwood, Killarney, South West Ireland, with a thick bryophyte mantle on the scree boulders. The oaks are heavily infected with ivy.

patterns by virtue of the dense shade it casts. These observations were presented in 1939 and it is obvious that *Rhododendron* has since spread greatly, often forming impenetrable masses (Fig. 8.15). Kelly notes that tree regeneration, herb layers and bryophyte communities are either completely suppressed or much impoverished in such areas. Beech trees and several other alien shrub species are also present in these woodlands. How these developments will influence woodland successions in the region is not yet fully known.

The successions which have been established are of two broad types: those on acidic rocks of the Old Red Sandstone, occurring well above the Killarney lakeside, and that described by Shimwell for limestones and sandstones found at lower levels around the lakes (Fig. 8.16). Shimwell thinks the presence of a *Taxus baccata* (yew) stage in the lakeside woodland sere may reflect microclimatic rather than edaphic differences. He also notes that *Arbutus* grows particularly well around the lake shore and this could be for the same microclimatic reasons.

OAKWOOD DYNAMICS

So far oakwoods have been examined mainly in terms of their structure and floristic composition. But they are dynamic entities and may also be viewed in terms of nutrient cycling, faunal influences and tree regeneration patterns. These aspects are not divorced from those previously outlined since, for example, it can easily be demonstrated that the structure and composition of

Fig. 8.15 Heavy invasion by *Rhododendron ponticum* of old oakwood (*Quercus petraea*) around the lake shores of Killarney, leading to the suppression of all other shrub and ground flora species. Two self-sown seedlings of spruce (*Picea* sp) are also present as 'escapes' from a nearby plantation.

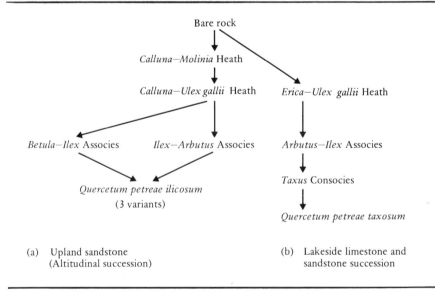

Fig. 8.16 Woodland successions in the Killarney Oakwoods (a) upland sites, (b) lakeside sites. (Modified from Shimwell, 1971.)

an oakwood often reflects the strong influence of the herbivores present. Recent research on oakwood dynamics owes much to the *International Biological Programme* (IBP) initiated in 1965 to study the basic processes of life-supporting environmental systems. The programme of research was designed to run for about 10 years and one area of intense research has been the woodland ecosystem.

NUTRIENT CYCLES

Meathop Wood, near Morecambe Bay in north-west England, was a main woodland site within the British section of the IBP. Brown has summarized the results of nutrient cycle studies at this site, noting a picture broadly similar to that found in other temperate deciduous woodlands. Meathop Wood is about 45 m above sea-level on Carboniferous Limestone covered with a variable layer of glacial drift. The brown earth soils and mull humus have pH values running from 4·1 to 7·5. The site supports a mixed deciduous woodland of sessile oak, ash and birch in the upper canopy and smaller proportions of sycamore, wych elm and other tree species. Hazel, hawthorn, ash saplings and calcicole shrub species form a well-developed under-storey and the ground flora is species-rich.

   Fig. 8.17 shows the quantities of macronutrient chemicals involved in the above-ground cycle at Meathop Wood and Fig. 8.18 shows one of the methods of measuring some of these components. The main source of plant nutrients is obviously that contained in the soil. Of those taken up by the root system some will be retained in the tree to form new tissues, i.e. the tree increment. Large quantities are released again, returning to the soil in the form of litter (leaves, catkins, bud-scales, bark, etc.), a major component in the nutrient cycle. However, a small quantity of chemicals is re-absorbed from the leaves before they fall. This is particularly pronounced for nitrogen and phosphorus, as Fig. 8.17 shows. This translocation back into the tree before leaf-fall is described by Brown as an 'internal cycle' to distinguish it from the 'external cycle' of litter fall.

   Another input to the system is the small, but ecologically significant, content of plant nutrients dissolved in the rainfall before it reaches the canopy level. As this rainwater drips through the crown it becomes further enriched in the bases potassium, calcium and magnesium, which are leached out of the foliage or from the microorganisms living on the canopy leaves. Inorganic nitrogen, however, is regularly removed from this rainwater supply as it passes through the canopy and in some months total nitrogen, both inorganic and organic, and phosphorus are also reduced. This is probably due to the microorganisms of the canopy zone but some absorption directly by the leaves may also be involved. The vegetation also filters out from the atmosphere small chemical particles (aerosols and dust) not dissolved in the rain. This input is then washed from leaves and twigs to join the stem flow or leaf drip to the soil surface. These leached chemicals can be quickly and directly re-used by the plants and soil microorganisms, whereas the far greater amounts in the litter

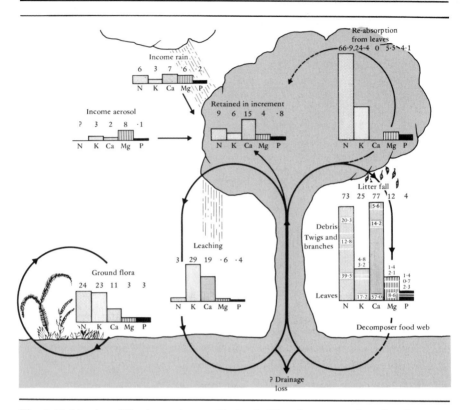

Fig. 8.17 Meathop Wood, north-west England. The above-ground cycle of macro-nutrients in the trees and ground flora. Quantities are kg/ha/year. The broad pattern of the nutrient cycle is similar to that found in other temperate deciduous woodlands. (Modified from Brown, 1974.)

must await mineralization, i.e. breakdown by the soil microorganisms such as bacteria and fungi.

From time to time, herbivore activity may cause severe defoliation as when, for example, outbreaks of *Tortrix viridana* (green oak tortrix) caterpillars occur. When this happens for several years in succession, it makes large demands on the nutrient reserves of the tree for the new growth of replacement leaves. Much of the leaf material eaten by the caterpillar will nevertheless remain in the ecosystem, being returned to the soil through death and decay of the herbivore. And Brown quotes the point made by Rafes that some partially compensating mechanisms may be associated with defoliation: the addition of readily mineralized nutrients contained in caterpillar excrement and the extra insolation resulting from a reduced canopy may both act to speed up decomposition of soil surface litter. However, Packham and Harding state that there is no evidence to support the first of these points.

Some 80–90 per cent of litter decomposition is carried out by microorganisms (bacteria, fungi, actinomycetes). The first stages of breakdown, however,

Fig. 8.18 Interior of Meathop Wood (see Fig. 8.17): an arrangement of collecting vessels (plastic bins, modified milk churns, etc.) for receiving materials from the canopy, e.g. litter fall, and by stem flow. (Photograph by permission of Dr E. J. White.)

are usually done by soil animals such as earthworms, springtails, nematodes and millipedes. These carry out the initial fragmentation of litter, providing a suitable substrate for subsequent microfloral and microfaunal growth. Edwards and Heath show that leaves buried in soil, in nylon bags with very small mesh to keep out the soil animals, will not change their physical appearance for several years. The microorganisms are further aided in their vitally important role by having available a rapidly utilizable nutrient source in the form of mineralized chemicals in the rain drip from canopy and ground flora.

Brown reports that losses of nutrients by soil drainage are small and are usually more than made up by the various inputs. However, important losses have occurred due to centuries of coppicing. This has meant removal from the site of all tree or shrub parts, including foliage, twigs and debris. This may lead to deficiencies in certain macronutrients (e.g. phosphorus) because cropping losses are not matched by inputs. A similar pattern of nutrient cycling could be demonstrated for most British oakwoods; any differences displayed probably being due to variations in the initial site conditions or management history. Details of specific nutrient cycles for British woodlands may be found in Heal *et al.* (nitrogen cycling) and Harrison (phosphorus cycling). These works demonstrate that nitrogen and phosphorus are normally tightly cycled with minimal loss through leaching from the soil or canopy. For example, uptake of phosphorus by plants depends largely on the recycling of phosphorus returned to the soil in organic matter – a finding based on studies at Meathop Wood. A useful summary of energy flow and nutrient cycling in woodlands is also provided by Packham and Harding, where again data from Meathop Wood are included.

FAUNAL INFLUENCES

Herbivores have the most direct influence on oakwood dynamics because they feed entirely on plant materials. They can be divided into two main groups: the invertebrates which live and feed on the oak itself, occupying a large range of habitats from the roots to the canopy, and those mammals and birds which freely move within the oakwood, operating at either one level (e.g. deer) or several (e.g. squirrels).

The oak has the richest insect fauna of any British plant. Several thousand species with many complex interrelationships are involved. Many of the galls on oaks are caused by insect attack, though other forms of parasite, including bacteria, fungi and mites, may also cause the plant to form a gall. One oak leaf may have between 50 to 100 spangle galls on it and the grubs therein plus their parasitic 'guests' can number nearly 200 creatures supported by the single leaf. As might be expected, more species are found in the oakwoods of southern Britain than in those of Scotland. Gradwell, writing on the effect of *defoliators*, reports that over one hundred species of moth feed on oak leaves in Britain. However, the two main defoliators are the winter moth and the green oak tortrix. Severe defoliation greatly decreases the annual growth of the tree: if repeated frequently and combined with other stress situations, such as drought, frost or fungal attack, mortality may follow. However, Gradwell shows that the oak is not passive in this relationship with defoliators. By producing 'lammas shoots' after defoliation (as in Wistman's Wood), the tree can build up food reserves again and withstand subsequent removal of primary leaves. There is also evidence that the high tannin content of older oak leaves upsets the digestive processes in the caterpillar larvae. Selective pressures exerted by the herbivores mean that only trees with this defence mechanism survive to produce acorns.

When the oak tree is at the seedling or sapling stage, the main defoliators may not be invertebrates but herbivorous mammals. Deer frequently produce a 'browse line' in woodlands, clearly showing the level to which they reach for leaves. Other mammals, such as squirrels, wood mice and voles, devour many of the acorns produced and if these are in short supply bark, buds or twigs may be substituted. Ashby has shown that mice and voles may destroy the vast majority of acorns. Tanton thinks squirrels and wood pigeons are perhaps more important, the latter sometimes removing up to 92 per cent of the acorn crop. But Mellanby sees their role somewhat differently. To him, these small animals are the main means by which the heavy acorns are dispersed. Pigeons carry acorns across barriers (walls, rivers) and deposit them on disturbed ground or clearings, good sites for germination. Jays and other members of the crow family, also effectively disperse acorns: a single jay may take and bury several thousand acorns each year.

In addition to the native fauna, many oakwoods are grazed by domesticated stock. When these animals are numerous they have a large bearing on the type of woodland which arises, i.e. the floristic composition and spatial structure reflects their selective feeding and prevention of tree regeneration. For example, the continuation of many upland oakwoods is severely threatened by sheep grazing. If we add these faunal influences to those of the much smaller invertebrates, the great importance of the faunal component in the oakwood ecosystem becomes apparent: there are, of course, far more animal species than plant species in any woodland.

OAKWOOD REGENERATION

The following summary on oakwood regeneration is based on Shaw's review of the problem. He notes that most ecologists and foresters are pessimistic, assuming that regeneration of our native oakwoods is failing. This view is based on many erroneous observations and conclusions: most studies have concentrated on the high losses of acorns known to occur regularly and not on the significance of the low number that survive. The oak is very long-lived and during a 300-year span each tree may produce about a million acorns. Only one seedling per parent tree needs to mature for successful continuance of the oakwood. In fact, if more survived we would soon be swamped with oak trees.

Acorn production is frequently high and lack of initial seed cannot be invoked as a cause of regeneration failure. Although there is very high acorn predation by small mammals and birds, a few acorns will be missed in the search and of the many taken by animals a few will always survive, some being buried and subsequently forgotten. These activities should not be regarded as a loss but as a very effective means of dispersal. Acorns germinate well on a variety of sites. It is the next stage which is the critical part of the reproductive process, i.e. seedling growth and survival.

Climatic, edaphic and biotic factors influence this stage. The main climatic variable of importance seems to be the light factor. At low levels, for example under dense parental shade, mildew growth is encouraged and insect attacks are higher. Adequate seedling growth occurs at about 10 per cent full daylight and maximum rates are found at about 28 per cent. This optimum regime is found in open woods or in small clearings and occurs commonly in present-day oakwoods. While we have not yet got a full picture of the edaphic requirements (nutrients, water, root competition) of oak seedlings, the conclusion of Shaw is that these are seldom limiting factors except at a few sites. The tolerance of the parent tree to a range of soils suggests that in most situations edaphic conditions are not of prime importance for successful seedling growth.

Grazing and browsing in an oakwood eliminate many seedlings. Both native fauna and domestic animals may be involved and much will depend on the stocking rates for these, i.e. the management of the large herbivores. This alone can prevent regeneration of all the tree species present, including the oak. But as Shaw notes, a certain amount of browsing is to be expected in a woodland. The oak seedling is well adapted to survive grazing in woodlands, provided these are not overstocked. The extensive root system functions to store most of the dry weight and nutrients of the plant below ground level beyond the reach of browsing animals. This is, perhaps, a response to browsing pressure. If grazing ceases, many heavily browsed seedlings quickly put on new leaf growth, using reserves in the swollen tap-root.

However, Shaw describes another form of grazing which he feels is extremely important for seedling growth. The damage here is caused by defoliating caterpillars which drop from parent trees. Only seedlings growing well away from an infected oak escape this defoliation completely. Many are eaten back so much that they have not even been recognized as oak seedlings in the past. Seedlings under large established oaks seldom survive more than three to five years. The replacement leaves which follow defoliation will only have about half the growing season left for carbon assimilation before they are shed in autumn. Hence they may need light intensities higher than those found under the canopy of the oakwood if they are to survive. This explains why field observations suggest light intensities of 50 per cent or more are necessary for good seedling growth. Once defoliation becomes important, the lower light levels previously suggested (28 per cent for maximum growth) are inadequate. If the site also has low levels of soil nutrients, then the uptake of nutrients to balance losses through leaf-fall (whether natural or caused by defoliators) may be insufficient and reserves in the plant are soon exhausted. At these particular sites the compensation point for certain nutrients may be more limiting than light conditions.

In large clearings and beyond the woodland, seedlings often do well. Shaw sees the biotic pressures preventing regeneration in the immediate vicinity of other oaks as entirely natural. As a mechanism, it prevents the sapling from being in direct competition with the parent plant for the same resources. The threat to regeneration comes when grazing and browsing are well above the natural levels. Other factors may also be important, such as competition with ground flora species or with the rooting systems of the parent oaks. Mellanby even suggests that root exudates from the mature oaks may be involved, as in the case of *Grevillea robusta* (see page 21). But all these aspects have not yet been fully investigated. Whitmore's review reminds us that tree regeneration in all woodlands can often be expressed in terms of varying sizes of gaps within the community: past, present and future gaps (related to senile or diseased or windthrown trees) are important in determining the patterns and processes that we see. Bormann and Likens also stress the importance of gaps and add that 'parasites, predators, symbionts, and decomposers may speed or direct changes in plant populations, or they themselves may decline or disappear as a result of changes in plant-community dynamics'.

PINE- AND BIRCHWOODS

Palynological studies show that pine has dominated the Northern Coniferous Forest formation in the eastern Highlands throughout most of the Flandrian period. This is particularly the case north of a line running approximately along the Grampian-Tayside county boundary. South of this birch and, to a lesser extent, alder have been equally or more important trees. Small woodlands of birch (of both *Betula pubescens* and *Betula pendula*) are the commonest natural woodland type remaining today in the Scottish Highlands. These species are found as pure stands or as mixtures with pine or oak. Birch grows well under the oceanic climate of western districts and on good soils at lower elevations it is sometimes seen mixed with oak or in competition with that species. Pine predominates on less favourable soils and under a more continental climatic regime. But heavily exploited pinewoods will be readily invaded by birch, as Steven and Carlisle noted. There are documented cases where devastated pinewood has been replaced by birchwood which, in time, has then reverted back to pinewood. Birch may also dominate open sites within or above pinewoods, such as scree slopes or around former bothies and crofts where it often plays a *pioneer* role. Pure birchwoods also occur in the north-west Highlands: Yapp sees these as the true climax for this area, similar to those of Scandinavia where this species occurs at higher latitudes than pine. The main species is *Betula pubescens* ssp *odorata*, which some consider the same variety as the Scandinavian ssp *tortuosa*.

The extent of the former Caledonian Forest, a term covering all the major woodland types in the Scottish Highlands, can never be exactly determined. It is clear, however, that most terrains were heavily wooded before the impact of man. The pine- and birchwoods, which now extend to about 610 m in only a few localities, formerly reached altitudes of 790 m or more. This is certainly the case in the Cairngorm Mountains where the highest pine stumps buried in the peat have been radiocarbon-dated by the author to 4,040 ± 120 BP

Table 8.4 Radiocarbon dates of peat macrofossils (buried tree stumps) from the Cairngorm Mountains, Scotland

Site	Grid ref.	Altitude (m)	Sample position	Date*
Carn Mor	028·907	610	(*a*) basal	6,700 ± 300
			(*b*) upper horizon	2,880 ± 220
Meall a'Bhuchaille	989·116	701	(*a*) basal	6,150 ± 150
			(*b*) upper horizon	4,400 ± 120
Jean's Hut	995·052	768	basal	4,630 ± 210
Sgor Mor	004·908	716	basal	4,140 ± 120
Coire Laogh Mor	004·068	793	basal	4,040 ± 120
Barns of Bynack	050·055	716	basal	5,110 ± 150

* Age BP (years before 1950, radiocarbon half-life 5570 years). Gakushuin University Dating Centre, ref. Gak. 2003–6 and Gak. 2538–41. A check date for the Sgor Mor site by Birmingham University (ref. Birm. 134) gave a date of 4130 ± 110 BP.

(Table 8.4). Under natural conditions these upper woodlands probably gave way to a *sub-alpine scrub zone* before open moorland was reached. Few examples of the transition exist now because of the repeated practice of moorland burning. The scrub zone was essentially a fern-rich juniper community. A good example of a sub-alpine scrub transition to open moorland occurs today at 647 m on Creag Fhiaclach in the Western Cairngorms (see map, Fig. 8.20). Poore and McVean report that these scrub zones are a common feature in Scandinavian forests.

Details of human interference in the Caledonian Forest are given in Chapter 10 but reference now to Fig. 8.19 will convey something of the extent of pinewood reduction. Most woodlands on this map are quite small. Many are not entirely composed of native pine (*Pinus sylvestris* var. *scotica*) but include plantations of conifers and areas of birchwood. This is clearly brought out on the more detailed map (Fig. 8.20) for the Rothiemurchus and Glenmore pinewoods (numbers 5 and 6 on the previous map). For the rest of this section we will concentrate mainly on the ecology of native pinewoods. Much of what is said applies equally to Scottish birchwoods. Indeed, in many cases the woods contain both pine and birch as virtual co-dominants.

WOODLAND COMPOSITION

McVean and Ratcliffe recognized two basic types of pinewood. Both have many species in common but the relative abundance and growth-forms differ in each type. In the moderately dense, native pinewood there is usually a slight admixture of birch (*Betula pubescens*) and perhaps a few rowan trees (*Sorbus aucuparia*). The shade cast does not encourage tall shrub growth. *Calluna vulgaris*, *Vaccinium myrtillus* and *V. vitis-ideae* form a well-developed dwarf shrub-layer except where mosses (*Hylocomium splendens* or *Rhytidiadelphus triquetrus*) replace shrubs when the tree canopy is particularly dense. Small patches of *Deschampsia flexuosa* grass also occur. This type of pinewood may be found from sea-level to over 475 m in the Central and Northern Highlands.

The second type is much more open in form and occurs throughout the Scottish Highlands. It is well seen in the Grampians where it replaces the first type with increasing altitude, occurring up to about 610 m on some slopes (Fig. 8.21). Because of its open form, the woodland contains more birch and rowan; in the West Highlands holly may also feature. The heather and whortleberry are much taller and juniper grows as a tall shrub, often columnar in form. *Sphagnum* tussocks, other mosses, several hepatics and the grass *Deschampsia flexuosa* may form the ground flora. These species are well seen in the higher rainfall areas of the west or on damp, north-facing slopes in the east where heather moor does not do well.

In like manner, two main forms of birchwood are recognized by McVean and Ratcliffe. There is a *Vaccinium*-rich birchwood with a ground flora of several species of fern, grass, rush and small flowering herbs. A tall shrub-layer is usually absent but juniper, hazel and honeysuckle may be locally abundant (see Fig. 2.4). The second type of birchwood is herb-rich and may occur in

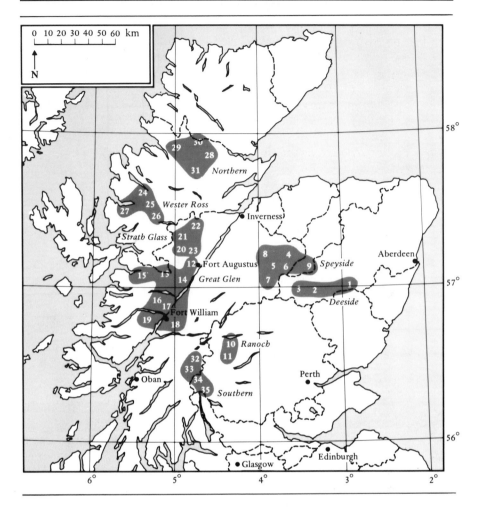

Fig. 8.19 The natural distribution of Scots pinewoods in Scotland: the remaining fragments of a once extensive forest cover. *Deeside*: 1. Glentanar 2. Ballochbuie 3. Mar. *Speyside*: 4. Abernethy 5. Rothiemurchus 6. Glenmore 7. Glen Feshie 8. Dulnan 9. Glen Avon. *Rannoch*: 10. Black Wood of Rannoch 11. Old Wood of Meggernie, Glen Lyon. *Great Glen*: 12. Glen Moriston 13. Glen Loyne 14. Glengarry 15. Barisdale 16. Loch Arkaig and Glen Mallie 17. Glen Loy 18. Glen Nevis 19. Ardgour. *Strath Glass*: 20. Glen Affric 21. Glen Cannich 22. Glen Strathfarrar 23. Guisachan and Cougie. *Wester Ross*: 24. Loch Maree 25. Coulin 26. Achnashellach 27. Shieldaig. *Northern*: 28. Amat 29. Rhidorroch 30. Glen Einig 31. Strath Vaich. *Southern*: 32. Black Mount 33. Glen Orchy 34. Tyndrum 35. Glen Falloch. (Modified from Steven and Carlisle, 1959.)

the same localities as the first. However, *Vaccinium* species are not usually present and grasses are more prominent. This community is often heavily grazed by either native or domesticated animals and this accounts for the lack of tall shrubs. More flowering herbs are found in the ground flora which is similar to that of mixed deciduous woodland. Many of these birchwoods, of

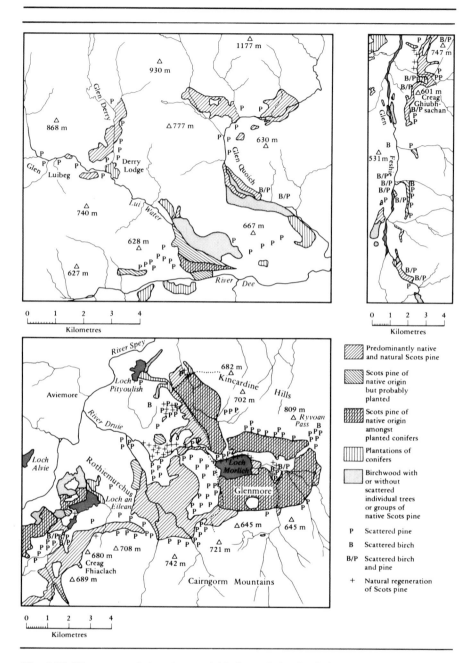

Fig. 8.20 The status of the pine- and birchwoods in the Cairngorm Mountains. The woodlands shown are numbers 3, 5, 6 and 7 Fig. 8.19. (Modified from Steven and Carlisle, 1959.)

both types, contain large numbers of boulders which often carry a rich bryophyte community. The soils of pine and birch woods are discussed in the next chapter.

Fig. 8.21 A natural fragment of Scottish pine and birch woodland at 400 m, Ryvoan, Cairngorm Mountains. The open nature of the woodland encourages the development of tall juniper (with a columnar growth-form) and dense heather and whortleberry.

NUTRIENT CYCLING

The same *major* components of the nutrient cycle demonstrated for Meathop oakwood will also be found in pine- and birchwoods. However, the values for each chemical in the cycle will vary because of differences in the vegetation itself and in the site characteristics under which these woodlands grow. Most of the available data for Scots pine woodlands come from studies of plantations of this species. Although there are some striking differences between plantations and natural pinewood, it can be assumed that broadly similar patterns of nutrient cycling exist in both. Ovington has assembled much of the available information on these woodland processes and the data presented here is mainly from that source. *Heal et al.* and Harrison also contain relevent data as does Chapter 9 of Packham and Harding.

Table 8.5 shows the annual organic balance-sheet for a Scots pine plantation. The evergreen pine photosynthesises over a greater part of the year than the oak: this partly accounts for greater total productivity of temperate coniferous woodland compared with temperate deciduous woodland. While much organic material is contained in the standing vegetation as leaves, branches and cones, a large quantity enters into the *litter fall component*. Packham and Harding report that the concentration of nutrients in conifer needles is lower than in leaves of temperate species and also a greater proportion of most elements is translocated from the needles before they drop. Thus the litter which accumulates is of low nutrient status. Compared with an oakwood, the litter decom-

Table 8.5 Annual organic balance-sheet for a Scots pine plantation, 32 years old (From Ovington, 1965)

Component of plantation	Productivity*
Primary production of photosynthetic plants	
By trees as	
cones	500
branches	4,170
leaves	3,750
trunks	9,500
roots	6,280
By field-layer	1,760
Litter-fall	
From trees	7,860
From field-layer	1,760
As caterpillar droppings	14
Accumulated	
In trees	9,040
In A_0 horizon	1,290
Removed	
As harvested tree trunks	5,200
By litter decomposition	8,330
Left in soil as roots of harvested trees	2,100

* Oven dry weight, kilograms per hectare.

position is much slower under pine trees. This may be partly due to the climate (lower temperatures) in upland areas. There is also a greater abundance and activity of earthworms, millipedes and woodlice in oakwoods and the important part they play in initial breakdown of leaf material, making it more suitable for subsequent attack by microorganisms, is well recognized. Leaching also occurs more easily from deciduous leaves than it does from pine needles. The thick A_0 horizon in pinewoods is thus less biologically active and mineralization of organic matter is achieved less readily. Under these conditions, Harrison quotes immobilization of phosphorus in the litter layer which may sometimes be in excess of 40 kg P ha^{-1}. Such large accumulations reduce the rates of cycling of phosphorus and may have long-term effects on soil fertility. However, nutrients that are released become tightly cycled through the soil mycorrhizae.

Ovington compares the nutrient circulation in respect of potassium for two woodlands of the same age, but of different tree species growing on the same soil type (Fig. 8.22). In the pedunculate oakwood there is a much greater uptake of potassium by the trees, whereas in the pinewood relatively large amounts of potassium circulate through the bracken ground flora. Thus these two wood-lands differ significantly in their patterns of nutrient circulation for potassium and presumably also for other macronutrients.

O'Sullivan's palynological studies of Abernethy Forest (woodland number 4 in Fig. 8.19), the largest remaining tract of semi-natural woodland in the

Quercus robur aged 47 years

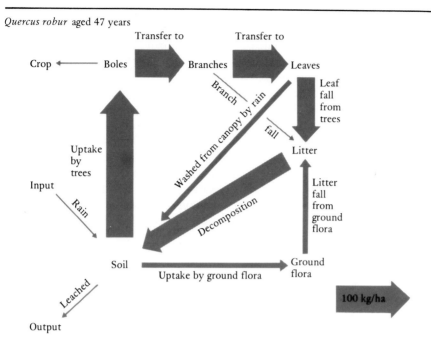

Pinus sylvestris aged 47 years

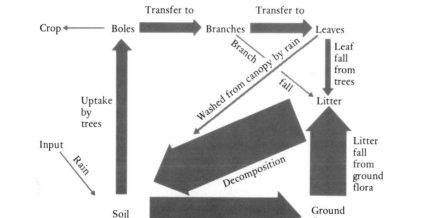

Fig. 8.22 A comparison of the potassium circulation in two adjacent woodlands growing under similar conditions. The oak trees take up much more potassium than the pines. This situation is reversed with the respective ground floras, where the bracken in the pine plantation has a strong uptake of this macronutrient. Arrow thickness indicates the magnitude of flow. (From Ovington, 1965.)

British Isles, have shown that the forest formerly contained more birch and other *deciduous* species such as willow, alder, poplar, wild cherry and rowan. Today, it is largely pine-dominated with large expanses of heather. The initial expansion of heathland in the forest began about AD 400 and was due to the felling, grazing and burning activities of the Pictish Dark Age settlements. This interference continued throughout the Medieval period, eliminating many pioneer deciduous species. It encouraged a change towards a *pine and dwarf-shrub* (largely *Calluna vulgaris*) *community* and soil impoverishment (e.g. the build-up of thick A_0 horizons of raw humus (mor)). While the forest was more mixed in composition and contained deciduous trees and certain herbaceous field-plants, whose litter had 'soil improving' properties, the soil degradation and regeneration failure did not occur. Although these changes in composition are now seen mainly in terms of soil alteration they also imply strong changes in many other aspects of the former nutrient cycle for Abernethy Forest. Unfortunately, we can now do no more than surmise on the exact nature of these changes.

ASSOCIATED FAUNA

Steven and Carlisle noted that the prehistoric fauna of Scotland was principally a *forest* fauna. Because many of the remnants of the Caledonian pinewoods are remote from human settlement, their fauna has been altered to a lesser extent than is the case with southern oakwoods. The distinctive nature of the pine-woods is made more so by the occurrence of a number of uncommon species in both the flora and fauna. Because of this and because many of the woodlands have enjoyed uninterrupted occupation of these landscapes since early Flan-drian times, a strong case can be made for their *conservation*. For convenience, the fauna which influence pinewood ecology can be considered under three broad headings: mammals, birds and insects.

MAMMALS.   The small native mammals include a group of carnivores: the pine marten (*Martes martes*), the wild cat (*Felis silvestris*), the fox (*Vulpes vulpes*), the stoat (*Mustela erminea*) and the weasel (*Mustela nivalis*). These do not cause any appreciable damage to the forest vegetation. It is the *larger* herbivores which have the main faunal impact on the forest ecology: the smaller herbi-vores, mountain hare, rabbits, red squirrel, mouse and vole, seldom reach sufficient numbers to be more than locally significant. The pine forest is the traditional winter shelter and grazing habitat for the native red deer (*Cervus elaphus*). These exist in quite large numbers where shooting and culling are not practised regularly. In the summer, the red deer seek higher, open ground but when winter returns they readily browse on young trees well within the main forest. The much smaller roe deer (*Capreolus capreolus*) lives in small herds for most of the year within the forest where they browse on young trees. However, domestic herbivores (sheep, cattle and, to a lesser extent, goats) have done considerable damage in the past and are still very important today in the grazing ecology of many woods.

BIRDS.   The principal birds influencing the ecology of the pine are the *game-birds*: the red grouse (*Lagopus scotica*), ptarmigan (*Lagopus mutus*), black grouse (*Lyrurus tetrix*) and capercaillie (*Tetrao urogallus*). The first two live on open ground but will enter woodland in severe weather. They are not so important as the last two which often frequent the forest and may do considerable damage to the buds of tree seedlings and saplings. The capercaillie became extinct in Scotland in the eighteenth century but has since been successfully reintroduced from Sweden. It is now found in most native pinewoods in eastern and central Scotland.

INSECTS.   Unlike the oakwoods, native pinewood is seldom seriously damaged by insect outbreaks. Many species of insect exist but the populations are usually well balanced. Some damage is done by the larger pine-shoot beetle (*Myelophilus piniperda*), the pine weevil (*Hylobius abietis*) and the pine looper caterpillar (*Bupalus piniarius*). However, these insects are far more important in pine plantations where, presumably, monoculture favours the rapid spread of pests. Pine plantations sometimes fail to recover from defoliation by the caterpillars of the *Bupalus* moth. According to Cousens, if the number of *Bupalus* pupae in the pinewood litter each winter does not exceed 5 per square metre the subsequent looper population will not reach epidemic proportions. Steven and Carlisle report the suggestion of Thalenhorst that the great numbers of wood ants in native pinewoods reduce insect populations by consuming many potential pests. Ants spend much of the day in the litter or on the trees and, together with the flocks of small insectivorous birds in the forest, probably prevent massive pest outbreaks.

REGENERATION OF PINE AND BIRCH WOODS

Little *natural* regeneration takes place in most Scottish pine and birch forests today. Ample seed is produced about every three to six years, but of that which germinates only a few reach the sapling stage. In the cooler, wetter climates of upland Scotland, pine litter often accumulates on the forest floor to form a thick carpet of raw humus. Seeds deposited on this layer may die of *drought* because they are unable to develop a rooting system quickly enough to reach the moisture reserves in the mineral soil beneath. In many pinewoods, the open canopy encourages the growth of tall field-layer species and these may also suppress seedlings. Likewise, Kinnaird shows that birch seed germination and seedling growth are suppressed by the shade of parent trees and undergrowth species. Other factors, however, may also be involved.

Birch seedlings are susceptible to competition from the mature birches and field-layer vegetation for moisture and nutrients. Phytophagous insects, which are numerous in the birch canopy, may also defoliate birch seedlings. Kinnaird notes that the best sites for birch germination are where 'compact' *Sphagnum* or bare soil occur: two situations where moisture stress is less likely to be encountered. Bare or disturbed soil seems equally vital for pine seedling development (except at higher altitudes when seeds in bare soil patches may suffer great losses by frost-heave). Any disturbance producing this soil condition, such as a fire or extensive wind-throw, is usually followed by a wave of pine regeneration. This can lead to the development of *even-aged* forest, influencing the structure and spatial pattern of vegetation for several hundred years hence. Ford has concluded that conifers in even-aged stands are increasingly likely to suffer from stress and when this happens the stand becomes vulnerable to catastrophe (e.g. wind throw, disease).

Although germination is difficult at certain sites, enough seedlings arise to maintain the forest. But very few of these survive because, for both pine and birch woods, *grazing* is the main cause of regeneration failure. Kinnaird gives a 99 per cent mortality rate for birch seedlings. Miller, investigating regeneration in Glen Feshie (woodland number 7 in Fig. 8.19), where some 2,000 red deer and a few sheep graze about 1,300 ha of range land below 450 m, found that 90 per cent of the birch and juniper saplings in the field were grazed at least once between October and June. In 1969, he planted seedlings of birch, pine, broom and gorse at 30 different sites in Glen Feshie. Within a year, about 57 per cent had died or disappeared, 32 per cent were alive but damaged by grazing or trampling, and only 11 per cent were undamaged. More recent studies by Miller and Cummins and Miller *et al.* have confirmed the importance of repeated deer browsing. In addition to deer grazing, some regeneration is prevented by game-birds and in many Scottish woods domesticated animals, particularly sheep, play a vital role.

In a study of pine at its altitudinal limits in the Cairngorm Mountains, the author investigated tree seedling distribution above the tree-line on the slopes of Creag an Leth-choin (see Fig. 6.6). The summit of this mountain is at 1,051 m, and an area between 550 m and 915 m, measuring approximately 2 km long by 400 m wide, was examined. Only 102 seedlings (one birch, the rest pine) were recorded. All but 11 were found below 730 m although 50 per cent of the area investigated occurred above this height. Numbers did not decline steadily with altitude, as might be expected if the distributional pattern was solely under climatic control. Whereas 55 seedlings showed moderate to heavy die-back – a climatically induced phenomenon – 87 seedlings had lost their terminal growth-point due to grazing. Much of this grazing is due to deer (there are few sheep on these particular hills) or game-birds, probably when they are forced down to the forest edge by severe weather on the summits. Where a deer fence exists, grazing is concentrated along the upper forest margins. Under these conditions, we cannot expect any regeneration of native pine at its altitudinal limits. In time, the forest at these levels must become more open and eventually shrink in area.

UPLAND MOORS

It is difficult to define precisely what is meant by the term 'moor' because, as Gimingham noted, 'heath' and 'moor' are terms which have been applied indiscriminately to one and the same area. Heath is a more general term and implies a community dominated by *evergreen dwarf shrubs*. A few trees or tall shrubs may be present but quite often the landscape is completely devoid of tall vegetation. Such heaths can occur at any altitude but most found in lowland areas of southern Britain only persist because they occupy infertile, sandy or stony soils unsuitable for agriculture. Moorland has come to mean *open* heathland lying *above* the level of enclosed or improved agricultural land. As such, it is mainly confined to the wetter, colder uplands of western and northern Britain and is often associated with peaty soils (Fig. 8.23). Many species are common to both lowland heaths and upland moors and the predominance of common heather (*Calluna vulgaris*) is a characteristic feature of both types. In this country, nearly all of the evergreen dwarf shrubs, including heather, belong to the family Ericaceae. Within a moorland area, these dwarf shrubs may be replaced locally by other low-growing vegetation dominated by species of grass, sedge, moss or lichen. Thus, although much moorland vegetation superficially may look monotonous there is frequently a mosaic of communities present, reflecting the subtleties of environmental variation.

Fig. 8.23 *Calluna*-dominated moorland communities (Callunetum) developed on the upper slopes of Glen Tanar, east Scotland. The strips in the heather are muirburns of different ages.

Whilst moorland would exist naturally above the tree-line, much in our uplands is obviously *derived* from former woodland. Vast stretches of land in Scotland below 610 m now carry a *Calluna*-dominated vegetation where once woods of pine, birch or oak existed. These dwarf shrubs are sometimes a natural component of the ground flora of such woodlands, especially on acidic soils. Tree destruction, burning and grazing, have allowed them to spread and dominate the ground vegetation. McVean and Ratcliffe describe three main types of heather moorland derived from former forest. The dry heather moors are more or less pure *Calluna* and they form the famous purple heather moors of

Scotland. A moss layer is present, consisting of *Hypnum cupressiforme, Dicranum scoparium* and *Hylocomium splendens* and, after fire, *Erica cinerea* (bell heather) and *Vaccinium) spp* may reach co-dominance locally with the *Calluna*. This is the most important vegetation type for grouse moors. A variant of this pure heather community is the *Arctostaphylos*-rich moor which, in addition to *Arctostaphylos uva-ursi* (bearberry), contains a richer herb and dwarf-shrub flora. The third type is damp heather moor in which *Vaccinium myrtillus* may co-dominate with *Calluna*. Mosses (including *Sphagnum* spp) and hepatics are prominent, especially in western oceanic districts. The community often reaches its best development on north-facing slopes.

Outside Scotland, similar types of moor have come into existence on formerly well-wooded land. For example, large parts of the North York Moors, although of no great altitude, are now heather-clad. Dimbleby has shown that extensive woodland covered this region until clearance by man began in the early Sub-boreal period. Atherden has traced the impact of prehistoric cultures on this area, while Tinsley has done likewise for the Nidderdale area of the Pennines. Similarly, much of Dartmoor was previously wooded to about 460 m. *Pteridium aquilinum* (bracken), a component of woodland ground floras, has since spread rapidly on some lower moorlands, particularly in milder,

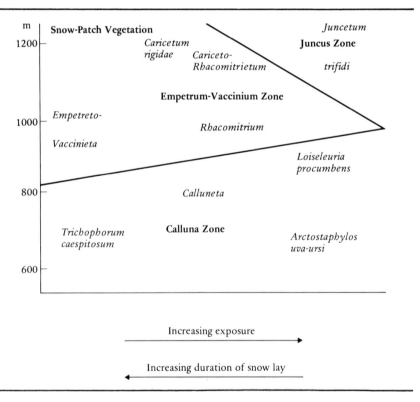

Fig. 8.24 Main plant communities above the present tree-line in the Cairngorm Mountains: the interaction of altitude, exposure and snow cover on their altitudinal distribution pattern. (From Watt and Jones, 1948.)

wetter, western districts. It may form a monoculture not unlike that of *Calluna* and, as such, it is a serious problem in hill farming.

Much moorland above the natural tree-line has experienced a similar long history of burning and grazing to that of the forest zone. Despite this, a simple division of the vegetation can usually be made on the basis of *increasing altitude*. This is well seen in the granitic Cairngorm Mountains (the most important mountain system in Britain, according to Ratcliffe's survey) where Watt and Jones recognized three main altitudinal zones above the forest edge (Fig. 8.24): the *Calluna* zone, the *Empetrum–Vaccinium* zone and the *Juncus* zone (and snow-patch vegetation). The topographic complex influences this altitudinal zonation to give interdigitation of the vegetation zones where suitable habitats come in. Within each zone, edaphic variation produces further diversity.

THE CALLUNA ZONE

The passage from pinewoods to *Calluna* moor is usually abrupt, there seldom being any transitional scrub zone left on British mountains. *Calluna* dominates up to 850 m and may be found beyond to a limit of 915–75 m. Two facies are recognized: a wet *Calluna* moor with *Trichophorum ceaspitosum* (deer grass) is found on steeper, sheltered slopes (the surface vegetation in Figs. 8.1 and 8.5 is of this type) and on more exposed ground heather is associated with *Arctostaphylos uva-ursi* at lower levels and *Loiseleuria procumbens* (Loiseleuria) at higher levels.

THE EMPETRUM-VACCINIUM ZONE

On exposed slopes the *Calluna* community passes directly to the *Juncus* (rush) zone. On more sheltered slopes it grades into the *Empetrum-Vaccinium* zone, which reaches its main development between 915–1,066 m. This may extend to the summit if the locality is not too exposed. The main species are *Empetrum hermaphroditum* (crowberry), *Vaccinium myrtillus* and *Vaccinium uliginosum* (bog whortleberry). *Rhacomitrium lanuginosum* (a lichen) will replace these species locally on exposed ridges in the Cairngorms.

THE JUNCUS ZONE

Juncus trifidus (three-leaved rush) can tolerate snow cover and exposure. It is the characteristic species of exposed slopes above 975 m and on the summit plateaux. Associated species are *Deschampsia flexuosa*, the sedges *Carex rigida* and *Carex bigelowii*, and *Rhacomitrium lanuginosum*. The last two are more characteristic of exposed sites, sometimes appearing as 'islands' within the *Juncus* sward. At these heights, there is much bare ground and *Juncus trifidus* has a patchy distribution with a typical tussocky growth-form.

The snow patches usually show an outer rim of *Nardus stricta* (mat grass) with an inner zone in which *Nardus* is still prominent together with *Carex bigelowii*, *Deschampsia flexuosa*, *Salix herbacea* (least willow) and *Gnaphalium supinum*. The core is occupied by bryophytes, particularly *Dicranium* spp and *Webera ludwigii*.

This sequence will not hold for all Scottish mountains. Where soils are less acidic a richer floral mixture comes in. Some uplands have a much higher proportion of grasses in their flora. As McVean and Ratcliffe note, natural grass heaths belong mainly to a zone above the dwarf-shrub-dominated vegetation but, due to the activities of man, grassland may now occur at any level on our mountainsides. Within these grasslands, two main directions of variation can be demonstrated. First, there is a series which reflects soil moisture differences. *Agrostis-Festuca* grasslands occupy drier soils (Perkins' energy model for such a grassland – the IBP Llyn Llydaw site in Snowdonia – is presented as Fig. 8.25). *Nardus stricta* grasslands are found on wetter soils. The wettest soils carry *Juncus squarrosus* (heath rush) communities. Secondly, within each of these types a species-rich or species-poor facies occurs according to whether the soils are base-rich or base-deficient. Because *Calluna* grows less well under waterlogged conditions, *Molinia caerulea* (purple moor grass), *Eriophorum* spp (cotton grass) *or Trichophorum ceaspitosum* (deer grass) may be equally or more important species in western districts. The frequency of burning and the grazing intensity experienced also control floristic diversity to a large measure. In the Pennines, for example, *Nardus stricta* has greatly extended its range on wetter slopes where heavy grazing has eliminated more palatable species.

Many of our uplands contain extensive flat or gently sloping surfaces where *peat* has developed. Further comments on these upland peats will be left until the next chapter but we may note here that peats carry various surface vegetations according to whether they are still actively accumulating or drying out and undergoing erosion. Peat-bogs which are still growing do not have a uniform surface but show a pattern of very small ridges and hollows. Different sets of species occupy each microhabitat in this complex (Fig. 8.26). It was previously believed that peat surfaces grew upwards when the hollows filled in with plant growth and the ridges or hummocks then became flooded to form new pools (the cyclic regeneration theory of peat growth). However, Barber's detailed palaeoecological research has now shown that even small-scale features of bog stratigraphy are under climatic control (each phase shift in peat growth is the result of a climatic shift – the Phasic Theory of bog growth). In these situations, heather is confined to the drier hummocks but it will quickly invade the much drier surface of eroding peat to form extensive spreads. Within many moorlands, springs break the surface and the floristic diversity of these *flush communities* reflects the nutrient status of the water.

Variations in altitude, angle of slope, type of rock, past and present biotic influence, climatic parameters, etc. combine to produce a mosaic of communities in upland moors. Apparently uniform vegetation dominated by a few species, on closer inspection, often reveals considerable ecological variety. The intensification of climatic factors with elevation and the confinement of certain species to distinct zones present many interesting spatial patterns for the biogeographer to study.

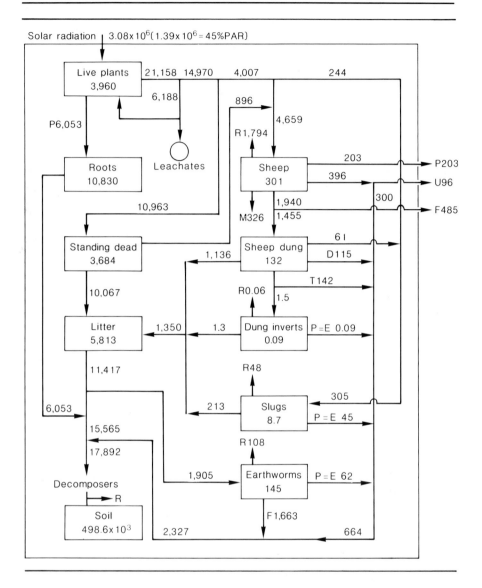

Fig. 8.25 Energy flow model for a montane *Agrostis-Festuca* grassland, the IBP site at Llyn Llydaw, Snowdonia, Wales at 488 m. Boxes represent the mean biomass of grassland components and reserves of soil (gm^{-2} or kJm^{-2}) throughout the year except sheep and sheep dung and invertebrates which are summer values. Transfers, on the arrows connecting boxes, represent the flows (gm^{-2} yr^{-1} or kJm^{-2} yr^{-1}) between components.

P: production; E: material eliminated; U: urine; F: faeces; M: methane (rumen fermentation); R: respiration; D: sheep dung decomposed in the soil surface; and T: sheep dung estimated to have been transferred to soil by dung beetles. (From Perkins, 1978.)

Fig. 8.26 Hollow-hummock pattern of an actively growing peat surface. 720 m on Upper Deeside. The pools contain *Sphagnum* species, *S. cuspidatum* growing under the water and *S. rubellum* on the edges. The drier hummocks support *Calluna vulgaris* and *Trichophorum caespitosum* (*Scirpus*), while *Eriophorum vaginatum* (cotton grass) grows at intermediate points.

THE ECOLOGY OF CALLUNA VULGARIS

Much research on upland moors has been concerned with the ecology of heather. A detailed review of this work is provided by Gimingham. The emphasis on *Calluna* is justified since this plant dominates many moorlands and is a *key food source* for all the large herbivores in which man is interested. Moorland management techniques aim for a continuous supply of heather in a suitably nutritious form for these herbivores.

Cyclic growth processes occur in several species of heath communities and *Calluna* itself has a four-phase life history which Barclay-Estrup described as follows:

1. *Pioneer phase*. This phase of establishment may last for 3–10 years and cover is about 10 per cent or slightly more. Mosses and other vascular plants are well represented and the biomass of *Calluna* is low. At this stage, heather has minimal influence on the other species present.
2. *Building phase*. Individual heather plants are 7–13 years old and cover values now reach 90 per cent. Most other plants are greatly suppressed and the ground-level microclimate is strongly influenced by the dense, closed *Calluna* canopy.

Table 8.6 Some of the main changes during the four-phase *Calluna vulgaris* development cycle (Modified from tables 14, 15 and 16 in Gimingham, 1972)

	Stage of development cycle			
	1 Pioneer	2 Building	3 Mature	4 Degenerate
Mean height (cm)	24·1	52·1	63·2	55·2
Mean age of individuals (years)	5·7	9·0	17·1	24·0
Mean cover (%)	12·0	93·8	78·4	41·3
Biomass (gm^{-2})				
(*a*) *Calluna* only	287·2	1,507·6	1,923·6	1,043·2
(*b*) Other dwarf shrubs and grasses	179·6	41·2	52·0	83·2
(*c*) Bryophytes	422·4	153·2	329·6	434·4
(*d*) Total, all plants	889·2	1,702·0	2,305·2	1,560·8
Net production of young *Calluna* shoots $(gm^{-2}$ in one year)	148·8	442·4	363·6	140·8
Depth of bryophyte layer (cm)	6·0	2·4	4·1	6·3
Surface illumination (% of that in open)	*c.* 100·0	2·0	20·0	57·0

3. *Mature phase*. By now, the individual plants will be 12–28 years old but cover value declines to about 75 per cent. The branches in the middle of each clump begin to spread out, allowing greater light penetration. This permits bryophytes and the lichen *Parmelia physodes* to colonize the centre, the lichen growing on the main branches of the heather. Although biomass is now at its maximum the net production of young shoots is less than in the preceding phase.

4. *Degenerate phase*. Heather plants will now be 16–29 years old and clumps are much more open. Cover values seldom exceed 40 per cent and other species, especially bryophytes, flourish. Net production of young shoots is at a minimum. The microclimate is once again much nearer to that of the pioneer phase. As the large, open centres of each clump spread, either other dwarf shrubs invade here or seedlings of *Calluna* may initiate a new pioneer phase in the middle of the old patch. Table 8.6 summarizes these developments.

Calluna produces copious seed which may remain viable for several years. Providing the site is not subject to desiccation, germination is marked. The high temperatures of a fire will destroy much seed but Gimingham reports that temperatures in the range 40–160 °C for a short period greatly increases germination. *Calluna* requires good light for germination and seedling growth and is well suited for colonizing bare ground. Seedlings are not usually found in the dense growth of the parent plant or other closed communities. The developing plant needs protection from water stress and is favoured by high relative humidities at the surface level. It will not tolerate long periods of waterlogging and on wet sites *Calluna* is replaced by species which are better adapted. Because of its water and light requirements for establishment and because the very small seeds have limited food reserves, new *Calluna* grows best in rather *open* habitats. These are today largely provided by moor burning and the extensive spread and dominance of *Calluna* is closely correlated with this practice, as Gimingham has shown. *Calluna* has well-developed mycorrhizae and these play an important role in the phosphorus and nitrogen nutrition of heather on the many sites of limited fertility where it grows.

MOORLAND FAUNAS AND THEIR MANAGEMENT

The invertebrate, herbivorous fauna of *Calluna* moorland has little pronounced influence on the plant ecology as far as man is concerned. Damage to plants is usually localized. Man's chief interest lies in the native and domesticated vertebrate herbivores which freely range our upland moors. The wild herbivores include red deer, mountain hares, red grouse and ptarmigan. Hares have not usually been cropped but the other species provide for sporting interests. More recently, the commercial market for red deer meat (venison) has expanded and a profitable trade now exists with the continent. Most of the natural predators of these wild herbivores have now been removed by man. Indeed, man is now the main predator and unless periodic culls are taken the carrying capacity of these *low*-production plant communities is soon exceeded. As Miller and Watson pointed out, the highest recorded production for *Calluna*, 442 g/m^2, is far less than that of pinewood (1,300 g/m^2) or oakwood (800 g/m^2). As we shall see in Chapter 10, there is a long history of cattle grazing on our uplands. Sheep and goats were also kept for local use. But by the nineteenth century large-scale sheep farming had more or less replaced cattle rearing in most districts. This replacement came earlier in Wales, the Pennines, the Lake District and the southern uplands of Scotland than it did in the Scottish Highlands.

For both wild and domesticated stock, *Calluna* is an important part of the diet, especially in winter when grasses are less readily available. However, some heather is probably grazed at all seasons by upland herbivores: in some seasons, grouse may have crop contents of 100 per cent *Calluna*. However, utilization of the primary production of heather on Scottish moors by grouse is low; no more than 1 to 2 per cent. Nevertheless, Miller and Watson show that grouse are very selective heather feeders, using plant material with the highest nutritive value. The nutrient content of heather declines with age as nutrient supply becomes locked in the increasing amount of unpalatable woody tissue. According to measurements by Thomas and Dougall, the maximum availability of nutrients for herbivores is found in heather about seven years old. The food value of heather also varies with the geology and it has been demonstrated by Picozzi that moors on granite soils carry smaller populations of some herbivores, for example red grouse, than moors overlying base-rich rocks.

Management of upland moorland is invariably through the use of fire. The chemicals in old vegetation are quickly returned to the soil in the ash and new *Calluna* growth is stimulated. This practice has a long history, extending back as far as the coppice management of oakwoods. However, only recently has it been regularized. Gimingham summarized the set of *principles* which should govern management by burning if site deterioration is to be avoided. The aim must be for a high production of nutritionally suitable *Calluna* and good germination conditions for regeneration. The essentials of good management are as follows:

1. Stands should be burnt in the late building phase, before they reach the mature or degenerate condition.
2. The fire should be hot enough to consume most of the above-ground heather. Canopy temperatures should not exceed 500 °C and soil surface temperatures should remain below 200 °C.
3. Heather should be burnt in numerous small patches, strips being preferred to square or circular areas.

Attention to these points will cut down the risk of soil or peat erosion, reduce losses of chemicals in the smoke, prevent plants such as *Nardus stricta* replacing heather on wetter sites, and not lead to the destruction of heather seed. It will also produce a mosaic of different-aged stands. This is important for grouse who have a strong territorial sense and need taller *Calluna* for shelter. Unfortunately, much burning has been along haphazard lines and environmental deterioration can be demonstrated over wide areas.

NUTRIENT CYCLING

On many moorlands the vegetation is rooted in peat and there is virtually no nutrient input to the ecosystem from the weathering of underlying bedrock. Heal and Smith report that many bog plants depend on recirculation of nutrients from decaying remains, but release by decomposition is slow and the availability of many ions declines markedly below c.pH 4·5. Even where mineral soil forms the surface it is heavily leached by high rainfalls and percolating, acidic soil water. Most dwarf shrubs and moorland grasses are relatively shallow-rooted and they only bring small amounts of leached chemicals back into circulation (*Eriophorum vaginatum* (cotton grass) appears to be an exception to this). There may be small additions to the ecosystem from lateral drainage and this is well seen in flush communities. But their occurrence is of limited extent. Another small addition comes from the impaction of airborne particles when the shrubs act as a filter to air-flow. Inputs in the form of dissolved chemicals in rainfall or snow are much more significant. Further small amounts of chemical may be added in the form of leachates from the vegetation itself. However, a major portion of the nutrients will be contained in the standing crop and in the litter accumulation. Fire, if of the right type and at the right time in the *Calluna* cycle, will quickly release relatively large quantities of these chemicals which then become a major input to the soil.

Coulson and Whittaker argue that many moorland animals (invertebrates and vertebrates) and their excreta represent high nutrient concentrations with rapid decay rates which may also be important sources of minerals for plant growth. They point out that many peat areas experience a vast 'spring' emergence of insects (as in the Tundra) which only live a few days. Their rapid death and decay puts back into circulation over a very short period quantities of minerals which are important for the vegetation.

Several attempts have been made to determine the *losses* of nutrients from upland moors. Because many moors are in heavy rainfall areas, loss by leaching will be a regular feature of their nutrient cycle. Gimingham noted that this was normal for the ecosystem and the rate of loss is assumed to be slow. The breakdown products of *Calluna*, however, readily mobilize cations and the heather monoculture itself may cause more leaching loss than hitherto suspected. There could be additional losses when ash deposits from burnt heather are partly dissolved and washed away by surface run-off.

Crisp has identified an important loss due to *peat erosion*. Working in the northern Pennines, he assessed the various inputs for a small stream catchment (83 ha) rising on the moors at 686 m. Table 8.7 shows the balance-sheet for this site which, apart from the areas of eroding peat, is dominated by heather and cotton grass. Sheep, which some have regarded as important in the impoverishment of moorlands, had a negligible effect in the direct draining of nutrients from the system. However, the sheep most certainly influence other aspects of the moorland ecology, e.g. by manuring, selective feeding and causing erosion. Two factors were responsible for a major loss of chemicals from the catchment: the down-stream movement of nutrients in solution and the loss of peat by erosion. Crisp showed that the most important losses were in respect of nitrogen and phosphorus. The annual phosphorus loss was two to four times that reported by Allen for a single heather burn and the nitrogen loss was about a quarter of the substantial loss occurring in a single heather burn (mainly in the smoke). Harrison has compared the phosphorus cycle at Meathop Wood with that for the IBP Llydaw montane grassland site (see also Fig. 8.25). He found a much faster rate of phosphorus cycling for the *Agrostis-Festuca* grassland but in both ecosystems the most important feature of the cycle was the dependence of plant uptake of phosphorus on the release of phosphorus from organic debris returned to the soil. He also concluded that sheep grazing tends to stimulate phosphorus cycling, leading to soil improvement (cf. conifer forest).

Most upland moor today is largely a *man-made* ecosystem. Gimingham summed up the position thus: '. . . destruction of native forest on so widespread a scale in an oceanic climate and on acid, generally infertile soil was a most unfortunate exploitation of a natural resource of great economic and ecological value. It set in train a series of consequences, both as regards land use and management and as regards developments in vegetation and soils, which have been inclined towards deterioration of the habitat. But at the same time, the spread of heath lands has provided an object lesson of the utmost value and a landscape of much delight.' Our understanding of this ecosystem, including the part man plays in the processes operating at these altitudes, is only just beginning.

Table 8.7 Water and mineral balance-sheet during the study year for a Pennine moorland site, Rough Sike, a stream catchment of 83 hectares rising at 686 m O.D. (Modified from Crisp, 1966)

	Water (thousands m³)	Sodium (kg/year)	Potassium (kg/year)	Calcium (kg/year)	Phosphorus (kg/year)	Nitrogen (kg/year)
1. Stream water output	1,368	3,755	744	4,461	33	244
2. Evaporation	403	–	–	–	–	–
3. Peat erosion	–	23	171	401	37	1,214
4. Drift of fauna in stream	–	0·004	0·011	0·003	0·010	0·118
5. Drift of fauna on stream	–	0·11	0·38	0·07	0·43	4·6
6. Sale of sheep and wool	–	0·16	0·44	1·58	0·98	4·4
7. Total output (1. to 6.)	1,771	3,778	916	4,864	71	1,467
8. Input in precipitation	1,771	2,120	255	745	38–57	681
Difference between 7. and 8. (Net loss for catchment)	–	1,658	661	4,119	14–33	786
Net loss per hectare	–	20·01	7·97	49·68	0·17–0·40	9·48

DISCUSSION SECTION

How certain can we be of pollen-analysis findings?

It is true to say that the more we have probed into palynology the more sources of potential error have been discovered. Nevertheless, there is agreement on the main findings and most workers accept the vegetational changes indicated. Controversy largely focuses on the detail within this framework. Certain *basic assumptions* underlie the method. For example, we believe plant species showed the same response to ecological factors in the past as they do now. We also assume that plants in the past were grouped into assemblages (communities) broadly similar to those of today. However, the more the artificiality of present woodlands is appreciated the less certain we may be about interpreting past forest covers in terms of them.

Further difficulties may arise over the question of *pollen inputs* to a sediment and the problem of interpreting trends in diagrams. Some species, such as pine and hazel, are prolific pollen producers while others may be under-represented at the preservation site. While many plants are wind pollinated quite a few are insect pollinated or even self-fertilizing, exposing no pollen for dispersal. It is the pollen of the first group which will largely feature in pollen diagrams. The exact limits of the pollen catchment area for a sediment are impossible to determine but evidence suggests most pollen is locally derived. However, much can happen to pollen after release and Tauber has shown that the surrounding vegetation can have a strong filtering effect. Pollen can also be disturbed or re-sorted in the sediment by water currents, burrowing animals and so on. Palynologists are now well aware of these potential sources of error. Much work is being done on the study of modern pollen inputs where some of these variables can be accurately determined. The results of such studies will put us in a much better position to interpret past pollen assemblages.

Perhaps the greatest controvery in palynology centres on the *interpretation* of sudden changes in pollen frequencies of certain species in pollen diagrams. For example, the marked decline in elm pollen, found in most diagrams from north-west Europe, was originally taken to indicate a climatic change, the beginning of a cooler Sub-boreal period. Further research has suggested other causes. Garbett has recently concluded that the decline was due to over-exploitation of a resource, initially by Mesolithic cultures who were changing from a hunter-gatherer to a fodder-gathering economy. The practice was continued and expanded as Neolithic cultures developed: lopping off young elm branches and leaves to feed to stalled domestic animals would greatly reduce pollen outputs. Another suggestion has been the onset of disease, similar to the outbreak of Dutch elm disease now ravaging much of southern Britain. In much the same way, Turner has argued for an anthropogenic cause rather than a climatic one for the decline in lime pollen characterizing the Sub-boreal period. It is true to say that with the arrival of man the interpretation of pollen diagrams becomes much more difficult. Also, it is possible that some trends in pollen diagrams result from normal seral developments and are not a response to strong environmental changes.

Disagreement is not confined just to the floral evidence. Coope, studying Coleoptera (beetles) remains in the peat, finds these *faunal indicators* point to a much higher summer temperature for the Late-glacial climatic amelioration (the Alleröd or Windermere) than that suggested by the pollen assemblages. He suggests a July average temperature of 17 °C by the end of the Lower Dryas whereas pollen studies indicate a July average of only 13 °C coming somewhat later, well within the Windermere zone. At the end of Zone III (the Upper Dryas) temperatures rose rapidly according to the Coleoptera findings but more gradually on the botanical evidence. Coope believes this can be explained by the fact that beetles respond to climatic improvement much faster than slow-growing trees. Hence, they would colonize areas more quickly. They are, therefore, better indicator species than the plants.

In spite of all the potential sources of error and problems of interpretation, there is widespread confidence in the Late-glacial and Flandrian climatic and vegetation history derived from peat and pollen studies. Because many researchers now use palynological techniques any findings are subject to close scrutiny. Palynological interpretations have a great relevance for many other field sciences and this ensures that they are closely checked against other evidence before being accepted.

Before the advent of man, was Britain clothed in vast areas of monotonously similar climax woodland or were there marked regional variations?

Although this is a small country with only two climax forest types (the Summer Deciduous Forest and Northern Coniferous Forest), it is reasonable to suppose there would have been marked *regional variations* in this woodland cover. These variations might have involved the associated trees, shrubs or ground flora species rather than the dominant trees. The strong climatic gradients across the British Isles (from the mild maritime regime of south-west Ireland to the cold, tundra-like climate of the High Cairngorms: from the wet hills of western Wales to the relative dryness of East Anglia) and the very varied geology also would lead to regional differences. Godwin's review of the forest differentiation of the late Atlantic shows birch forests in far northern Scotland, pine-birch-alder in the Central Highlands and several types of mixed-oak forest in England and Wales (e.g. alder-lime-oak-elm forest in East Anglia and south-eastern England, and oak-alder-birch forest in south-west England). Ireland was noteworthy for the strong presence of hazel in the mixed deciduous forests and the absence of beech, hornbeam and lime. Greig has recently summarized evidence which establishes that lime woods were quite common in southern Britain and many so-called oakwoods probably had lime as their main species.

In many parts of Scotland, woods of oak, birch and pine existed in close proximity, no more than a few kilometres apart due to the 'concertina effect' of elevation. Woodlands in wetter, western districts would have had a higher number of bryophytes and epiphytes, if the Killarney Oakwoods and Wistman's Wood are any guide. In south-west Ireland further variation was provided by the Lusitanian floral element. An important point, not obvious from the present landscape, is that many low-lying areas were much more liable to *flooding* in the past. Most rivers in this country have been dredged, cleared or had their flow improved. Before this work was initiated, most valley floors experienced seasonal flooding. This contrast between well-drained upper slopes and wet, waterlogged valley bottoms would be reflected in the vegetation. Extensive tracts of alder, willow and hydrosere scrub development would exist.

Natural fires occurred, even in damp western Scotland, before the arrival of man and these might have spread unchecked. These, and other disturbances, would lead to *seral* stages in the landscape, adding further variation to the plant cover. In some areas soil peculiarities probably prevented full colonization by the appropriate climatic climax type. While pinewood generally receded northwards as the Summer Deciduous Forest advanced, we cannot be absolutely certain that some pinewood did not persist in the south. On very sandy, infertile soils in the south-east several small patches of pine woodland are found today which appear quite natural. Most believe that these are subspontaneous communities established in the last few centuries by seed from nearby plantations. But there remains the possibility of survival on these poor soils from early Flandrian times. Other areas of the British Isles also have unexpected floras where edaphic influences outweigh climatic factors. The area known as The Burren (Co. Clare) is a good example. Here, on extensive pavements of Carboniferous Limestone just above sea-level, a peculiar mixture of lime-tolerant species (e.g. *Festuca ovina*, *Sesleria caerulea*, *Koeleria cristata*), plants of acid heathland (e.g. *Empetrum nigrum*, *Calluna vulgaris*) and plants usually found on mountains (e.g. *Dryas octopetala*, *Saxifraga hypnoides*) exists together.

Care is needed, however, in arguing from what we see today about variations in former plant cover. Some present woodland differences reflect different past histories of human interference (variations in grazing pressures, coppicing, burning frequencies, planting policies, etc). For example, it was long held that the almost pure ashwoods found today on steep dale slopes in Derbyshire (e.g. Dovedale Woods) were a natural climax on these limestone rocks. But recent studies, including palynological investigations, now show that a very mixed deciduous woodland previously existed here. This was cleared for pasture land and pure ashwood has only arisen through the recent colonization of abandoned grazings.

How big a component in the present British flora are exotic tree species?

The simple answer is very big indeed. Table 8.8 sets out the relevant statistics. All the conifers are exotic species except for the Scots pine. Much pinewood is, of course, planted and not natural. Of the broad-leaf species, sycamore is an introduction (fifteenth century) and so is sweet chestnut. Some poplar

Table 8.8 The composition of British woodlands in 1972 (Modified from Rooke, 1974)

Conifer	a	b	Broad-leaf	a	b
Sitka spruce	364	19·8	Oak	204	11·1
Scots pine	260	14·1	Birch	172	9·3
Larch	156	8·5	Beech	64	3·5
Norway spruce	128	7·0	Ash	52	2·8
Lodgepole pine	92	5·0	Hazel	32	1·7
Corsican pine	48	2·4	Sycamore	28	1·5
Douglas fir	48	2·4	Sweet chestnut	16+	0·9
Others	61·8	3·3	Alder	16	0·8
			Poplar	8	0·4
			Elm	8	0·4
			Others	104	5·8
Totals:	1,157·8	61·8		704	38·2

a = area in thousands of hectares   *b* = percentage composition (approximate)

species are native and some are introduced. We may distinguish *three* types of *introduced* tree species. First, those that arrived early and have since spread throughout many woodland types, often unaided, and now behave in much the same way as native trees (e.g. sycamore). Secondly, many exotic trees introduced in the eighteenth century for ornamental purposes when parklands were laid out for the nobility. In some districts, these trees now contribute a highly valued and distinctive element to the landscape. Thirdly, those trees, almost entirely conifers, brought in by practical foresters as crop species. In the first instance, this was done by private landowners, particularly in northern England and Scotland. More recently, the Forestry Commission has become the major agent in their spread. In terms of the previous question, exotic trees may add variety but, unfortunately, in some areas they may also produce very monotonous tree landscapes. Peterken states that by 2025 AD 76 per cent of all British woodland will consist of plantations on land which was not wooded at the start of the twentieth century, if current plans for afforestation are fully implemented. Such developments would have strong implications for our uplands, leading to a great loss of moor and other 'open' land.

Since many upland moors are derived from other vegetations by the activities of man, just how stable are these heather communities?

The man-made *Calluna* moors are stable only while the activities which created them continue. If burning ceases and grazing by large herbivores is prevented, then many are readily invaded by trees. Pine, birch and oak have little difficulty in this respect, as long as adequate sources of seed are nearby and suitable conditions for germination prevail. Many upland moors are extensive and nearly all trees have been eliminated over wide areas. Thus a suitable source of seed is frequently not available. Spruce trees (*Picea sitchensis*) seldom grow well on *Calluna* moors, as the Forestry Commission have found to their cost. This is the most frequently planted exotic conifer, and Gimingham reports evidence that *Calluna* may release a root toxin to which spruce is sensitive.

Calluna is also susceptible to competition from other dwarf-shrub species of the moorland ecosystem. Two species very similar to *Calluna* in many respects, *Erica tetralix* (cross-leaved heather) and *Erica cinerea* (bell heather), replace it according to soil hydrology conditions. *Erica tetralix* tolerates much wetter sites and *Erica cinerea* does better on drier sites. Mismanagement of moorland can change the hydrology in either direction, leading to *Calluna* replacement.

An interesting case arises when bracken competes with *Calluna*. Bracken passes through a four-phase growth cycle similar to that of heather (pioneer, building, mature and degenerate). It is well adapted to fire, and damage by trampling, cutting or burning stimulates a dormant bud to grow from the underground perennating rhizome. Bracken is seldom grazed and casts a deep shade. When established it forms a virtually monospecific community. Like *Calluna*, it does not tolerate waterlogged soils but it does less well than heather on thinner drier soils. In the past, bracken was valued by the rural community and Rymer lists its uses for compost, thatching, animal bedding, fuel, as a source of potash for glass and soap making, and even as a food or medicine. This generally kept it under control but depopulation and the change from cattle to sheep economies in our uplands may well be responsible for its very rapid spread on moorland. As Gimingham noted, in the competition between *Calluna* and *Pteridium* much will depend upon which of the two is in the more aggresive growth phase. *Calluna* will succumb to *Pteridium* when its competitive vigour is reduced. This may occur when the plant is in the degenerative phase, when heavily grazed by sheep, or after a recent burn. Once again, incorrect management may greatly influence the stability of the heather ecosystem.

REFERENCES

Allen, S. E., 1964. Chemical aspects of heather burning, *J. appl. Ecol.*, **1**, 347–67.

Ashby, K. R., 1959. 'Prevention of regeneration of woodland by field mice (*Apodemus sylvaticus* L.) and voles (*Clethrionomys glareolus* Schreber and *Microtus agrestis* L.)', *Q. Jl. For.* **53**, 148–58.

Atherden, M. A., 1976. 'The impact of late prehistoric cultures on the vegetation of the North York Moors', *Trans. Inst. Br. Geogrs.*, **1**(3), 284–300.

Barber, K. E., 1981. *Peat stratigraphy and climatic change. A palaeoecological test of the theory of cyclic peat bog regeneration*, Balkema, Rotterdam.

Barclay-Estrup, P., 1971. 'The description and interpretation of cyclical processes in a heath community. III Microclimate in relation to the *Calluna* cycle', *J. Ecol.*, **59**, 143–66.

Barkham, J. P., 1978. 'Pedunculate oak woodland in a severe environment: Black Tor Copse, Dartmoor', *J. Ecol.*, **66**, 707–40.

Birks, H. J. B. and Birks, H. H., 1980. *Quaternary Palaeoecology*, Arnold, London.

Bormann, F. H. and Likens, G. E., 1979. *Pattern and Process in a Forested Ecosystem*, Springer-Verlag, New York, Heidelberg, Berlin.

Brown, A. H. F., 1974. 'Nutrient cycles in oakwood ecosystems in N.W. England', in *The British Oak: its history and natural history* (eds Morris, M. G. and Perring, F. H.), Botanical Soc. Br. Isles, E. W. Classey, Faringdon, pp. 141–61.

Brown, A. H. F., 1981. 'The role of buried seed in coppicewoods', *Biol. Conserv.* **21**, 19–38.

Coope, G. R., 1970. 'Climatic interpretations of Late Weichselian Coleoptera from the British Isles', *Revue de Géographie Physique et Géologie Dynamique*, **12**(2), 149–55.

Coope, G. R. and Pennington, W., 1977. 'The Windermere Interstadial of the Late-Devensian', *Phil. Trans. R. Soc. Lond. B.*, **280**, 337–9.

Coulson, J. C. and Whittaker, J. B., 1978. 'Ecology of Moorland Animals', in *Production Ecology of British Moors and Montane Grasslands* (eds Heal, O. W. and Perkins, D. F.), Springer-Verlag, Berlin, Heidelberg, New York, pp. 52–93.

Cousens, J. 1974. *An introduction to Woodland Ecology*, Oliver and Boyd, Edinburgh.

Crisp, D. T. 1966. 'Input and output of minerals for an area of Pennine moorland: the importance of precipitation, drainage, peat erosion and animals', *J. appl. Ecol.*, **3**, 327–48.

Cushing, E. J., 1967. 'Late-Wisconsin pollen stratigraphy and the glacial sequence in Minnesota', in *Quaternary Palaeoecology* (eds Cushing, E. J. and Wright, H. E.), Yale University Press.

Dimbleby, G. W., 1952, 'The historical status of moorland in north-east Yorkshire', *New Phytol.*, **51**, 349–54.

Edwards, C. A. and Heath, G. W., 1963. 'The role of soil animals in breakdown of leaf material', in *Soil organisms* (eds Doeksen, J. and van der Drift, J.), North Holland, Amsterdam, pp. 76–84.

Ford, E. D., 1982. 'Catastrophe and disruption in forest ecosystems and their implications for plantation forestry', *Scottish Forestry*, **36**, 9–24.

Garbett, G. G., 1981. 'The Elm decline: the depletion of a resource', *New Phytol.*, **88**, 573–85.

Gimingham, C. H., 1972. *Ecology of Heathlands*, Chapman and Hall, London. .

Godwin, H., 1975. *The History of the British Flora* (2nd edn.), CUP, Cambridge.

Gradwell, G. R., 1974. 'The effect of defoliators on tree growth', in *The British Oak: its history and natural history* (eds Morris, M. G. and Perring, F. H.), Botanical Soc. Br. Isles, E. W. Classey, Faringdon, pp. 182–93.

Greig, J., 1982. 'Past and present lime woods of Europe', in *Archaeological Aspects of Woodland Ecology* (eds Limbrey, S. and Bell, M.) British Archaeological Reports, S146, Oxford, pp. 23–55.

Harrison, A. F., 1978. 'Phosphorus cycles of forest and upland grassland ecosystems and some effects of land management practices', in *Phosphorus in the environment: its chemistry and biochemistry*, CIBA Foundation Symposium No. 57, Elsevier, pp. 175–99.

Heal, O. W. and Smith, R. A. H., 1978. 'Introduction and site description', in *Production Ecology of British Moors and Montane Grasslands* (eds Heal, O. W. and Perkins, D. F.), Springer-Verlag, Berlin, Heidelberg, New York, pp. 3–16.

Heal, O. W., Swift, M. J. and Anderson, J. M., 1981. 'Nitrogen cycling in United Kingdom forests: the relevance of basic ecological research', in *The Nitrogen Cycle* (a discussion organised by Stewart, W. D. P. and Rosswall, T.), *Phil. Trans. R. Soc. Lond. B.*, **296**, pp. 427–44.

Holmes, G. D., Wareing, P. F. and Harley, J. L. (eds) 1975. 'A discussion on forests and forestry in Britain', *Phil. Trans. R. Soc. Lond. B.*, **271**, 45–232.

Jacobson, Jr., G. L. and Bradshaw, R. H. W., 1981. 'The selection of sites for paleovegetational studies', *Quaternary Res.*, **16**(1), 80–96.

Kelly, D. L., 1981. 'The native forest vegetation of Killarney, south-west Ireland: an ecological account', *J. Ecol.*, **69**, 437–72.

Kinnaird, J. W., 1971. 'Birch regeneration in relation to site characteristics' (pp. 19–23). 'Effects of shade on the growth of birch' (pp. 24–5), *Range Ecology Research: 1st Progress Report*, Nature Conservancy, Edinburgh.

Longman, K. A. and Coutts, M. P., 1974. 'Physiology of the oak tree' in *The British Oak: its history and natural history* (eds Morris, M. G. and Perring, F. H.), Botanical Soc. Br. Isles, E. W. Classey, Faringdon, pp. 194–221.

McVean, D. N., 1963. 'Ecology of Scots pine in the Scottish Highlands', *J. Ecol.*, **51**, 671–86.

McVean, D. N. and Ratcliffe, D. A., 1962. *Plant communities of the Scottish Highlands*, Nature Conservancy Monograph No. 1, HMSO, London.

Mellanby, K., 1968. 'The effects of some mammals and birds on regeneration of oak', *J. appl. Ecol.*, **5**, 359–66.

Miller, G. R., 1971. 'Grazing and the regeneration of shrubs and trees', *Range Ecology Research: 1st Progress Report*, Nature Conservancy, Edinburgh, pp. 27–40.

Miller, G. R. and Cummins, R. P., 1982. 'Regeneration of Scots pine *Pinus sylvestris* at a natural tree-line in the Cairngorm Mountains, Scotland', *Holarctic Ecology*, **5**, 27–34.

Miller, G. R., Kinnaird, J. W. and Cummins, R. P., 1982. 'Liability of saplings to browsing on a red deer range in the Scottish Highlands, *J. appl. Ecol.*, **19**, 941–51.

Miller, G. R. and Watson, A., 1974. 'Heather moorland: a man-made ecosystem', in *Conservation in Practice* (eds Warren, A. and Goldsmith, F. B.), Wiley, London. pp. 145–66.

Miller, G. R. and Watson, A., 1978. 'Heather productivity and its relevance to the regulation of red grouse populations', in *Production Ecology of British Moors and Montane Grasslands* (eds Heal, O. W. and Perkins, D. F.), Springer-Verlag, Berlin, Heidelberg, New York, pp. 277–85.

O'Sullivan, P. E., 1970. 'The ecological history of the forest of Abernethy, Inverness-shire', Ph.D. thesis, New University of Ulster.

Ovington, J. D., 1965. *Woodlands*, English University Press, London.

Packham, J. R. and Harding, D. J. L., 1982. *Ecology of Woodland Processes*, Arnold, London.

Pears, N. V., 1968 'The natural altitudinal limit of forest in the Scottish Grampians', *Oikos*, **19**, 71–80.

Pears, N. V., 1975. 'Radiocarbon dating of peat macrofossils in the Cairngorm Mountains, Scotland', *Trans. Bot. Soc. Edinb.*, **42**, 255–60.

Pennington, W., 1974. *The History of British Vegetation* (2nd. edn.), English University Press, London.

Perkins, D. F., 1978. 'The distribution and transfer of energy and nutrients in the *Agrostis-Festuca* grassland ecosystem', in *Production Ecology of British Moors and Montane Grasslands* (eds Heal, O. W. and Perkins, D. F.), Springer-Verlag, Berlin, Heidelberg, New York, pp. 375–95.

Peterken, G. F., 1981. *Woodland Conservation and Management*, Chapman and Hall, London and New York.

Picozzi, N., 1968. 'Grouse bags in relation to the management and geology of heather moors', *J. appl. Ecol.*, **5**, 483–8.

Poore, M E. D. and McVean, D. N., 1957. 'A new approach to Scottish mountain vegetation', *J. Ecol.*, **45**, 401–39.

Proctor, M. C. F., Spooner, G. M. and Spooner, M. F., 1980. 'Changes in Wistman's Wood, Dartmoor: photographic and other evidence', *Rep. Trans. Devon Ass. Advmt. Sci.*, **112**, 43–79.

Rackham, O., 1967. 'The history and effects of coppicing as a woodland practice', in *The biotic effects of public pressures on the environment* (ed. Duffey, E.), Monks Wood Expt. Station Symposium, Abbots Ripton, **3**, 82–93.

Rackham, O., 1974. 'The oak tree in historic times', in *The British Oak: its history and natural history* (eds Morris, M. G. and Perring, F. H.), Botanical Soc. Br. Isles, E. W. Classey, Faringdon, pp. 62–79.

Rackham, O., 1980. *Ancient woodland: its history, vegetation and uses in England*, Arnold, London.

Rafes, P. M., 1971. 'Pests and the damage which they cause to forests', in *Productivity of Forest Ecosystems* (ed. Duvigneaud, P.), UNESCO, Paris.

Ratcliffe, D. A. (ed.), 1977. *A Nature Conservation Review*, vol. I and II, Cambridge University Press, London, New York.

Rooke, D. B. (ed.), 1974. *British Forestry*, Forestry Commission, HMSO, London.

Rymer, L., 1976. 'The history and ethnobotany of bracken', *Bot. J. Linnean Soc.*, **73**, 151–76.

Shaw, M. W., 1974. 'The reproductive characteristics of oak', in *The British Oak: its history and natural history* (eds Morris, M. G. and Perring, F. H.), Botanical Soc. Br. Isles, E. W. Classey, Faringdon, pp. 162–81.

Shimwell, D. W., 1971. *The description and classification of vegetation*, Sidgwick and Jackson, London.

Simmons, I. G., 1962. 'The development of the vegetation of Dartmoor', Ph.D. thesis, University of London.

Simmons, I. G., 1965. 'The Dartmoor oak copses: observations and speculations', *Field Studies*, **2**, 225–35.

Steele, R. C., 1974. 'Variation in oakwoods in Britain' in *The British Oak: its history and natural history* (eds Morris, M. G. and Perring, F. H.), Botanical Soc. Br. Isles, E. W. Classey, Faringdon, pp. 130–40.

Steven, H. M. and Carlisle, A., 1959. *The Native Pinewoods of Scotland*, Oliver and Boyd, Edinburgh.

Tansley, A. G., 1949. *The British Islands and their Vegetation*, Vols. 1 and 2, CUP, Cambridge.

Tanton, M. T., 1965. 'Acorn destruction potential of small mammals and birds in British woodlands', *Q. Jl. For.*, **59**, 230–4.

Tauber, H., 1967. 'Investigations on the mode of pollen transfer in forested areas', *Rev. Palaeobotan. Palynol.*, **3**, 277–86.

Thomas, B. and Dougall, H. W., 1947. 'Yield of edible material from common heather', *Scott. Agric.*, **27**, 35–8.

Tinsley, H., 1976. 'Cultural influences on Pennine vegetation with particular reference to North Yorkshire', *Trans. Inst. Br. Geogrs.*, **1**(3), 310–22.

Tittensor, R. M., 1970. 'History of the Loch Lomond oakwoods', *Scott. For.*, **24**, 100–18.

Turner, J., 1962. 'The *Tilia* decline: an anthropogenic interpretation', *New Phytol.*, **61**, 328–41.

Turner, J. S. and Watt, A. S., 1939. 'The oakwoods (*Quercetum sessiliflorae*) of Killarney, Ireland', *J. Ecol.*, **27**, 202–33.

Turner, C. and West, R. G., 1968. 'The subdivision and zonation of interglacial periods', *Eiszeitalter Genenw.*, **19**, 93–101.

Varley, G. C., and Gradwell, G. R., 1965. 'Interaction between oaks and winter moth', paper delivered at Winter Meeting, British Ecological Society, Oxford University, reported in *J. appl. Ecol.*, **2**, 441.

Warren, A. and Goldsmith, F. B., 1974. *Conservation in Practice*, Wiley, London.

Watt, A. S. and Jones, E. W., 1948. 'The ecology of the Cairngorms. I. The environment and the altitudinal zonation of the vegetation', *J. Ecol.*, **36**, 283–304.

West, R. G., 1968. *Pleistocene Geology and Biology*, Longman, London.

Whitmore, T. C., 1982. 'On pattern and process in Forests', in *The Plant Community as a Working Mechanism* (ed. Newman, E. I.), Special publication No. 1 of The British Ecological Society, Blackwell, Oxford, London, Boston, pp. 45–59.

Yapp, W. B., 1974. 'Birds of the north-west Highland birchwoods', *Scott. Birds*, **8**, 16–31.

9

THE SOILS

INTRODUCTION

Virtually all the truly natural vegetation of the British Isles has been replaced by a plant cover which, to a greater or lesser extent, reflects the presence of man. In much the same way, many British soils have been modified by the activities of man. Nevertheless, centuries of agriculture, whether arable or pastoral, have not completely blurred the original soil characteristics of each region. Certain fundamental differences still persist in our soils to permit, on the one hand, a broad division into main types for the British Isles and, on the other hand, to allow considerable subdivision at the local (or large-scale) level of mapping (see, for example, Fig. 9.7).

The interplay of climate, geology, biota, topography and time produces a mosaic of soil types. This interplay is expressed in terms of differences in soil depth, texture, structure, horizon development, associated humus, propensity to gleying, etc. These criteria are used for classifying soils into 'types', distinguished mainly on a profile study. But Fitzpatrick reminds us that the profile is simply a two-dimensional section through soil which really exists as a *three-dimensional continuum*. Soil horizons intergrade with each other through lateral and vertical changes to give a continuous spatial variation. The erection of a soil classification based on division of this continuum into 'types' is a descriptive convenience, the arbitrary nature of which must be recognized. However, strong variations in soils do occur in the British Isles and there is close agreement on the main types present. These pronounced changes reflect the marked geological variations across the country (cf. the soft, young limestones of the south-east and the resistant, ancient igneous and metamorphic masses of the Scottish Highlands. Catt also reminds us that over two-thirds of England has its solid geology mantled by Quaternary deposits which provide very variable parent soil materials), the contrasting topography of lowland and upland Britain and the climatic differences which range from a wet, exposed maritime western fringe to a drier, more continental eastern margin. This chapter can only attempt a summary of British soils and the reader should consult Curtis *et al.* for further descriptive examples, mainly derived from the work of the Soil Surveys of Great Britain and Ireland.

MAIN SOIL TYPES

The classification of soils into major groups presented below largely follows the scheme used in reports of the Soil Survey of Great Britain, such as that produced by Mackney and Burnham. The notation for the soil horizons is that given in Chapter 3 (see page 38). In describing humus types or organic horizons, three terms are frequently used: mull, mor and moder. Each refers to organic matter which has reached a particular state of decomposition, giving the humus distinctive *physical* properties.

MULL.   Mull is a humus which frequently develops under deciduous woodland. The pH value is usually above 4·5, the soils often being neutral or slightly alkaline. With a mull humus, plant debris decomposes rapidly and the litter layer of undecomposed, recognizable plant remains is thin. Organic material becomes quickly incorporated with the mineral fraction of the A_1 horizon, forming relatively stable clay-humus complexes. Earthworms play a major role in this rapid decomposition and Kubiena reports evidence for the formation of the clay-humus complexes in the intestines of earthworms. The humus particles are very small and widely dispersed throughout the top soil which tends to have a loose, crumb-like structure.

MOR.   In contrast, mor (or raw humus) is characterized by much less mixing of the organic and mineral horizons. A much slower rate of decomposition is suggested by the presence of three distinguishable organic layers: the litter layer of largely unaltered plant debris; the fermentation layer, where some breakdown has occurred but many plant structures are still visible; and the blackish humified layer where decomposition is more marked and much of the material consists of the droppings of small soil animals. The litter and fermentation layers are always well represented but the humified layer is sometimes less so. Mor is usually associated with cooler, wet climates or nutrient-deficient, acidic parent materials or vegetation rich in crude fibre. It is best developed where all three occur together, as in parts of upland Britain. Mor soils always have strong acidic reactions and much less earthworm activity. Plants rich in crude fibre are referred to by Kubiena as raw humus plants and examples are *Calluna*, *Erica* spp, *Empetrum nigrum* and most of the conifer trees, especially pines and spruces. Grieve demonstrates that development of a discrete fibrous mor can occur rapidly even in lowland Britain. It is well seen in the Forest of Dean where 50-year-old spruce plantations now exist on sites previously carrying mixed deciduous woodland.

MODER.   Moder is transitional between mor and mull. It has a well-defined humified layer, better developed and thicker than the combined litter and fermentation layers of a mor profile. It may appear like mull superficially, but the humus and mineral fractions are separable and do not form the closely bound clay-humus complexes of a mull. Variations in humus properties may be virtually continuous and mull, moder and mor provide convenient reference points in this spectrum.

RAW SOILS AND RANKERS

Raw soils are soils in the very earliest stages of formation, showing poor differentiation into horizons. At best, a thin upper horizon, with little or no humus, directly overlies relatively unaltered rock fragments. They dry out rapidly, having little ability to retain water. Those on solid rock or scree deposits are *lithosols*, while those on sand dunes or other small, loose fragments are known as *regosols*. They may be linked with the early stages in development of the lithosere or psammosere and often carry an incomplete vegetation cover (for typical sites see Figs. 4.1, 4.2 and 4.14). Today, these soils are largely confined to steep upland slopes, frequently undergoing active erosion, or coastal dune systems and hence have a rather limited distribution. In the Late-glacial period, however, such raw soils were very common. These freshly weathered surfaces would have been initially base-rich in many cases. Only after the development of a fuller vegetation cover and a period of leaching would pH values have begun to fall.

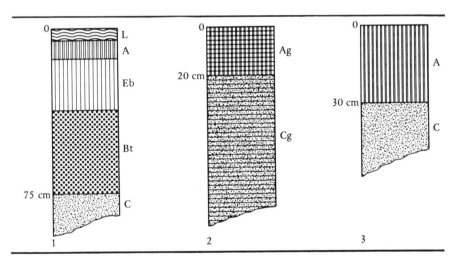

Fig. 9.1 Diagrammatic soil profiles: 1. leached brown soil, a sub-group of the brown earths; 2. ground-water gley soil; and 3. rendzina. The thickness of horizons are only approximate. Gley morphology (g) is shown by horizontal lines superimposed on other patterns. The horizon nomenclature is that used by the Soil Survey of England and Wales, see p. 38 (Modified from Burnham and Mackney, 1964).

Fitzpatrick regards *ranker soils* as the second stage in soil evolution. They are characterized by a fuller development of the upper horizon, which is usually humus-rich, lying directly on a more or less unaltered C horizon of parent material. Such soils are well seen on the higher Scottish mountains at levels above about 800 m. which have never been forested. Romans, Stevens and Robertson describe them as *alpine humus soils* (Fig. 9.2 and Fig. 3.9). They are usually deficient in nutrients and strongly acidic. At some sites the poorly decomposed humus layer can be rolled back like a carpet. The surface

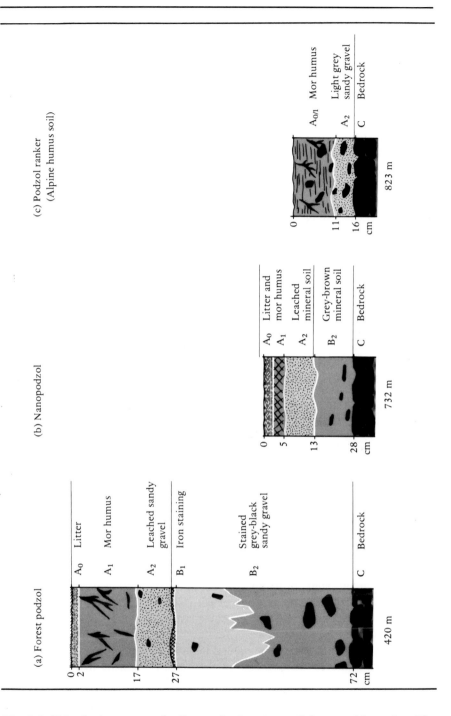

Fig. 9.2 Altitudinal sequence of soil types in the western Cairngorm Mountains. The forest podzol, the typical soil within the main forest, is replaced by a nanopodzol (nano = dwarf) on mid slopes above the forest edge. On higher slopes, a thin podzol ranker develops, largely consisting of raw humus directly overlying the parent material.

vegetation consists mainly of dwarf moorland shrubs, lichens and mosses. Raw soils and rankers have no agricultural value apart from limited grazing as the vegetation cover develops.

CALCAREOUS SOILS

As the name implies, these soils show the strong influence of parent materials rich in calcium carbonate or lime. Calcium carbonate is present in all horizons or at least those below the upper A horizon, which may be slightly acid in reaction. In many areas in southern Britain characterized by a scarp-and-vale topography (e.g. the Cotswolds, the Chilterns and the North and South Downs) the steep slopes of the limestone scarps have developed *rendzina soils*. These are always very shallow and consist of a dark brown or black A horizon resting directly on a fragmented surface of limestone. The A horizon is medium to fine-textured and has a granular or fine subangular blocky structure. Water drains away rapidly, leading to drought conditions during dry spells. In their earliest stages of formation, rendzinas are not unlike raw soils since they are characterized by shallowness, stoniness and a thin humus layer. On very steep slopes this skeletal nature can be maintained for many years. Surface material tends to be lost downslope and fresh limestone fragments become incorporated from below into the surface zone. As they develop, they become similar to rankers in profile, i.e. an A horizon directly over a C horizon. But the strong chemical influence of the parent material determines many of the properties of this soil (Figs. 9.1 and 9.3). The humus is usually a black calcareous mull, neutral or alkaline in reaction. It often contains numerous small fragments of limestone.

Under similar geology but on much less steep slopes, a *brown calcareous soil* may develop. This type may also be found on boulder clays rich in chalk fragments. These soils are thicker and show better horizon differentiation. Their greater stability has led to more leaching from the topsoil and the A horizon has a slightly acid or neutral or calcareous mull humus. The brownish B horizon, of coarser blocky peds, is calcareous and rests directly on the weathering limestone. These soils may show gleying features in the lower profile where drainage is not good. Whereas rendzinas have little agricultural use, these thicker calcareous soils, providing the slopes are not too steep, are widely used in farming. *Gleyed calcareous soils*, usually derived from calcareous shale or clay, have impeded drainage. The coarse-structured, deep profile may show zones of secondary carbonate deposition and frequent mottling.

There is some debate as to the naturalness of rendzina soils on the chalklands of England. Limbrey has argued that these areas originally had brown forest soils with high loess (wind-blown silt) content. Forest clearance, from the Neolithic onwards, led to the process of lessivage (mechanical eluviation of clay and iron oxides from the A horizon) and to soil erosion. Upper soil horizons were lost on steep slopes (accumulating as colluvium in the valley bottoms) and the rendzina profile resulted. However, Barker and Webley present evidence pointing to the rendzina as the natural soil type on these slopes.

Fig. 9.3 A rendzina profile developed on the scarp slope of the Chilterns. The humus enriched A horizon lies directly above the weathered C horizon, which is a mixture of chalk fragments and clay-with-flints. The surface vegetation is chalk grassland with beechwood in the background. The lens cover provides a scale.

BROWN EARTHS

There is a very widespread occurrence of *brown earths* in the British Isles. They represent the main but not the only soil type associated with the former climax deciduous woodlands of lowland Britain. Brown earths are slightly to strongly acid soils with a mull or moder humus: surface pH values of 5–6·5 increase with depth to about neutrality. The profile does not show distinct mineral horizons with sharp boundaries and there is no tendency for sub-surface layers to form which show signs of bleaching or accumulation of humus or iron oxides. Indeed, Fitzpatrick notes that one of the main differentiating criteria for brown earths is that the free Fe_2O_3 is uniformly distributed throughout the upper and middle horizons. The soils are usually of medium texture and the clay content is highest near the surface, only gradually decreasing with depth. This allows some profile differentiation on a textural basis. Kubiena describes them as being formed by deep-reaching chemical weathering under moist conditions and good aeration. They are not confined to any particular geology but hard siliceous rocks are unfavourable for their formation. Mackney and Burnham describe four main subdivisions of the brown earths in their study of West Midlands soils:

1. *Leached brown soils* with acid mull or moder humus and a distinctive Bt horizon rich in clay eluviated from the Eb horizon above (Fig. 9.1). These

soils are usually of medium to fine texture and the Bt horizon may not be highly permeable. This can cause some gleying.

2. *Acid brown soils* with acid mull or moder humus and a B horizon. Moderate to strong acidic reactions occur throughout the profile. They tend to be sandy or silty in texture with a low clay content. These soils were formerly known as 'brown earths of low base status'.

3. *Ferritic brown earths* with acid mull humus, a B horizon with a loamy texture and a well-developed subangular blocky structure. The profile has a moderately acidic reaction and a characteristic feature is the bright red-brown colouration due to high iron content. In examples from the West Midlands,the C horizon is a decalcified ferruginous limestone which often contains over 50 per cent Fe_2O_3

4. *Brown warp soils* which are deep soils formed on alluvium or reclaimed estuarine silts. They are slightly or moderately acidic and usually have a high base status. Although they may be flooded occasionally, they are not poorly drained.

PODZOLS

The profile described in Chapter 3 (page 36) was essentially that of a Podzol. These soils have clearly differentiated horizons due to the process known as *podzolization*. This leads to heavy leaching of iron, aluminium and humus compounds to lower profile horizons. Organic material accumulates at the surface to give a mor humus divisible into litter, fermentation and humified layers. The black humus zone is fairly sharply divided from a dark grey zone where the organic material becomes mixed with mineral fragments. Beneath this comes the pale, bleached zone which has had most of its colouring substances, the iron compounds mainly, removed by percolating water and organic acids, leaving behind a concentration of the more chemically resistant quartz and felspars. Removed humus, iron and aluminium compounds are deposited in the lower horizons, just above the bedrock, mainly as coatings on the mineral particles. The clay content in the upper profile is low but shows some increase with depth due to redeposition at lower levels. Similarly, humus products may increase slightly the organic content of the B horizons or zones of illuviation. The surface mor is strongly acid, pH 4 or less, but the lower layers show an increase in value towards about pH 5·5. The process of podzolization is chemically very complex and Anderson *et al.* have recently carried out a detailed reassessment of podzol formation.

Podzols will readily form under cool, wet climates, especially where the vegetation produces an acid litter. Podzols, of one form or another, are typically associated with the Northern Coniferous Forest formation and much of our derived upland moors. They will even form on intermediate or basic parent materials though, of course, acidic rocks are more readily podzolized. Even in areas of relatively low rainfall, an easily leached coarse-textured parent material will frequently develop a good podzol profile. Good examples are found in Norfolk on the sandy surface materials of the Breckland.

A typical *forest podzol* examined in the pine and birch wood flanking the Cairngorms is shown in Fig. 9.2a. These mountains have relatively low temperatures, high rainfalls and a short growing season. These conditions intensify with altitude, favouring heavy leaching and the accumulation of raw humus. Further, the parent soil materials are derived from acidic rocks whose content of available plant nutrients is low. The forest podzol described below carried a surface vegetation of open pinewood with a thick ground layer of *Calluna vulgaris* and *Vaccinium myrtillus* on a slope of 5 degrees at 420 m.

0–2 cm:	fresh and slightly decomposed litter.
2–17 cm:	dark brown to black peaty humus (mor) with numerous plant roots of the surface vegetation.
17–27 cm:	light brown to grey leached sandy gravel with moderate iron staining, concentrated at about 25–27 cm (B_1 horizon).
27–72 cm:	dark grey to black sandy gravel with iron staining. Rock fragments present, increasing in size towards base.
72 cm:	parent material; drift, mainly of Moine Schist boulders.

Nearby, on a gently sloping site at 732 m, a *nanopodzol* (dwarf podzol) could be demonstrated (Fig. 9.2b). The surface vegetation was wind-clipped *Calluna vulgaris*, *Rhacomitrium lanuginosum* (reindeer moss) and *Trichophorum caespitosum* (deer grass).

0–5 cm:	dark brown fibrous raw humus, many plant structures visible.
5–13 cm:	ash grey mineral layer.
13–23 cm:	grey to medium brown sandy mineral layer with quartz crystals and mica flecks prominent. Some iron staining.
23 cm:	parent material of grey Moine Schist boulders.

At higher levels, 800 m and above, the vegetation cover is often less complete and at the *podzol ranker* site (Fig. 9.2c) wind-clipped *Calluna* mounds only 4 cm high and thin spreads of *Rhacomitrium lanuginosum* alternate with patches of loose mineral fragments, each set among collections of large granite boulders on the surface. These thin soils really consist of the upper part of a podzol, the A horizons, largely of raw humus, lying directly on the granite parent material.

Where the podzolization process is advanced, we find *humus-iron podzols* and *iron podzols* with strongly developed Bh and/or Bfe horizons. When not so pronounced, the soils are more properly described as *podzolized acid brown soils* and they usually have a moder humus. Mackney and Burnham note that in this case the Ea horizon still contains some iron content and is often brown or yellow in colour. Very similar soils have been described by Ball. He has drawn attention to a group of freely drained soils, transitional in their morphology and distribution between lowland brown earths and mountain peaty podzols. They occur widely throughout the British Isles, especially on steep stopes, and are suitable for forestry utilization. Although they may approach the acid brown soils (the brown earths of low-base status) in character, they nevertheless have distinctive differences in their morphology and chemistry, such as lower values for pH, percentage base saturation and exchangeable Ca; increases in surface organic matter and free iron oxide in the B horizon; a moder rather than a mull humus. Unlike the brown earths, they may readily develop an A_0, A_2, B, C profile. For these reasons, Ball considers them to be a sub-group of the podzols rather than the brown earths and refers to them as *brown podzolic soils* (Table 9.1).

Table 9.1 Mean values for some chemical properties of North Caernarvonshire soils (From Ball, 1966)

Chemical property	Brown earths (7 profiles)		Brown podzolic soils (9 profiles)		Peaty podzol (5 profiles)	
	A horizon	B horizon	A horizon	B horizon	A_2 horizon	B horizon
pH	5·8	5·8	4·8	5·1	4·2	5·0
% loss on ignition	11	5	26	7	13	13
% base saturation	40	–	10	–	6	–

Another variation is found where water accumulates in the lower profile of a podzol. This gives rise to the soil type known as a *gley podzol*. If drainage near or at the surface is impeded, then *peaty gleyed podzols* may result.

It was previously stated that oakwoods were widely associated with brown earths, although it was noted that this was not always invariably so. Mackney has studied an old oakwood at Sutton Park (Warwickshire) where the soils are podzols and podzol intergrades (a type showing a blend of podzolic and brown earth features but with closer affinity to the podzol). Decomposition does not keep pace with leaf-fall under the dense canopy shade and a raw humus forms on the moist surface. The soil profiles show the typical horizon differentiation of a true podzol. A pollen analysis of the organic and mineral layers here shows the vegetation has always been largely dominated by oaks, the influence of other species being minor. This, and other studies, have now established that podzolization can occur in oakwoods when conditions favour the accumulation of thick, moist, strongly acidic organic layers. Thus, while certain types of soil and vegetation may readily be linked, several important exceptions may arise to these general rules.

GLEY SOILS

Many brown earths or podzols may show a limited development of gleying which is given weight when such soils are mapped. Thus Mackney and

Burnham further divide the brown earths by recognizing 'wet analogues' for most of the sub-groups, e.g. leached brown soils with gleying, acid brown soils with gleying and brown warp soils with gleying. However, they also recognize a distinct category of gley soils whose morphological characteristics largely result from pronounced waterlogging and reduction and re-oxidation of the iron compounds. They show the strong influence of anaerobic conditions and can be divided into two sub-groups.

SURFACE-WATER GLEYS.   These soils have an acid mull or moder. A greyish, eluvial, coarse-textured Ebg horizon allows water to pass through fairly easily but it is held up by a clay-rich, plastic Btg horizon of prismatic peds present at about 15–30 cm below the surface. This causes waterlogging of the upper horizons and ground surface in wet seasons. Characteristic rusty mottlings are seen in the upper zones and the sub-soil layers show grey or orange blotches. These soils are also known as *stagnogleys*.

GROUND-WATER GLEY SOILS.   The humus type is an acid mull. In some cases, the profile may show little development, the humus lying directly on the C horizon. Elsewhere, Ag and Bg horizons may be present and texture and structure can be quite variable. But the feature of overriding importance is the permanent waterlogging of the lower profile. These soils develop in basins, estuaries and on flood plains where water readily accumulates (Fig. 9.1). Fitzpatrick refers to the surface-water gleys as *supragleysols* and the ground-water gley soils as *subgleysols*.

ORGANIC SOILS

The Soil Survey defines organic soils as those with a surface layer more than 38 cm (15 in) thick containing at least 30 per cent organic matter. Soils which just fall short of this criterion are described by the use of the prefix 'peaty', e.g. peaty gleyed podzol. Quite obviously, an organic soil will only form where the accumulation of plant remains greatly exceeds the decomposition rate. A high water table together with a cold wet climate will impede the thorough breakdown of plant debris. Under these conditions, the organic accumulation may reach several metres in depth. These *peats* are sub-divided into three main types:

FEN PEATS.   These form in hollows where basic (lime-rich) or only slightly acidic waters draining from surrounding limestones collect. The peat is usually a black, well-humified, structureless mass derived from a relatively rich floristic community. Small examples of fen peats exist throughout the British Isles but extensive spreads occur in low-lying areas of eastern England, such as the Fens. Most of these spreads have now been reclaimed and form excellent arable land.

RAISED BOG PEATS.   In basin and valley sites receiving drainage water from acidic rocks, brownish peat composed of poorly decomposed plant remains may develop. The central part of the bog grows somewhat more rapidly than do the edges and this gives the bog a slightly convex surface. The surface vegetation does not have the floristic luxuriance of a fen peat and is often dominated by a few species. An actively growing peat surface is characterized by the hollow-and-hummock complex described in Chapter 8. The peat is very acidic with pH values about 4·0. Under the anaerobic conditions of standing water, however, the pH value of the peat surface may reach about 5. But several workers have demonstrated that oxidation of the peat leads to higher acidities, pH values of 3–4 being recorded. This will occur where the peat stands above the water table (on the hummocks) or has dried out (around the margins of the bog) or has undergone erosion. Like fen peats, these sites may be drained and reclaimed for agriculture purposes but they will generally be far less productive. In the wetter uplands they have been largely left in a natural state, but are sometimes used for grazing.

BLANKET (HILL) PEATS.   As the name suggests, these peats lie like a blanket, masking flat surfaces and minor irregularities across the landscape. They owe their origin to the direct effects of atmospheric moisture rather than groundwater influence and consequently occur extensively in our northern and western uplands and may even be found near sea-level in Ireland and western Scotland. Depths of 6 m or more are known but the peat thins out to a few centimetres against steeper slopes. The floristic diversity is even less than that of the raised bog and pH values between 3 and 4 are common. Whereas the raised bog can grow rapidly and may receive inwashed mineral particles from the surrounding slopes, the blanket peat grows slowly and usually consists of well-decomposed plant remains free of mineral matter. In some areas, such as Wales, hill peats have formed over a previous soil and the reasons for this are discussed later in this chapter. Blanket peats have limited agricultural value though they do form a considerable part of many upland grazings. When drained, they are frequently used for coniferous plantations, providing the site is not too exposed. Many areas of peat today are not actively growing and have probably developed under a previously wetter climate.

OTHER CLASSIFICATION SCHEMES.   There are several further ways of classifying peats. These arise because each organic increment at the site may have experienced a somewhat different history of accumulation and decomposition although perhaps composed of the same set of contributing plant species. We may thus look at peats in terms of their state of decomposition (or humification). Long ago, von Post proposed a 1 to 10 scale of humification for peats where, for example:

H1   = completely undecomposed peat, free of amorphous material. Clear
          colourless water exuded.
H6   = fairly decomposed, fair amount of amorphous material. Indistinct

plant structures. About one-third of the peat passes through the fingers. Residue strongly pasty but showing plant structures more distinctly than before.

H10 = completely decomposed peat without discernible plant structures. All peat passes between the fingers on squeezing as a 'slime-like' material.

Peats may also be viewed in terms of the water table position during their formation. This gives us a division into three basic types:

(a) Limnic – formed in zone entirely below the water table.
(b) Terrestrial – formed in zone entirely above the water table.
(c) Telmatic – formed in zone of fluctuating water table.

Equally, peats may be classified into three types according to their base and mineral contents:

(a) Eutrophic – relatively rich content (e.g. many fen peats).
(b) Oligotrophic – relatively poor content (e.g. many hill peats).
(c) Mesotrophic – midway between (a) and (b) in properties.

PLAGGEN SOILS.   The organic soils discussed above are of natural occurrence. However, the *plaggen soil* is essentially man-made and its characterized by a thick organic A horizon. They usually develop in areas where the original soil was infertile (e.g. very sandy). Over the centuries, man has added to the profile considerable amounts of farm manure together with either forest litter, heather or grass sods. This long-continued manuring leads to a thick man-made surface and plaggen soils are reasonably common in north-west Europe. Conry describes coastal examples from Ireland which involve additions of dung and calcareous seasand initially applied in pre-Christian times and well authenticated during the Medieval period.

THE REGIONAL PATTERN

Burnham has drawn attention to the close correlation between the broad climatic divisions within the British Isles and the regional distribution of major soil groups. A review of his findings will serve as a summary of the preceding sections of this chapter and help to place them in a geographical context. While not denying the strong influence of parent materials and topography on soils, Burnham points out that significant climatic gradients exist across the British Isles. A difference in mean annual temperature of 10 °C occurs between the higher Scottish mountains and the lowlands in southern England. This leads to greater chemical and biological activity within soils of the south. He cites the differences to be found in the peat bogs of Caithness and the mineral soils of south-west England where both have developed under similar landscapes and rainfall regimes. Another important climatic variable within this country is precipitation, ranging from 500–4,000 mm (20–160 in) p.a. Evapotranspiration is usually higher in low rainfall areas and this causes drainage through the soil to reach a total of less than 100 mm (4 in) around the Thames Estuary during the short winter period. In contrast, in parts of north-west Scotland drainage amounts to about 3,000 mm (120 in) spread throughout the whole year. Burnham declares this variation to be the most important factor explaining British soil variations on a regional scale.

Consideration of these gradients allows him to divide Great Britain into five main climatic regions, each with a characteristic soil or soil group (Table 9.2). Obviously, the soil changes are gradual (remembering that the soil is essentially a continuum) and local factors may lead to considerable variations within any given areas. Nevertheless, it is possible to map these regions (Fig. 9.4) and soil types (Fig. 9.5) and a broad but useful correlation emerges when the two maps are compared.

Table 9.2 The climatic regions of the British Isles and their characteristic soil types (Modified from Burnham, 1970)

Climatic region	Mean annual temperature	Rainfall	Characteristic soils
1. Warm, dry	Over 8·3 °C	Under	Leached brown soil
2. Cold, dry	4·0–8·3 °C	1,000 mm	Semipodzol/Podzol
3. Warm, wet	Over 8·3 °C	Over	Acid brown soil
4. Cold, wet	4·0–8·3 °C	1,000 mm	Peaty gley podzol/Blanket peat
5. Very cold, wet	Under 4 °C		Alpine humus soil

An important factor producing noticeable soil differences within small areas is topography. Taylor has described this influence on British soils by considering a theoretical transect from an exposed high plateau above 300 m with rainfall above 1,500 mm (60 in) p.a. through upper, middle and lower valley-side slopes to a valley bottom at about 76 m with about 750 m (30 in) rainfall p.a. The drainage characteristics and water table levels of this slope sequence give rise to a hydrologically based pattern of soils. These range from hill peats and peaty gley podzols (on the gently sloping high plateau sites) to podzols and shallow brown earths (on the steep upper slopes) to deeper brown earths and seasonal gley soils (on the middle and lower slopes) and to permanent gley soils and saturated basin peats (along the valley bottoms). This approach represents an extension of the catena concept (Chapter 3) to the regional level. The emphasis on topographical relationships again makes it an appropriate theme for the geographer to develop. Gerrard's recent book, although not concentrating on the British Isles, can be seen as a further step towards integrating soils and landforms along the lines of this theme.

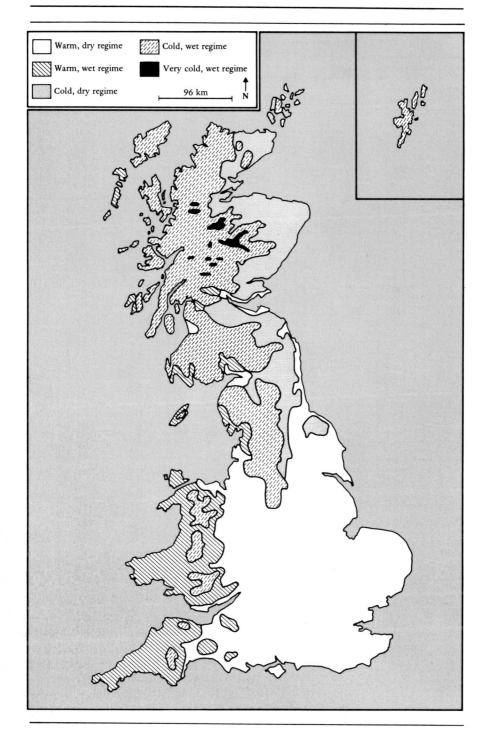

Fig. 9.4 Climatic Regions of Great Britain. (From Burnham, 1970.)

Fig. 9.5 Soil Regions of Great Britain. (From Burnham, 1970.)

SOIL SURVEY AND MAPPING

The most comprehensive accounts of British soils are contained in the *Regional Memoirs* produced by the Soil Survey of England and Wales. They describe soils and their properties from the viewpoint of the pedologist and the agriculturist. Because of this, each Memoir now contains two maps; one showing the *soils* and the other the *land-use capability*. The latter displays the soil's suitability or limitations for various forms of land use. The basic unit of classification and mapping used by the Soil Survey is the *soil series*, a group of soils with similar profiles developed from lithologically similar parent materials. Each soil series is named after the area or site where it was first described or is commonly found. These names do not apply just to the region where first used but are retained for any appropriate soil profile wherever it may occur. For example, the *Banbury Series* was first recognized in north Oxfordshire but the same soils occur extensively in northern Leicestershire and are mapped as such. Each series is determined from an examination of the intrinsic soil profile properties, emphasis being placed on the more permanent soil characteristics such as soil texture, structure, organic matter content, moisture regime and degree of horizon differentiation.

After detailed examination of the field profile, samples from each horizon are analyzed in the laboratory to supplement field observations. *Soil phases* are subdivisions of a soil series and are thought to be of agricultural significance. They are based on features like stoniness or depth and are too detailed to appear on the final map. The term *soil complex* is used for an area where the boundaries between two or more soil series are indefinite and the delimitation of individual series becomes impossible at the mapping scale used for publication (in the past usually at a scale of 1: 63,360, though in England and Wales areas of particular interest are now mapped at 1: 25,000). Beckett and Webster have tested the reliability of the soil series as mapped and have shown that the percentage of the area actually belonging to the soil series depicted on the map may be as low as 50 per cent. However, it is usually much higher (up to 85 per cent) and Courtney's study of soil variability in mapping units has generally confirmed the correctness of the surveyor's classification.

Although a group of soils may be very different in their profiles, nevertheless, they may occur in a characteristic pattern in the landscape, the soil types being linked through the topographical relationships of distinctive landscape units. The term *soil association* is used to describe this pattern. It is a small-scale mapping unit and an example in the form of a block diagram is presented as Fig. 9.6. This association is related to the development of a scarp-and-vale landscape unit based on alternating hard and soft geological outcrops. At least eight different soil series are recognized in the association, reflecting the differences between sites in terms of relief and parent materials.

The detail possible when an area is mapped in terms of soil series and complexes is illustrated in Fig. 9.7. This shows an extract from a soil survey map of the Melton Mowbray district by Thomasson. In all, ten soil series and complexes are found in this small area and these are classified into Soil Groups

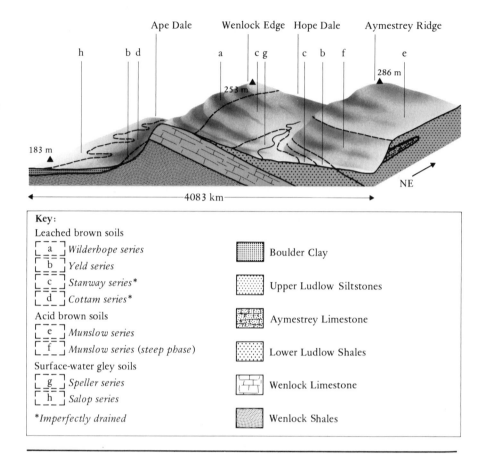

Fig. 9.6 The Wenlock Edge Soil Association: a group of soils related to the development of a scarp-and-vale landscape unit, reflecting differences in relief and/or parent material between each site. (Modified from Mackney and Burnham, 1964.)

(Table 9.3). This simple landscape (Fig 9.8) consists of a dissected boulder clay plateau, rising to the north to nearly 150 m and sloping south-eastwards towards the Wreake Valley at 58 m. A series of small tributaries of the River Wreake have cut back into the edge of this plateau to give a sequence of valley-bottom, valley-side and spur-top sites. The most widespread soil is the *Ragdale Series*, which is the typical soil on Chalky Boulder Clay throughout much of the East Midlands (Fig. 9.9). A dense grey clay or silty clay, containing many erratics, forms the parent material on the plateau. Between 15 and 30 per cent calcium carbonate may be present in the fine-earth fraction but these soils are poorly drained and often waterlogged in winter. They are classified as surface-water gleys and are almost impermeable below 60 cm. Excess rainfall is usually shed laterally through natural fissures between peds in the upper horizons.

The Hanslope Series, also on Chalky Boulder Clay, forms a discontinuous

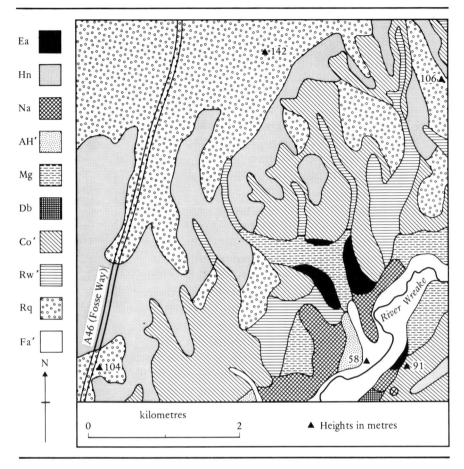

Ea

Hn

Na

AH'

Mg

Db

Co'

Rw'

Rq

Fa'

N

A46 (Fosse Way)

River Wreake

▲142

106▲

▲104

58▲

▲91

kilometres

0   2   ▲ Heights in metres

Fig. 9.7 Soil-mapping at the scale of 1:63,360: an extract from Sheet 142 (Melton Mowbray), Soil Survey of England and Wales (Thomasson, 1971). See Table 9.3 for a description of the soil types. Fig. 9.8, a general view of this area, was taken looking north-westwards from the point marked ⊗.

periphery to the Ragdale soils, occurring mainly on valley sides or slopes. Though similar to the Ragdale soils in clay content, the calcium carbonate percentage is higher and the drainage is better. *The Cottam Complex* occurs on sandy reddish boulder clay derived from Triassic rocks. This heterogeneous clay outcrops on valley sides from beneath the Chalky Boulder Clay of the plateau. The profile shows a loamy A horizon and a strong increase in clay content from A to B horizons. The calcium carbonate content increases with depth so that the C horizon is often calcareous.

These three soil series make up more than 60 per cent of the area depicted in Fig. 9.7. The other seven soil series on this map are of limited extent and they are described in Table 9.3. The properties of some are strongly associated with particular parent materials (e.g. the *Evesham Series* on Lower Lias or Rhaetic clay-calcareous shales where the influence of drift material is slight).

Table 9.3 The mapping units and soil types for the area shown in Fig. 9.7 (From the Soil Survey of England and Wales: Melton Mowbray, sheet 142)

Soil mapping units (soil series and complexes)	Texture and geology	Soil group
Ea – Evesham Series	Clayey; over clay-shale (Lower Lias and Rhaetic)	Gleyed calcareous soils
Hn – Hanslope Series	Clayey; Chalky Boulder Clay	
Na – Newport Series	Coarse loamy; fluvio-glacial deposits	Brown earth
AH[1] – Astley Hall Complex	Coarse loamy; fluvio-glacial deposits over clay	Gleyed brown earth
Mg – Melton Series	Coarse loamy; mixed head and fluvio-glacial deposits over Jurassic clay	
Db – Dunnington Heath Series	Coarse loamy over clayey; head over reddish marl (Keuper)	
Co[1] – Cottam Complex	Loamy over clayey; reddish boulder clay	
Rw[1] – Rowsham Complex	Fine loamy over clayey; head over Jurassic clay	Surface-water gley soils
Rq – Ragdale Series	Clayey; Chalky Boulder Clay	
Fa[1] – Fladbury Complex	Clayey; riverine alluvium	Ground-water gley soils

Fig. 9.8 A view north-westwards from the point marked ⊗ in Fig. 9.7, looking towards the Chalky Boulder Clay plateau. The Wreake Valley forms the foreground. This simple landscape, with no striking topographical or pronounced geological boundaries, can be divided into ten different soil types (see text).

Fig. 9.9 A Ragdale Series soil profile, near Ragdale Wood, Melton Mowbray. The parent material is Chalky Boulder Clay. The B and C horizons contain more than 35 per cent clay and the A horizon has more than 30 per cent clay. The C horizon shows limestone and chalk-with-flints fragments. The thin A horizon has a fine subangular blocky structure while the mottled B horizon has coarse angular blocky to prismatic peds. Low permeability throughout makes this a surface-water gley soil. The surface vegetation is mixed oakwood. The white marker indicates the approximate junction of the A and B horizons.

With others, their characteristics largely reflect the low-lying nature of the site (e.g. the poorly drained *Fladbury Complex* of gleyed brown warp soils and groundwater gleys developed on the alluvium of the River Wreake flood-plain).

A landscape like that depicted in Fig. 9.8 on first examination may appear to hold little interest for the biogeographer. There are no striking topographical changes and the geological boundaries are not obvious on the ground. Nevertheless, it is possible to map many soil series in these simple landscapes by paying careful attention to a few basic soil features such as texture, calcium carbonate content and drainage.

LATE-GLACIAL AND FLANDRIAN CHANGES

It is obvious that agriculture has modified many British soils, particularly in their upper horizons. However, in some regions, other changes of a less obvious nature have to be taken into account for a full understanding of the evolution of soil patterns. Pearsall pointed out long ago that the newly formed soils of the Late-glacial period would often be basic in reaction, being young soils with a predominance of freshly weathered material and only slightly leached. However, these soils would soon undergo progressive leaching. In our cold, wet uplands soil acidification would set in and poorly decomposed humus might accumulate on the surface. It was thus envisaged that soils, by an entirely natural process, might pass from a skeletal (raw soils) stage through an intermediate brown earth stage (associated with the mixed-oak climax forests) towards pronounced podzolization – i.e. the concept of soil as a gradually acidifying medium approaching a degraded state. Similar views were also expressed by Dimbleby (who pioneered soil-pollen analysis in this country) and, more recently, Bridges has reasoned along these lines but has emphasized the role of man in such processes (see below).

Iversen has traced these changes in some detail for Danish forests, using palynological techniques. He termed the vegetational phenomenon *retrogressive succession* and showed how podzolization of the soil began and proceeded independently of later human activity in these forests. These changes may be caused naturally by either progressive soil leaching or a local rise in the ground water table.

Smith and Taylor have since argued similarly for the Flandrian development of vegetation and soils in northern Cardiganshire. Their evidence indicates changing acidities in upland soils between 300 and 600 m due to progressive leaching. This led to a degeneration of the damp, climax oak forests which consequently became much more open in character. This occurred long before man began to have an impact at this altitude. When Bronze Age cultures entered these areas they encountered secondary forest, naturally derived. Data from Cardiganshire show that while climatic changes have general significance for the formation of peats (organic soils) they may often operate through their influence on soil processes in such a way that any original relationship with climate becomes obscured. Pedogenic peats may form when surface drainage is impeded because of textural changes in the lower soil profile which may lead to pan development. They may also form when raw humus begins to accumulate. This restricts the plant rooting and eventually leads to increased water retention at the surface. Both these processes can be brought about either by natural changes or through practices associated with man such as burning, deforestation, grazing and alteration of ground floras.

However, the evidence increasingly points to the vital role of man in the initiation of these developments. Moore, and Merryfield and Moore have argued this for the creation of many of our upland peats and peaty soils in Wales and Scotland and on Exmoor. The development of peats may reflect a climatic change, the view generally held in the past, or it may be the end result of certain soil changes, as outlined above. But Moore, while not rejecting the importance of these two natural processes, believes that man was the major initiator. An increasing number of radio-carbon dates now link commencement of upland peat growth with the presence of early Neolithic farming cultures in the same area. Although population levels were probably low during Neolithic times, from about 3,000 BP, extensive areas of grazing would have been created for their cattle herds simply because natural grazing within climax forests would be limited. It was about this time that the marked decline in elm pollen is seen in pollen diagrams and, as noted in Chapter 8, this may well have been due to the feeding of young elm growth to stalled cattle. This decline in elm also coincides with the first appearance of peat layers in many uplands (though the author has produced data from the Scottish Highlands which show that the age of upland peats can be very variable: a finding that Chambers confirms for the uplands of South Wales). Recent studies show that the clearance of forest can increase the likelihood of ground waterlogging. Following Moore's interpretation, Neolithic clearance, perhaps coinciding with a climatic shift towards higher rainfalls, would have led to the replacement of woodland soils by wet peats and peaty soils.

Several workers have reported the occurrence of loess contributions to soil profiles (e.g. Perrin *et al.*, Findlay) and Catt has mapped the wide distribution of these aeolian deposits in southern Britain. The silty material usually occurs now as thin spreads (> 0·3 m) and while it is still of great agricultural significance there has no doubt been a pronounced loss of this material downslope in the past in many areas due to man's mis-use of the terrain. Greig has suggested that the marked decrease in lime forest since the Climatic Optimum may well be associated with this exploitation of loess-rich soils by early man. This represents yet another way in which soils have changed during the Flandrian.

In Chapter 8, O'Sullivan demonstrated a change from mull to mor humus in the soils of Abernethy Forest brought about by a strong reduction in deciduous woodland species. However, soils may be changed in the opposite direction by human activities and the picture would not be complete without some reference to this. O'Sullivan noted the possibilities of reversing the mull to mor change in the forests studied and McVean and Ratcliffe, sampling in the Rothiemurchus Forest (number 5, Fig. 8.19), describe a site where this has occurred. A mor humus has been changed to a mull by a single generation of birches colonizing the open ground left by the removal of large pines. The two humus types and their associated woodlands are now separated by only a stone wall. This soil change is at 305 m on a parent material of intermediate base status. Details of the two soil profiles are shown in Table 9.4. A similar change of woodland at sites in the vicinity with a more acidic parent material, such as granite or quartzite, failed to convert the mor into a mull.

Despite the small size of the British Isles, we have an impressive variety of soil types. This range is explicable not only in terms of the climatical, geological and topographical variations of this country but also in terms of 5,000 years or more of human interference. Bridges has presented a very readable summary of this interaction between man and soils in Britain. He argues that the raw parent materials of the Late Devensian developed initially into immature (skeletal) soils which then gave way to brown soils during the early Flandrian development of a forest cover. From Mesolithic times onwards (the late Boreal-early Atlantic), these soils showed three lines of evolution. First, on *coarse*-textured materials, acid brown soils formed and then, through progressive podzolization, became podzols which, during the Sub-atlantic underwent

Table 9.4 Soil profiles from the Rothiemurchus Forest illustrating the change of a mor humus to a mull formation (Modified from McVean and Ratcliffe, 1962)

Horizon	Birch (mull)	Pine (mor)
A_0L	0·5 cm birch leaves, etc.	1·0 cm pine needles, etc.
A_0F	0·5 cm dark brown rooty mull	2·0 cm moist black rooty mor
A_1	13·0 cm brown sandy loam with schist pebbles	2·0 cm black mor
A_2	———	1·5 cm white sand (leached)
B_1	22·0 cm bright red-brown sandy loam with schist pebbles	1·0 cm indurated coffee-coloured sand
B_2	———	36·0 cm light red-brown sandy loam
C	Grey brown sandy drift with schist and granite pebbles	

gleying to form stagnopodzols. On *fine*-textured materials, the brown soils experienced eluviation to become argillic brown earths (clayey). These then underwent gleying to form gleyic argillic brown earths. In only a few areas have the brown soils of the Boreal period continued through to the present day essentially unchanged.

Initially, the modifications of man were mainly in terms of changes to the vegetation cover and drainage pattern. More recently, to these have been added the massive applications of chemical fertilizers and chemicals to control weeds and pests. Undoubtedly, these too will produce further changes in our soils and this has already been noted in some areas. It is important to stress again the dynamic nature of soils and how this may be changed by man's use of the land. Soil, together with the vegetation cover it carries, is, after all, one of the most, if not *the* most, important basic resources of any country.

DISCUSSION SECTION

There seem to be many ways of classifying soils and this gives rise to a multiplicity of terms. Where do British soils fit into the picture?

Unfortunately, it is true that there are several systems in use, developed by particular researchers or applied to specific countries. These have been well reviewed by Clayden. As yet there is no single agreed system for soil classi-fication and perhaps there may never be. Russian pedologists, especially Dokuchaev, had a profound influence on early soil classification, emphasizing the importance of the soil profile and the relationship between soils, vegetation and climate. This led to a division into three major groups: *zonal* soils, mature types with well-developed profiles reflecting the strong influence of vegetation and climate; *intrazonal* soils, where the profile is well developed but some local factor, such as relief or parent material, has had a greater influence than climate on its properties; *azonal* soils, skeletal forms with poor profile differentiation.

Another influential early system was that of Marbut in the USA. He divided soils into two broad groups, the *pedalfers* and the *pedocals*. The former were soils developed under moderate to high rainfalls where many chemicals (especially calcium or lime) were readily leached because the predominant water movement was down through the profile. The latter were found in light rainfall regions which often experienced a pronounced dry season. This resulted in less leaching and an accumulation of lime in the profile. Marbut's system is seldom used today and its limitations have been spelt out by Buol *et al.*

A major advance was the work of Kubiena on European soils. He produced a detailed hierarchical system based on many more soil properties than hitherto considered. These included macro- and micro-morphological features, microscopic examinations of humus and of thin sections of minerals. This enabled him, for example, to recognize 10 different types of rendzina. In his major work soils are described in great detail and often illustrated with colour plates. Another influential European pedologist has been Duchaufour. He stressed the importance of *lessivage* (clay translocation) and clay enrichment (argillic horizons) in profile development.

Other approaches include those of Northcote for Australian soils and the system of the Soil Survey staff of the United States Department of Agriculture, the so-called *7th Approximation* (1975). Both are bifurcating systems based on the presence or absence of soil properties. The Northcote system (1960) divided all soils into one or other of four groups based on profile features, particularly textures. The groups are: soils with uniform texture profiles (U), soils with gradational profiles (G), soils with contrasting (Duplex) texture profiles (D) and soils with organic profiles (O). Each is then further subdivided on the basis of other profile properties.

In the 7th Approximation, which is probably the most widely used classification in the world today, soils are divided into ten *Orders* which are subdivided into *Sub-orders*. Each sub-order is further divided into *Great Soil Groups* which, in turn, may contain *Sub-groups* (Table 9.5). The system also takes into account the fact that some soils are no longer in a virgin state but have been cultivated to varying extents. Mitchell has considered the criticism levelled at this approach but comes down firmly in favour of the 7th Approximation. (Indeed, the *Soil Map of the World* at 1:5,000,000 published by FAO/UNESCO (1974–8) is mainly derived from the 7th Approximation. Unfortunately, however, the FAO/UNESCO scheme has developed its own somewhat different terminology. Northcote *et al.* have recently adopted a revised approach to Australian soils that follows this FAO/UNESCO scheme.)

These systems have an overall orderly appearance, as Fitzpatrick notes, but he is critical of both. He proposes his own scheme which, although very detailed, is exceptionally well illustrated. It is intended for international use and is based on an ascending system using *soil formulae*. These allow almost unlimited versatility since they can be built up in various ways to suit specific situations. Unfortunately, the system involves a new and complex terminology which may deter many would-be users.

Table 9.5 The Soil Orders of the 7th Approximation Scheme, United States Department of Agriculture

Order	Meaning	Soil characteristics
1. Entisol	Recent soil	Profile development very weak or absent
2. Vertisol	Inverted soil	Expanding clays, self-mulching. Found in subhumid to arid climates
3. Inceptisol	Young soil	Weak profile development, no strong illuvial horizon (as in a podzol) but a cambic horizon present (an altered subsoil horizon similar to a spodic horizon but altered to a far lesser extent)
4. Aridisol	Arid soil	Arid region soils, often with salt horizons of calcium, gypsum or sodium
5. Mollisol	Soft soil	A_1 horizon thick and dark; best development under grass vegetation
6. Spodosol	Ashy soil	Illuvial horizon accumulation of iron and organic colloids (spodic horizon). Cemented pan development
7. Alfisol	Pedalfer soil	Clay horizon with high base saturation (> 35%)
8. Ultisol	Ultimate (leaching)	Clay horizon of low base saturation (< 35%)
9. Oxisol	Oxide soils	Clay horizon very high in iron and aluminium oxides
10. Histosol	Organic soils	Organic surface horizon at least 6 inches thick and with more than 30% organic matter

A system now used as the basis for soil mapping by the Soil Survey of England and Wales is the one proposed by Avery in 1973 and outlined in Appendix 9.1. This system recognizes *Major Groups*, *Groups*, *Sub-groups* and *Soil Series*, but only the first two divisions are shown in the appendix. Many names will already be familiar as they have much in common with the terminology used in early Soil Survey Memoirs which have already been referred to several times in this chapter. The *pelosols* are clayey soils which expand and contract markedly with wetting and drying. The rankers and rendzinas are grouped together as *lithomorphic* soils and those soils retaining water at or near the surface are designated *stagnogley* soils. Texture remains the most important criterion for separating many of the groups. The man-made soils include, in addition to plaggen soils, the increasing acreage resulting from modern attempts to reclaim land made derelict by mineral exploitation (e.g. colliery spoil, slag heaps, china clay wastes, fly ash, sand and gravel). Bradshaw has written on the detailed reconstruction work that goes into producing this category of soil.

What is the cause of the extensive erosion of upland peats in this country?

Erosion is taking place extensively throughout blanket peats, not only in our uplands but also in some northern lowland areas. Most of the early work on

Fig. 9.10 Peat erosion at 700 m on Upper Deeside, east Scotland. The channels, about 1 m deep, have exposed numerous pine stumps in the upper peat showing that the area was once well wooded.

this problem has been done by Bower, mainly in the Pennines. A developing peat bog contains massive amounts of water and occasionally the system can become unstable, giving rise to a bog burst. Most peat erosion, however, is not of this spectacular type; rather it consists of a complex network of channels eating back or down into the peat (Figs. 9.10 and 9.11). Controversy exists as to the cause of this type of erosion. One suggestion sees it as a natural end-point to peat-bog growth, since accumulation cannot go on for ever. Erosion may also be a concomitant of accumulation, occurring intermittently during the build-up phase (though there is little evidence to show this happens other than on a very local scale). Both these processes could operate under constant environmental conditions.

Some field workers see the cause as due to a recent change in environment. This could be a climatic change but it is not clear what direction of change would be needed to trigger-off erosion. Some argue for a drier phase, causing accumulation to cease and erosion to begin. Others have proposed a wetter phase, making the surface and upper layers saturated, less compact and hence less stable. Another change of great influence has been a marked increase in biotic avtivities. Most of these areas have a long history of grazing and heather burning; practices which could breach the surface vegetation and expose the soft peat to wind and water erosion. Small erosion scars can be healed fairly

Fig. 9.11 Advanced peat erosion at 690 m, flanks of Carrantuohill (1,040 m), south-west Ireland. Only two small 'islands' of peat remain, with a pine stump exposed between them (beneath the ranging pole).

quickly by the growth of a fresh *Sphagnum* cover. But extensive or repeated breaching or anything that prevents *Sphagnum* growth could cause serious peat erosion. More recently, atmospheric pollutants have been cited as a cause for the severe reduction in the *Sphagnum* cover now seen over much of the southern Pennines and this has probably led to a pronounced extension of erosion channels in this region.

The answer might lie in a combination of all or several of these factors. It used to be argued that climate caused peat accumulation but we now know that other factors may be involved. So while we can envisage multiple causes for peat accumulation we also have the same possibility for peat erosion.

Which has the greater influence, soils on vegetation or vegetation on soils?

Undoubtedly, most pedologists would come down in favour of the latter. Some soils formed on old erosion surfaces have experienced several cycles of development, as with the polycyclic soils of the Tropics. In these cases, it is often difficult to determine the exact influence of each pedogenic factor. Other soils have formed on a more recent surface under a climatic phase showing only minor fluctuations. This is certainly the case for most British soils whose formation dates from the Late-glacial period. Where the natural vegetation has persisted for a long period, it will have made a profound contribution to the soil profile. Geological influences are strong in the early stages of soil formation but, with time, they become less and less important, or are confined to the lowest soil horizons. The upper and middle horizons increasingly come under the influence of the plant cover.

Plants are responsible for much of the biological and chemical characteristics of the profile: they largely control the cycling of minerals. At the soil surface, the vegetation is a major influence on microclimate and microbe activity. It contributes the bulk of the humus and plays a large role in determining the hydrological regime of the profile through its influence on run-off and infiltration rates. On many slopes, the plant cover has the key role of preventing soil loss through erosion. And, in the extreme case with peats, the vegetation is essentially the soil.

The vegetation is of fundamental importance for humification processes and humus types. Some species are 'acidifying' and always give rise to a mor humus (e.g. *Calluna*, *Picea* (see the study of spruce plantations by Grieve on page 271 of this chapter), *Pinus*). Others 'improve' the soil and encourage mull formation (e.g. *Fraxinus*, *Alnus* and many grasses). The importance of these controls is seen when it is realized that on different parent materials the same vegetation often leads to a convergent evolution of humus types and upper horizons. Where man has disturbed vegetation and encouraged 'acidifying' species to spread, soil changes can be rapid indeed, again illustrating the profound importance of the plant cover. Dimbleby has detailed the podzolization of brown forest soils with mull that followed the oakwood clearances on the North York Moors by Bronze Age cultures. He also has shown how recent growth of birch can alter these derived moorland podzols by rooting to some depth and renewing the biogeochemical cycles, changing the mor to mull and encouraging earthworm activity. Grubb, Green and Merrifield have examined the 'acidifying' effects of *Calluna* litter, noting strong correlations between the size of heather bushes and the soil pH underneath their centres. These pH values and the pattern in the surrounding soil are shown in Fig. 9.12. This paper considers in some detail the reciprocal nature of soil-plant relationships. It must also be remembered that in derelict land reclamation plants play a key role in soil development: Palaniappan *et al.* detail the remarkable effects of *Lupinus arboreus* (tree lupin) on china clay wastes as an example.

This contribution of vegetation was recognized many years ago and is seen, of course, in the zonal concept of Dokuchaev. However, there are cases where the soil determines the vegetation cover. At some sites, the parent material influence predominates. An extreme case is where serpentine rocks are found. These outcrop to form parent materials at only a few localities in Britain, such as the Lizard Peninsula and in the Shetland Islands. The soils carry an excess of magnesium, chromium and nickel but are deficient in some major nutrients. Most plants are unable to adapt to these conditions and the sites on the Lizard Peninsula support four distinctive types of heath. These are (a) *Festuca ovina-Calluna* heath; (b) *Erica vagans-Ulex europaeus* heath; (c) *Erica vagans-Schoenus* heath; and (d) *Agrostis setacea* heath. Coombe and Frost also note the presence of *Juncus pygmaeus*, a species of rush confined exclusively to these soils in Cornwall.

While the influence of vegetation on soils is the greater factor in Britain we should remember that this may not always be true of other parts of the world.

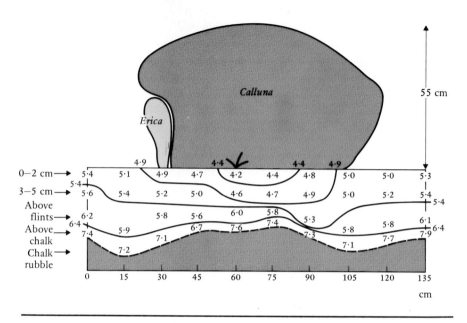

Fig. 9.12 The pH variations in the mineral soil beneath an isolated bush of *Calluna* and a suppressed plant of *Erica cinerea* on chalk heath in south-east Sussex. No exaggeration of scale above ground but within the soil the vertical scale is 2·5 times the horizontal scale. pH determinations are 15 cm apart. (From Grubb, Green and Merrifield, 1969.)

For example, extensive areas of saline soil exist under semi-arid and arid climates and the accumulation of salts only allows certain species to persist. Quite clearly, the species present do not determine the salt content of the profile though, nevertheless, they may influence some of the other profile properties.

Modern agriculture relies heavily on the use of chemical fertilizers – but in what other ways do present-day farming practices modify the soil?

Modern farming methods in Britain also rely on the use of heavy machinery and have created much larger fields for cultivation (one is largely a consequence of the other). These two developments have caused concern about possible damage to the inherent fertility and fundamental structure of the soil. These matters were brought to a head by the very wet period in 1968–9. As a result, the Agricultural Advisory Council undertook an urgent inquiry and some of the main conclusions from Strutt's report were:

1. There is little cause for concern over the nutrient fertility of our soils. The replacement of ley-farming and farmyard manure by chemical fertilizers does not appear to have caused any fertility loss (though we may add that

excessive loss of chemicals from farmland in the drainage waters can lead to ecological disturbance of rivers, ponds and lakes by the process of eutrophication).

2. There is much greater concern over the decline in soil structure. Many soils now suffer from a lack of organic matter and their structure has been badly altered by the use of heavy machinery in unsuitable conditions. This has accentuated to a marked degree the drainage problems on many farms.

3. Whereas many weeds and pests are adequately and suitably controlled by chemicals some have spread alarmingly (e.g. wild oats) and the move towards crop specialization has increased the likelihood of serious outbreaks. As yet, there is no evidence for deleterious effects on the soil from these chemicals.

4. High stocking densities on grasslands in areas of high rainfall have led to structural damage to the soil and deterioration of the sward. Compaction of the soil, by machines and cattle, and subsequent surface waterlogging have led to a decline in pasture productivity, particularly in the main dairy farming areas.

Another concern is the increasing incidence of wind erosion on light friable soils. The year 1967 was a bad one for such erosion with cases being reported on the sandy soils of East Yorskhire (Radley and Simms) and Lincolnshire (Robinson). Pollard and Millar also describe the situation on the reclaimed Fen soils of East Anglia in 1968 when conditions were the worst in living memory.

Fig. 9.13 Recent severe soil erosion in a newly sown pasture on the rural fringe of Leicester. The main gullies, containing running water, are already 20 cm deep. The 12 ° slope has a thin covering of glacial drift material over Lias Clay.

Surface peat, fertilizers and even seeds were blown from the fields, blocking the important drainage dykes at these low levels. Standing crops also suffered from abrasion by grit-laden winds. Costs were estimated at £100 per ha. 1973 was another bad year for wind erosion and Evans and Nortcliff report spectacular erosion in north Norfolk in 1975. The blowing of soil has probably been made worse by the large-scale growing of sugar beet, which requires a fine seed-bed, and the removal of hedges to increase field size and cut down 'wastage' of land. Large-scale hedge removal has extended to much of eastern England in recent years and this, together with other modern farming practices, will increase the likelihood of further soil deterioration. Morgan's survey of the problem leads him to conclude that soil erosion is an increasing hazard in Britain (see Fig. 9.13). It is to be hoped that now we are aware of the situation a return to good farming techniques will do much to counter these adverse trends.

APPENDIX 9.1: SOIL CLASSIFICATION

The soil classification scheme now used as a basis for soil mapping by The Soil Survey of England and Wales (Modified from Avery, 1973.)

Major group	Group
1. **Terrestrial raw soils** Mineral soils with no diagnostic pedogenic horizons or disturbed fragments of such horizons, unless buried beneath a recent deposit more than 30 cm thick	1.1 *Raw sands* Non-alluvial, sandy (mainly dune sands) 1.2 *Raw alluvial soils* In recent alluvium, normally coarse textured 1.3 *Raw skeletal soils* With bedrock or non-alluvial fragmental material at 30 cm or less 1.4 *Raw earths* In naturally occurring, unconsolidated, non-alluvial loamy, clayey or marly material 1.5 *Man-made raw soils* In artifically disturbed material, e.g. mining spoil
2. **Hydric raw soils** (Raw gley soils) Gleyed mineral soils, normally in very recent marine or estuarine alluvium, with no distinct topsoil, and/or ripened no deeper than 20 cm	2.1 *Raw sandy gley soils* In sandy material 2.2 *Unripened gley soils* In loamy or clayey alluvium, with a ripened topsoil less than 20 cm thick

Major group	Group
3. Lithomorphic (A/C) soils With distinct, humose or organic topsoil over C horizon or bedrock at 40 cm or less, and no diagnostic B or gleyed horizon within that depth	3.1 *Rankers* With non-calcareous topsoil over bedrock (including massive limestone) or non-calcareous, non-alluvial C horizon (excluding sands) 3.2 *Sand-rankers* With non-calcareous, non-alluvial sandy C horizon 3.3 *Ranker-like alluvial soils* In non-calcareous recent alluvium (usually coarse textured) 3.4 *Rendzinas* Over extremely calcareous non-alluvial C horizon fragmentary limestone or chalk 3.5 *Pararendzinas* With moderately calcareous non-alluvial C horizon (excluding sands) 3.6 *Sand-pararendzinas* With calcareous sandy C horizon 3.7 *Rendzina-like alluvial soils* In recent alluvium
4. Pelosols Slowly permeable (when wet), non-alluvial clayey soils with B or BC horizon showing vertic features and no E, non-calcareous Bg or paleoargillic horizon	4.1 *Calcareous pelosols* Without argillic horizon 4.2 *Non-calcareous pelosols* Without argillic horizon 4.3 *Argillic pelosols* With argillic horizon
5. Brown soils Soils, excluding pelosols, with weathered, argillic or paeloargillic B and no diagnostic gleyed horizon at 40 cm or less	5.1 *Brown calcareous earths* Non-alluvial, loamy or clayey without argillic horizon 5.2 *Brown calcareous sands* Non-alluvial, sandy, without argillic horizon 5.3 *Brown calcareous alluvial soils* In recent alluvium 5.4 *Brown earths* (sensu stricto) Non-alluvial, noncalcareous, loamy or clayey, without argillic horizon 5.5 *Brown sands* Non-alluvial, sandy or sandy gravelly 5.6 *Brown alluvial soils* Non-calcareous, in recent alluvium 5.7 *Argillic brown earths* Loamy or clayey, with ordinary argillic B 5.8 *Paleo-argillic brown earths* Loamy or clayey, with paleo-argillic B

Major group	Group
6. Podzolic soils With podzolic B	6.1 *Brown podzolic soils* (podzolic brown earths) With Bs below an Ap or 15 cm, and no continuous albic E, thin ironpan, distinct Bhs with coated grains or gleyed horizon at 40 cm or less 6.2 *Humic crytopodzols* (humic podzolic rankers) With very dark humose Bhs more than 10 cm thick and no peaty topsoil, thin ironpan, continuous albic E, Bs, or gleyed horizon 6.3 *Podzols* (sensu stricto) With continuous albic E and/or distinct Bh or Bhs with coated grains and no peaty topsoil, bleached hardpan or gleyed horizon above, in or directly below the podzolic B or at less than 50 cm 6.4 *Gley-podzols* With continuous albic E and/or distinct Bh or Bhs, gleyed horizon directly below the podzolic B or at less than 50 cm, and no continuous thin ironpan or bleached hardpan 6.5 *Stagnopodzols* With peaty topsoil and/or gleyed E or bleached hardpan over thin ironpan or Bs horizon (wet above a podzolic B)
7. Surface-water gley soils (Stagnogley *sensu lato*) Non-alluvial soils with distinct, humose or peaty topsoil, non-calcareous Eg and/or Bg or Btg horizon, and no G or relatively pervious Cg horizon affected by free groundwater	7.1 *Stagnogley soils* (sensu stricto) ($\approx$ *Pseudogley*) With distinct topsoil 7.2 *Stagnohumic gley soils* With humose or peaty topsoil
8. Ground-water gley soils With distinct, humose or peaty topsoil and diagnostic gleyed horizon at less than 40 cm, in recent alluvium ripened to more than 20 cm, and/or with G or relatively pervious Cg horizon affected by free groundwater	8.1 *Alluvial gley soils* With distinct topsoil, in loamy or clayey recent alluvium 8.2 *Sandy gley soils* Sandy, with distinct topsoil and without argillic horizon 8.3 *Cambic gley soils* Non-alluvial, with distinct topsoil, loamy or clayey Bg horizon and relatively pervious Cg or G horizon 8.4 *Argillic gley soils* With distinct topsoil and argillic (Btg) horizon over relatively pervious Cg

Major group	Group
	8.5 *Humic-alluvial gley soils* With humose or peaty topsoil, in loamy or clayey recent alluvium 8.6 *Humic-sandy gley soils* Sandy, with humose or peaty topsoil and no argillic horizon 8.7 *Humic gley soils* (sensu stricto) Non-alluvial, loamy or clayey, with humose or peaty topsoil
9. Man-made soils With thick man-made A horizon or disturbed soil (including material recognizably derived from pedogenic horizons) more than 40 cm thick	9.1 *Man-made humus soils* With thick man-made A horizon, including Plaggen soils 9.2 *Disturbed soils* Without thick man-made A horizon
10. Peat (organic) soils	10.1 *Raw peat soils* Without earthy topsoil or ripened mineral surface layer 10.2 *Earthy peat soils* With earthy topsoil or ripened mineral surface layer

REFERENCES

Anderson, H. A., Berrow, M. L., Farmer, V. C., Hepburn, A., Russell, J. D. and Walker, A. D. 1982. 'A reassessment of podzol formation processes', *J. Soil Sci.*, **33**, 125–36.

Avery, B. W., 1973. 'Soil classification in the Soil Survey of England and Wales', *J. Soil Sci.*, **24** (3), 324–38.

Ball, D. F., 1966. 'Brown podzolic soils and their status in Britain', *J. Soil Sci.*, **17** (1), 148–58.

Barker, G. and Webley, D., 1978. 'Causeway camps and early Neolithic economies in central southern England', *Proceedings of the Prehistoric Society*, **44**, 161–86.

Beckett, P. H. T. and Webster, R. 1971. 'Soil variability: a review', *Soils and Fertilizers*, **34**, 1–15.

Bower, M. M., 1960. 'The erosion of blanket peat in the Southern Pennines', *East Midland Geographer*, **13**, 22–33.

Bower, M. M., 1962. 'The cause of erosion in blanket peat bogs', *Scott. Geogr. Mag.*, **78**, 33–43.

Bradshaw, A. D., 1983. 'The reconstruction of ecosystems', *J. appl. Ecol.*, **20**, 1–17.

Bridges. E. M., 1978. 'Interaction of soils and mankind in Britain', *J. Soil Sci.*, **29**, 125–39.

Buol, S. W., Hole, F. D. and McCracken, R. J., 1980. *Soil Genesis and Classification*, Iowa State University Press, Ames, Iowa.

Burnham, C. P., 1970. 'The regional pattern of soil formation in Great Britain', *Scott. Geogr. Mag.*, **86**, 25–34.

Burnham, C. P. and Mackney, D., 1964. 'Soils of Shropshire', *Field Studies*, **2** (1), 83–113.

Catt, J. A., 1978. 'The contribution of loess to soils in lowland Britain', in *The effect of man on the landscape: the Lowland Zone* (eds Limbrey, S. and Evans, J. G.), Council for British Archaeology Research, Report 21, pp. 12–20.

Catt, J. A., 1979. 'Soils and Quaternary Geology in Britain', *J. Soil Sci.*, **30**, 607–42.

Chambers, F. H., 1982. 'Two radiocarbon-dated pollen diagrams from high-altitude blanket peats in South Wales', *J. Ecol.*, **70**, 445–59.

Clayden, B., 1982. 'Soil classification', in *Principles and Applications of Soil Geography* (eds Bridges, E. M. and Davidson, D. A.), Longman, London and New York, pp. 58–96.

Conry, M. J., 1971. 'Irish plaggen soils: their distribution, origin and properties', *J. Soil Sci.*, **22**, 401–416.

Coombe, D. E. and Frost, L. C., 1956. 'The heaths of the Cornish Serpentine', *J. Ecol.*, **44**, 226–56.

Courtney, F. M., 1973. 'A taxonometric study of the Sherborne soil mapping unit', *Trans. Inst. Brit. Geogr.*, **58**, 113–24.

Curtis, L. F., Courtney, F. M. and Trudgill, S. T., 1976. *Soils in the British Isles*, Longman, London and New York.

Dimbleby, G. W., 1961. 'Soil-pollen analysis', *J. Soil Sci.*, **12** (1), 1–11.

Dimbleby, G. W., 1962. *The Development of British Heathlands and the Soils*, Oxford Forestry Memoir 23, Clarendon Press, Oxford.

Duchaufour, Ph., 1977. *Pedologie. I Pedogenese et classification*. Masson, Paris.

Evans, R. and Nortcliff, S., 1978. 'Soil erosion in north Norfolk', *J. agric. Sci., Camb.*, **90**, 185–92.

Findlay, D. C., 1965. 'Soils of the Mendip District of Somerset', *Mem. Soil Surv. Gr. Brit.*, Soil Survey, Harpenden.

Fitzpatrick, E. A., 1971. *Pedology: a systematic approach to soil science*, Oliver and Boyd, Edinburgh.

Gerrard, J. A., 1981. *Soils and Landforms. An Integration of Geomorphology and Pedology*, Allen and Unwin, Hemel Hempstead.

Greig, J., 1982. 'Past and present lime woods of Europe', in *Archaeological Aspects of Woodland Ecology* (eds Limbrey, S. and Bell, M.), BAR International Series, 146, Oxford, pp. 23–55.

Grieve. I. C., 1978. 'Some effects of the plantation of conifers on a freely drained lowland soil, Forest of Dean, U.K.', *Forestry*, **51**, 21–8.

Grubb, P. J., Green, H. E. and Merrifield, R. C. J., 1969. 'The ecology of chalk heath: its relevance to the calcicole-calcifuge and soil acidification problems', *J. Ecol.*, **57**, 175–212.

Handley, W. R. C., 1954. 'Mull and mor formation in relation to forest soils', *For. Comm. Bull.*, **23**, HMSO, London.

Iversen, J., 1964. 'Retrogressive vegetational succession in the Post-glacial', *J. Ecol.*, **52** (Suppl), 59–70.

Kubiena, W. L., 1953. *The Soils of Europe*, Thos. Murby, London.

Limbrey, S., 1975. *Soil Science and Archaeology*, Academic Press, London.

Mackney, D., 1961. 'A podzol development sequence in oakwoods and heath in central England', *J. Soil Sci.*, **12** (1), 23–40.

Mackney, D. and Burnham, C. P., 1964. *The Soils of the West Midlands*, Soil Survey of Great Britain, Bull. 2, Harpenden.

McVean, D. N. and Ratcliffe, D. A., 1962, *Plant Communities of the Scottish High-lands*, Monographs of the Nature Conservancy, No. 1, London.

Merryfield, D. L. and Moore, P. D., 1974. 'Prehistoric human activity and blanket peat initiation on Exmoor', *Nature*, London, **250**, 439–41.

Mitchell, C. W., 1973. 'Soil classification with particular reference to The Seventh Approximation', *J. Soil Sci.*, **24**, 411–20.

Moore, P. D., 1968. 'Human influence upon vegetational history in North Cardiganshire', *Nature*, London, **217**, 1006–9.

Moore, P. D., 1973. 'The influence of prehistoric cultures upon the initiation and spread of blanket bog in upland Wales', *Nature*, London, **241**, 350–3.

Moore, P. D., 1975. 'Origin of blanket mires', *Nature*, London, **256**, 267–9.

Morgan, R. P. C., 1980. 'Soil erosion and conservation in Britain', *Progress in Physical Geography*, **4**, 24–47.

Northcote, K. H., 1960. *Factual Key for the Recognition of Australian Soils*, Aust. Soils Div. Report, CSIRO, Canberra.

Northcote, K. H., Hubble, G. D., Isbell, R. F., Thompson, C. H. and Bettenay, E., 1975. *A description of Australian Soils*, CSIRO, Australia.

Palaniappan, V. M., Marrs, R. H. and Bradshaw, A. D., 1979. 'The effect of *Lupinus arboreus* on the nitrogen status of china clay wastes', *J. appl. Ecol.*, **16**, 825–31.

Pears, N. V., 1975. 'Radiocarbon dating of peat macrofossils in the Cairngorm Mountains, Scotland', *Trans. Bot. Soc. Edinb.*, **42**, 255–60.

Pearsall, W. H., 1952 'The pH of natural soils and its ecological significance', *J. Soil Sci.*, **3** (1), 41–51.

Perrin, R. M. S., 1956, 'The nature of "Chalk Heath" soils', *Nature*, London, **178**, 31–2.

Pollard, E. and Millar, A., 1968, 'Wind erosion in the East Anglian Fens', *Weather*, **23** (10), 415–17.

Radley, J. and Simms,C., 1967. 'Wind erosion in East Yorkshire', *Nature*, London, **216**, 20–2.

Robinson, D. N., 1968. 'Soil erosion by wind in Lincolnshire, March 1968', *East Midland Geogr.*, **4**, 351–62.

Romans, J. C. C., Stevens, J. H. and Robertson, L., 1966. 'Alpine soils of north-east Scotland', *J. Soil Sci.*, **17** (2), 184–99.

Smith, R. T. and Taylor, J. A., 1969. 'The Post-glacial development of vegetation and soils in Northern Cardiganshire', *Trans. Inst. Br. Geogrs.*, **48**, 75–96.

Strutt, N., 1970. *Modern Farming and the Soil*, Ministry of Agriculture, Fisheries and Food, HMSO, London.

Taylor, J. A., 1960. 'Methods of soil study', *Geography*, **45**, 52–67.

Thomasson, A. J., 1971. *Soils of the Melton Mowbray District*, Soil Survey of Great Britain (sheet 142), Harpenden.

THE IMPACT OF MAN

INTRODUCTION

The vital role of man in ecology has been stressed within several topics considered so far. This last chapter summarizes man's past impact on British ecosystems and presents two detailed regional examples. The ecological activities of man are a continuing phenomenon and present and probable future patterns of these are also examined in ecosystem terms. The difficulties arising from multiple land use and mounting recreational pressures on our countryside are then considered, particularly with respect to the conservation problems they pose for environmental planners.

THE LEGACY OF HUMAN INTERFERENCE

From about 5000 BC onwards man has had an ever-increasing influence in determining the distribution and many of the properties of our flora, fauna and soils. The early cultures, Palaeolithic and Mesolithic, existing as hunters and gatherers at low population densities, were once thought to have produced only slight changes in their environment. However, for certain upland habitats, such as the North York Moors and Dartmoor, Simmons has argued for a clearly detectable Mesolithic impact and there is increasing evidence for this view. There was also impact on those marginal lowland sites which were ecologically susceptible to change by virtue of their pedological characteristics.

By about 3000 BC Neolithic man had arrived in this country and his influence was certainly pronounced. As a cultivator, he cleared more forest and to protect his own grazing animals and increase his food supplies he began to reduce the competing native fauna of the woodlands. We have strong evidence that selective clearance of some tree species was taking place at this time (young lime and elm trees to supplement scarce grass supplies for domesticated cattle) and some slopes were experiencing soil erosion as a result of use by man. The first Neolithic forest clearances were temporary plots which were subsequently abandoned, reverting by subseres towards the climax forest composition. Before long, however, more permanent clearances were established and by then much of the forest in some regions would have been essentially secondary in form. This was particularly so on the better drained soils and slopes. It was initially thought that a high concentration of prehistoric remains on limestone scarplands (e.g. the Chilterns and North Downs) and uplands (e.g. Dartmoor, Pennines and North York Moors) meant that early man avoided lowland sites. But recently many more prehistoric settlement sites have come to light in the lowlands causing a revision of this view. During the following Bronze Age and Iron Age the forest underwent further reduction – Bronze Age settlements, for example, reached quite high levels on Dartmoor.

It was these cultures that the Romans encountered during their invasion of Britain. In this conquest they set fire to woodland as part of their military stra-tegem against local populations. The Romans were also important in clearing lowland areas for cultivation. They were able to tackle some sites which would have proved unattractive to earlier cultures whose technology for dealing with the denser vegetation and poor drainage of the wettest valley bottoms was limited. Throughout this and earlier periods the British flora was receiving additional species and Godwin lists many introductions which are now largely accepted as native. These include *Castanea sativa* (sweet chestnut), *Papaver rhoeas* (poppy), *Juglans regia* (walnut) and *Sinapis arvensis* (wild mustard).

Forest reduction must have continued throughout the Roman period and into the Dark Ages but there would still have been vast tracts of woodland landscape. The next clear evidence of profound environmental alteration comes with the arrival of the Saxons. From about the sixth century onwards, these agricultural colonizers opened up much of the lowland forest and established perhaps 90 per cent of the existing pattern of English villages and hamlets. The process used was 'assarting' or grubbing up and burning trees and shrubs to prepare the land for tillage. About the same time, Norse, Danes and Vikings first raided and later settled many parts of northern Britain. They, too, used fire extensively to establish their rule. Under the Norman feudal system, the clearance and reclamation of land continued, but large areas were also set aside as royal preserves – open 'wastes' and woods suitable for game hunting, such as the New Forest. However, their introduction of the rabbit and their fostering of deer herds did much to restrict the regeneration of trees.

These destructive trends gathered pace during the Medieval period and locally timber supplies became very scarce. The Exchequer Rolls of Scotland show Baltic timber being imported to Edinburgh as early as AD 1329. By the seventeenth century, Samuel Pepys, John Evelyn and other prominent writers were bewailing the dearth of timber in Britain. Much in the south had been used for fuel, buildings, ship construction and charcoal production. Increas-ingly, merchants moved into remote regions for their supplies (e.g. the Lake District, the Loch Lomond Woods, see p. 227). On Roman maps the simple descriptive term *Caledonia Silva* – the Caledonian Forest – had been boldly written across Scotland, yet in AD 1617 Sir Anthony Weldon, touring lowland Scotland, was led to comment, 'there's scarce a tree to hang a Judas on', an indication of the completeness of tree removal in these parts.

REGIONAL EXAMPLES

The human impact on the natural environment can be followed more specifically if we take two detailed examples. The first concerns the pine and birch woods flanking the Cairngorm Mountains of Deeside and Speyside. They are remnants of the ancient Caledonian Forest and their ecology has already been referred to in Chapters 8 and 9. In complete contrast are the numerous small mixed-deciduous woodland plots lying a few km east of Leicester, many of which are fox coverts. Some of these are much altered in character but originally were part of the once extensive Summer Deciduous Forest. Others, however, were entirely created by man as planted woodland for fox-hunting. Whereas most present coniferous woodland in Britain is quite obviously planted, so much of the deciduous woodland is equally 'artificial'.

THE CAIRNGORM FORESTS

These forests formerly extended to altitudes of 620 m or more but today the upper limit seldom exceeds 490 m. Much of the forest is now open ground or forestry plantation. The gross disturbance of these woods began about AD 400 in Pictish times when large tracts of lower forest were cleared for primitive cultivation. The decline of forest on the higher slopes took place much later, mainly in the Medieval period, with the evolution of cattle-breeding, the mainstay of the Highland economy until the late eighteenth century. Gaffney has shown that by the sixteenth century most forested land between 300 m and 600 m had been turned into highly prized grazings (Fig. 10.1). Through the area ran several important drove roads for moving the cattle to summer pastures (shealings) or to lowland markets. These routes attracted reivers or thieves who were extremely active until the military set up control points in the region. In addition to the local cattle on the summer pastures, there were 'strangers' or gall cattle. For local tenants it was common practice to take in such cattle for a fee from farms as remote as 35 km away. We cannot be certain just how well wooded the higher ground originally was but there are numerous references in the Gordan Castle Estate Papers to unlawful grazing and heavy exploitation of timber at these levels. It is clear, however, that the trend was always towards continual reduction of the forest cover.

Other animals, in addition to cattle, featured in the economy and ecology of the pre-sheep period of this part of Highland Scotland. Deer were numerous and greatly encouraged by the Dukes of Gordon, the upper glens eventually becoming overstocked. In winter their traditional feeding grounds were the native pinewoods. Horses, goats and sheep were also numerous. Macfarlane

Fig. 10.1 Upper Dee Valley at 400 m. Grazings along the riverside (straths) and muirburns on the heather slopes. A formerly forested area which became important for the
Highland cattle economy by the Medieval period.

in 1748 described the upper Dee as richly stocked with these animals. In 1779
the Forest of Glenavon was 'eaten up like a sheep pasture' and in 1794 'the
forest contained many green spots and . . . affords pasture for a thousand head
of cattle'. It is obvious that these animals would have a marked effect on tree
seedling regeneration and ground flora composition.

Several activities associated with cattle husbandry also had a pronounced
influence on the woodlands. First, the construction of temporary bothies
('scalans') at the highest pastures required the use of 'divots and trees'.
Secondly, by law, cattle had to be confined or 'hained' at night and this often
necessitated the building of timber enclosures. Thirdly, crops needed
protection from grazing animals and this was often achieved by erecting timber
barriers ('garthing'): at one site 5,000 young trees were used to protect a corn-
yard. A measure of this destruction of native trees is given by the petition of
James Grant of Grant to the Parliament of James VI 'that fir wood in Aber-
nethie and Glencherneck are dayly cutt, stollen and carried away by the tenants
of Strathspey . . . and that the birkwoods are wholly destroyed by peiling of
the timber peiled standing rotting in the woods . . . and against the great hurt
and prejudice done to the fir woods of Strathspey by cutting and destroying
standing trees for to be candle fir to all the inhabitants'. (During times of hard-
ship peeled birch bark was used as a human food source. The 'fir' mentioned
is Scots pine, candle fir being a primitive form of home lighting.)

The practice of muirburning was of even greater ecological importance. This
was widespread and frequently the fires got out of control. It became so excessive that by 1695 severe penalties were enforced for burning too near the pinewoods – a third offence merited hanging.

By the end of the eighteenth century large-scale sheep farming had replaced cattle farming but grazing and muirburning continued as dominant activities working against the remaining forest cover. In the Cairngorms sheep were never so important as in other parts of the Highlands and eventually most sheep runs were converted to grouse moors. Again, this meant the continuance of regular muirburning.

The eighteenth century also saw the intensive commercial exploitation of the remaining woods by the timber companies (Fig. 10.2). The York Building Company of Hull was responsible for the main onslaught in Abernethy Forest from 1728 onwards. Similarly, Glenmore Forest was worked over in 1783 and the adjacent Rothiemurchus Forest at this time was providing £10,000–£20,000 annual timber revenue. Several severe forest fires occurred in these woodlands during the eighteenth and nineteenth centuries and there was further heavy timber exploitation in the World Wars of the twentieth century.

The nobility have acted unwittingly to preserve some parts of these forests through their strict ownership and sporting interests. In the nineteenth century there was extensive tree planting by the nobility and one Cairngorm parish could boast 14 million 'firs' planted by 1877. However, the main plantings seen today are due to the Forestry Commission who, from 1923 onwards, began a comprehensive re-afforestation programme using several exotic conifers (mainly spruce, lodgepole pine, larch and fir). Glenmore became a National Forest Park in 1948 and in 1954 the large Cairngorms National Nature Reserve was declared.

Many of these recent measures to repair the ravages of previous centuries are now placed in jeopardy by the current massive influx of tourists, a phenomenon which we shall examine shortly. Recreational and sporting activities, forestry, conservation, and grazing, all operating together at these altitudes must lead to a severe clash of interests which will have to be resolved successfully if these already much modified woodlands are to survive in an acceptable form.

THE WOODLANDS OF EAST LEICESTERSHIRE

The subdued, rolling landscape of East Leicestershire (which now includes the former county of Rutland) stands in sharp contrast to the grandeur of the Cairngorm Mountains. The East Midlands are an important agricultural region but, nevertheless, the area contains a surprising number of small scattered woodland plots: 130 are shown on the O.S. 1:25,000 Leicester (East) SK 60/70 sheet. Many of these are less than 200 years old. Nearly all owe their form and existence to the fox-hunting activities for which the county is famous: the natural Summer Deciduous Forest which clothed the region in prehistoric times has long since given way to farming.

In the twelfth century all of Leicestershire and neighbouring Rutland was declared 'forest', i.e. wooded and open wasteland reserved for hunting. Leicester became exempt in AD 1235 but Rutland did not until much later. Some of the present-day woodlands around the parish of Withcote (Fig. 10.3),

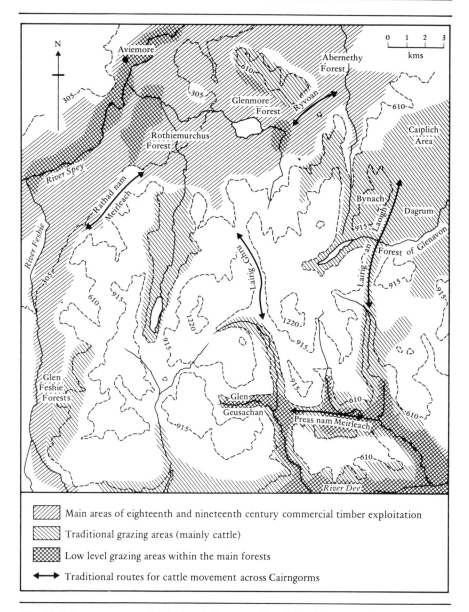

Fig. 10.2 Some aspects of the historical geography of the Cairngorm Mountains: areas of pronounced disturbance by man from early Medieval to recent times.

although now much altered by man, may be directly descended from the extensive early Medieval Leighfield Forest that was retained in this part of the Midlands until relatively late. The largest of them, Owston Wood, once grew very fine native oaks but these were clear-felled after the First World War. This wood was neglected for 35 years, acting simply as a source of stakes and fence material for the Cottesmore Hunt and local farmers. In 1960 the Forestry

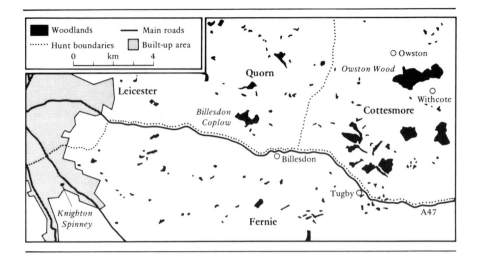

Fig. 10.3 The woodlands and hunt boundaries of part of East Leicestershire. (Map area based on O.S. 1:25,000 Leicester (East) SK 60/70 sheet.)

Commission took it over but have had continuing problems with weed species (willow herb, bracken, bramble and, particularly, willows) that became well established during the period of neglect. Another difficulty is that the heavy Lias Clay soils puddle badly in winter and crack deeply in summer, this condition being made much worse by the constant churning of the Cottesmore Hunt who draw these woods each winter.

Billesdon Coplow (Fig. 10.3) is even more of an artefact since much of the hill consists of mature planted beech reaching 30 m or more. The wood also has a larch plantation and open areas are planted out with a mixture of Scots pine and beech. A few of the original oaks remain (Fig. 10.4). Botany Bay Fox Covert, adjacent to Billesdon Coplow, was originally planted soon after the time of the Australian settlement, as the name suggests. It is mainly degenerate oak-ash wood with an open canopy which has encouraged profuse growth of hawthorn, hazel, blackthorn, privet and elder shrubs. The hazel was formerly coppiced but some shrubs are now 'chopped' to promote a tangled undergrowth suitable for fox cover (see Figs. 8.10 and 8.11).

Knighton Spinney (Fig. 10.3), now surrounded entirely by suburbs, lies 4.5 km from the centre of Leicester. It is an entirely man-made woodland and many other examples like it could be quoted from the East Midlands. Formerly the site was ploughed land but in the early nineteenth century it was planted with rows of oak. Ash, elm, sycamore and willow are now also present. The wood is rather neglected and has a poorly developed ground flora. It is no longer drawn by the Fernie Hunt and now forms the core of a municipal park in the Leicester suburbs.

In the late nineteenth century, at the height of fox-hunting as an acceptable and overt social phenomenon, over 300 horses might leave centres such as Market Harborough or Melton Mowbray for a day's sport in these woodlands.

Fig. 10.4 Billesdon Coplow (200 m), a small mixed woodland surviving in a rich agricultural region because of its strong association with the Quorn Fox Hunt.

Today, despite antagonism from some sections of society (a topic discussed later in this chapter), fox-hunting has not diminished and there are more hunts than at the turn of the century. In a landscape which is predominantly agricultural, it will be interesting to see what happens to these small but important woodlands in the future if rapidly rising costs and the present pace of social change leads to a decline in fox-hunting. Already many are under increasing pressure from tourists, and in several, vandalism is evident. Like the remote Cairngorms, these woodlands represent a dynamic situation. In both cases there is overwhelming evidence of man-induced changes during the past and a certainty for the future that man will become more not less important in the ecology of these areas. These two regional studies could be matched by similar examples drawn from virtually any part of the British Isles. As we shall see in a later section, for all British woodlands the basic problem is to so manage this resource that the pressures emanating from the various human uses do not undermine good conservation practices.

Much that we cherish in the countryside today is artificial and results from centuries of environmental modification by man. The open hillsides of the Lake District, the Chalk Downlands with their distinctive chalk floras, the expanses of heather moor and the 'typical' rolling English countryside of fields, hedges and small woodlands all fall into this category. Some species have proved resilient to change and others have adapted well, extending their range into man-made environments. The seagull, for example, appears to be just as much at home scavenging from inland urban refuse tips and frequenting reservoirs as it does in its more usual coastal habitat. Likewise, pigeons, closely related to rock doves, find suitable nesting sites on the 'cliff ledges' of office buildings. An even more interesting case is the collared dove (*Streptopelia decaocto*). This native of India reached Turkey in the sixteenth century but has now shown an explosive spread across Europe during this century. First reported in Lincolnshire in 1952 it had reached a population of 30–50,000 in Britain by 1970. Coombs *et al.* report that it is invariably associated with human settlement, particularly favouring town parks and suburban gardens. The introduced Oxford ragwort (*Senecio squalidus*) is an example from the plant world. It has spread rapidly in Britain along the cinder-strewn embankments of railways and into other disturbed stony sites where conditions approximate to the lava fields around Mount Etna, one of its original habitats in Sicily. An increasing number of species, however, are highly vulnerable to these changed landscapes and some have become extinct while many are now rare. These trends must be carefully considered as we contemplate present and future patterns of activity in rural Britain.

PRESENT AND FUTURE IMPACT

The increasing pressure of recreational pursuits on the countryside was previously mentioned and this will now be examined in more detail. The developments described mainly relate to the Cairngorm Mountains but most of the findings are of general application because many other parts of the British Isles are undergoing a similar experience, e.g. the Lake District, Dartmoor and the Peak District.

In the Cairngorms the hill-walker and mountain-climber, always keenly appreciative of what the region offers, have now been joined by rapidly growing numbers of motorised tourists. For some time tourists had limited access to the region but now there have been road improvements and a great extension of tourist facilities. The small settlement of Glenmore has been expanded and the road widened and continued to 760 m. Chair-lifts, ski-tows, snack bars and restaurants now exist at the terminus. These early investments soon bore fruit. On Easter Monday 1962, 5,000 people were estimated to be on Cairngorm itself and during the first seven months of operation 80,000 people used the chair-lift. Over a five-day period in 1965 the Cairngorm Board collected some £4,500 on lift and toll charges. Commercial success like this prompts more investment, an example being the nearby £2½ million Aviemore Centre which includes two large hotels (now frequently used as conference centres), chalets, a swimming pool, an ice-rink and artificial practice ski-slopes. In 1982 further proposals for expansion into adjacent corries were under consideration (including two ski-lifts, two tows, parking for 1,000 cars, cafeteria, toilets and snow fences). Lickorish, writing about planning for recreation and leisure, noted the point made by the US Department of Commerce as early as 1950: when a community can attract a couple of dozen tourists daily throughout the year, it is economically comparable with acquiring a new manufacturing industry whose annual payroll is $100,000. And business interests soon realise that an investment of £100,000 in a mountain that brings in £10,000 profit a year is just as useful as the same money made from more conventional investments. However, some might see the situation rather differently in that such schemes may not create as many jobs as the same money more conventionally invested and therefore the end result might be less socially acceptable.

An increasing number of visitors to the Cairngorms are people on day trips. Many will come in their own vehicle but there is a rapidly growing bus and coach tour element in these numbers (this expansion in the early years of these recreational developments can be seen from Scotland's bus and coach tour passenger numbers which increased from 3,900,000 in 1960 to 5,800,000 in 1963). Because of the relative remoteness of many recreational areas this implies a lot of travelling which leads to congestion on roads originally built to serve small, isolated Highland communities.

In the context of the United Kingdom, Dower, writing on the function of open countryside, has considered some of the main conclusions of the 28–volume report published in 1962 by the Outdoor Recreation Resources Review Commission of North America. The main American findings were:

1. The population of the United States will double by AD 2000 but the national demand for outdoor recreation will treble.
2. The recreational areas are remote from the centres of population.
3. Driving for pleasure is today the most popular outdoor activity.
4. Most people try to centre outdoor recreational activities around a body of water.

It is more than 20 years since Dower's paper and no such detailed comprehensive report on recreation exists for this country. However, an increasing number of studies of specific areas within the British Isles have since been produced (e.g. Moss; O'Riordan) and several general texts, often viewing the recreational problem in global terms, contain much that is relevant to this country (e.g. Simmons; O'Riordan and Turner). It is now quite clear that the pattern of leisure-time activities in Britain has moved strongly in a similar direction to that seen in the United States. We already know that the majority of motorised tourists seldom venture more than a few hundred yards from their car and this often gives rise to the 'honey-pot phenomenon' where large numbers congregate at popular vantage points. A good example in the Cairngorms is Loch Morlich (Fig. 10.5), where erosion of the shoreline is already a problem (similar shoreline erosion has been reported by Tivy and Rees for many Scottish lochs). There is a strong case for preserving extensive parts of the Cairngorms against massive penetration by tourists, but although a large Nature Reserve does exist the pleasure-seeking public have never shown great respect for lines on maps. Such a vast influx of tourists coupled with unplanned piecemeal development will do much to destroy the essential character of these mountains.

Fig. 10.5 Loch Morlich and the Cairngorm Plateau. The development of recreational facilities – water sports and hill-walking in summer, skiing in winter.

PROBLEMS OF MULTIPLE LAND USE

There is agreement that development in regions such as the Cairngorms must be planned and should consider all interests, including the larger national interest. The problem is to find a plan broadly acceptable to all and, at the same time, capable of implementation at reasonable cost. Difficulties arise because interests clash and priorities have to be determined. One interest, that of the tourist, has been mentioned but others include those of the water boards, the power stations, the military, mining, forestry, agriculture, sporting activities, transportation and last, but not least, conservation. Many of these uses are mutually exclusive and nearly all have their problems of operation greatly increased by the presence of large numbers of tourists.

Another factor adding to the difficulties of resolving clashes of interest in our countryside is the question of land ownership. Much of our landscape is still privately owned and the land-use pattern here will often, but not always, be determined by the necessity to make a profitable return on investment. Large areas are state-owned or state-controlled and motives other than profitable return may then decide land use. But, it should be remembered, many designated National Parks do include farms and private land within their boundaries. This makes a comprehensive development plan for the whole region that much harder to achieve.

It is beyond the scope of this book to deal fully with each interest in turn. Instead, a few comments under each heading will be made so as to convey something of the complexity involved in multiple land-use planning. However, fuller treatment will be given to conservation because of its central role in preserving as a resource these very landscapes that we wish to use and because effective conservation calls into consideration many of the ecological principles outlined in previous chapters. This makes it of special interest to biogeographers.

WATER BOARDS

The industrial and domestic demand for water has risen sharply in this century and it is usually argued that this will continue to be the case in the foreseeable future. For example, the installation of washing machines and dishwashers in the home and the rise in personal cleanliness have led to a dramatic increase in household consumption of water. Agriculture now uses more water and consumption is likely to expand even further in eastern England, where intensively grown crops are likely to benefit from the use of sprinklers and irrigation three years out of every five. Heavy demand, coupled with a very variable climate, already leads to local and regional shortages during spells of dry weather. However, forecasting future demand for water is not without its difficulties. Several large traditional industries which are heavy users of water are now in marked decline (e.g. steel). Many of the newer industries based on electronic technologies will not require massive inputs of water. The problem is well illustrated by the Kielder Reservoir. Since completion of this recent £167 million scheme the industrial demand in the Northumbrian Water Authority area has already declined by 5 per cent and full use of the water from Kielder is not now likely until the year 2000.

The creation of large reservoirs must take place at topographically suitable sites which, by their very nature and location, will tend to be attractive areas for recreation. Many reservoirs have been sited in our uplands but an increasing number are now in lowland areas. Wherever the site is, local opposition to the loss of land involved can be passionate: a group opposed to a new reservoir on farm land in the Roadford Valley, west Devon, even threatened deliberate pollution of the River Tamar if the scheme went ahead. Extensive areas of farm or forestry land have been flooded by schemes and in the past the fear of water pollution and bank erosion has usually meant the exclusion of people from these sites. Fortunately, this policy has been revised of late and more reservoirs now cater jointly for water and recreational needs. This is well seen at the new Rutland Water Reservoir which lies in the heart of an important agricultural region. The loss of farmland here will be partly compensated for by the provision of recreational facilities, particularly for fishing, boating and water-sport interests but also providing picnic areas and nature reserves (Figs. 10.6 and 10.7). However, at Cow Green, Upper Teesdale, plans to build a reservoir to serve the ICI industrial complex on Lower Teeside met great opposition. This was because the proposed site was the habitat of several rare plants, a unique arctic-alpine floral element containing species such as *Gentiana verna* and *Minuartia stricta* which are confined in Britain to the Upper Teesdale area. Despite a well organized national campaign and strong support from many internationally eminent botanists the plans were implemented.

Another activity by water boards that threatens wildlife is the drainage of wetlands. Halvergate Marshes in Norfolk is a classic case where controversy

Fig. 10.6 Rutland Water Reservoir, Leicestershire, in 1976: an early stage in the drowning of agricultural land along the valley of the River Gwash and its tributaries. Normanton Church, with the dam visible behind it, was built in the eighteenth century. It will remain on a reinforced base as an island surrounded by water and a reminder of former land use in this area. Trees planted in foreground to improve the visual impact.

Fig. 10.7 Rutland Water Reservoir, Leicestershire in 1983: What the people want! The same area as Fig. 10.6 but seven years later: car parking, eating and toilet facilities, leading to the 'honey-pot phenomenon'. Normanton Church appears just to the right of the conical toilet block.

between farming and conservation interests has raged for several years now. Drainage would allow profitable grain crops to be grown, but to the conservationist Halvergate is one of the last pieces of species-rich grazing marshland in Britain.

MILITARY TRAINING

The Ministry of Defence controls several large regions in the British Isles where the general public are excluded. Unfortunately, some of these regions are highly attractive landscapes with great recreational potential. Examples are western Dartmoor, parts of the Dorset coast and a portion of Breckland in Norfolk. Currently, these areas are under review with strong pressure for more to be released for enjoyment by the public. Some people have argued for an almost total reduction of military involvement in these regions. However, if we spend thousands of millions of pounds on defence it then seems illogical to deny proper training facilities. The importance of adequate preparation in the use of modern warfare technologies has been emphasized by the recent Falklands conflict.

In some cases the presence of the Military actually assists conservation. For example, access to most of the Pendine Sands of Carmarthenshire is prohibited because of a missile and gunnery testing range. Yet these dunes may well act as a refuge for native fauna and they do carry a fuller and more varied vegetation cover than adjacent sections which are regularly and heavily trampled by tourists (Fig. 8.3). Likewise, the Chemical Defence Establishment at Porton Down in Wiltshire controls about 2,000 ha of chalk grassland which is now one of our most important wildlife conservation areas. It is a diverse region which acts as a sanctuary for many interesting species that are rare or absent in the surrounding zone of intensive agriculture.

POWER STATIONS

By the very nature of their activities nuclear power stations must be sited at some distance from major settlements. The same applies to several of the associated Atomic Energy Authority Establishments (e.g. Dounreay, Windscale (now Sellafield) and Calder Hall). Because of this, many occupy areas that would normally be of prime conservational or recreational interest. This is particularly so since the vast majority of nuclear power stations also occupy coastal sites (e.g. the Central Electricity Generating Board's nuclear power stations at Dungeness, Bradwell, Sizewell, Hartlepool, Hunterston, Heysham, Wylfa and Hinckley Point). Not only are they massive structures in themselves but associated facilities and security considerations mean that a large area of coast is effectively prevented from being used for any other major purpose. Several of the possible sites for future pressurized water reactors (PWR's) of the Three Mile Island type are also coastal locations (e.g. Stakeness on the Moray Firth and Gwithian near Land's End).

MINING

Mining and quarrying are dirty, dangerous and unsightly operations and their intrusion into areas of natural beauty is to be deplored. They also generate a form of transport (the heavy truck) which is particularly damaging to the countryside and infuriating to other road users. Nevertheless, we need these raw materials for the very things we may value in other environments, such as good roads, important urban buildings, gravel driveways and garden rockeries. In the long term their worst effects on the landscape can be alleviated by insistence on proper soil restoration measures and the use of the flooded surface pits for recreational and conservational purposes. When carried out successfully, this can actually enhance the landscape. A good example is seen in the Cotswolds where some of the abandoned and now flooded, shallow, surface quarries are used as centres for fishing and boating, while others are reserved for wildlife. The complex will eventually cater for many needs in an area formerly lacking in surface water facilities. Another example of cooperation is seen at Thrislington, Co. Durham. Here, the Nature Conservancy Council and the Durham Conservation Trust are working in conjunction with a quarry company to move a 8 ha field complete with topsoil, rare plants (including several orchids) and animal life to a new site half a mile away. This will allow pure dolomite, an important ingredient in the steel industry, to continue to be mined.

Mining and quarrying are not always confined to hard rock areas, most of which occur as our uplands. Extensive lowland tracts have been worked for gravels, sands and brick clays, leaving numerous scars on the landscape. Only recently has the full recreational potential of some of these sites been appreciated. However, the picture is complicated because some lowland areas fall within National Parks. A case in point is the highly attractive Mawddach Estuary, which is near sea-level yet lies within the Snowdonia National Park. Because of this, plans to dredge for alluvial gold in the estuary aroused much protest.

FORESTRY

Many British woodlands are privately owned and contribute a very desirable element to our landscape. Whether they remain in their present form may well depend on government taxation policy as much as on anything else. In the past, governments have encouraged tree planting on private estates by a series of financial incentives. But any changes in capital transfer taxation following death of the owner could lead to a sharp decline in many of these woodlands. As Table 10.1 shows, the private sector still produces nearly 60 per cent of our cut timber and is responsible for the vast majority of our more attractive hard wood species.

The state owns vast areas of forest, mostly as coniferous plantations. Formed in 1919, the Forestry Commission was charged with the task of building up strategic reserves of timber following the depletion during the First World War. This has been a dificult task because of the long time-span for tree species

Table 10.1 Volume of timber cut annually in Britain: values in millions of cubic metres (Modified from Rooke, 1974)

Source	Softwoods	Hardwoods	Total
Forestry Commission	1·55	0·04	1·59
Private sector	0·82	1·30	2·12
Totals	2·37	1·34	3·71

to reach maturity and the poor growing conditions at many upland sites. The Second World War also caused further setbacks. However, the Forestry Commission is on target for its aim of 5 million acres (2 million ha) of afforested land (including both state-owned and privately owned). Nevertheless, we still import vast amounts of timber despite increasing home production. Wood and wood products are Britain's third largest import, costing about £2,370 million in 1978. Russia is a main supplier which could make our position vulnerable at times of international crisis. Net imports are unlikely to drop much below 90 per cent of our total timber consumption during this century (Table 10.2).

Table 10.2 Consumption, production and import of timber in Britain at the present and expected for the near future: values in millions of cubic metres (From Rooke, 1974)

Year	Expected consumption	Home production	Expected net imports	Net imports as % of total
1970	45	3·5	41·5	92
1975	51	4·5	46·5	91
1980	57	5·0	52·0	91
2000	82	8·7	73·0	89

One of the major difficulties with forestry as a land use is forecasting future trends within the industry because of the long-term nature of the crop. Raup has studied the basic concepts and assumptions underlying North American forest management and found them to be false. 'It looks as though a better assumption would be that predictions beyond one or two decades were more likely to be wrong than right' and 'uncertainties in the long run call for the greatest possible flexibility in resource use'. These lessons may have strong implications for British forestry. One of the major problems in Britain is the vast acreage now planted with fast-growing conifers, mainly Sitka spruce (*Picea sitchensis*). Sitka grown under our climate is largely a low-grade softwood timber and there is increasing difficulty in finding suitable and profitable outlets for it. While the need for more forest cover can be readily demonstrated, especially when given the forecasts of a massive rise in world timber consumption and severe demands being made on diminishing natural forests (see Bowman), technologies may change and so too may fashions for certain timbers. Thus, what we plant now may not be what is required some 50 or more years later. However, it is the tourist who is likely to cause the main changes in future forestry developments.

In early Forestry Commission plantations the public were discouraged, partly because of fire hazards and partly by the monotonous nature of the vegetation itself. But the key to present changes is contained in a declaration from the Seventh World Forestry Congress (1972): 'Foresters recognise that forestry is concerned not with trees but with how trees can serve people . . . His allegiance is not to the resource but to the rational management of that resource in the long-term interests of the community'. Forestry makes its greatest impact on the general public through the landscapes it creates. The Commission now realize that social and environmental demands must be harmonized with the need to produce timber. By 1974, the Commission had designated seven Forest Parks with a combined area of 176,600 ha and was investing over £1 million annually in recreational provisions. The Commission had about 14 million day visits to its forests in 1980. In Holland, 75 per cent of all state forests now cater for recreation and Britain is also moving rapidly in this direction.

AGRICULTURE

Farming has often been in competition with forestry, particularly in upland areas where clashes also exist with some of the other forms of land use discussed. Indeed, as we have seen, most agricultural land has resulted from the clearance of forest at some time in the past. In 1972 there were some 11,000 hill and upland farms in England and Wales. Much upland terrain is marginal for agriculture and may provide a better economic return and more rural employment under forestry. But the position is not simple, as James shows in his review; each case has to be carefully considered and conclusions applied to whole areas or even whole farms may be incorrect. Decision-makers have fallen into the trap of regarding agriculture in our uplands as a homogeneous activity when, in reality, some parts of an area or individual farm are more suited or less suited to agriculture than others.

We no longer live in a world of cheap food supplies (despite EEC surpluses) and a case can be made for either extending farming acreage or greatly intensifying productivity on land we already use. In lowland areas at least, intensive agriculture will increasingly involve large capital investments for complex technological systems in farming – an aspect which at best makes the presence of too many tourists a nuisance and at worst a positive danger, according to Bonham-Carter. In some areas, however, the farming community has adopted a more welcoming attitude to the tourist. Along certain stretches of coast and in some uplands, farmers were not slow to realize that a field of caravans can be more profitable than one full of cows. On the flanks of Dartmoor a very large component in the income of some farms is now derived from activities associated with the tourist industry.

Government policy (e.g. guaranteed prices or the addition or removal of subsidies) and EEC decisions may well tip the economic balance of marginal agricultural areas one way or another, sometimes strengthening sectional interests and sometimes weakening them. Schemes such as the drainage of wetlands may benefit the farmer, but seldom the taxpayer. Increased production is often in commodities already in surplus (e.g. milk, grains). This means that through the EEC's Common Agricultural Policy these surpluses are brought into intervention to maintain the price paid to the farmer (i.e. storage, until sold later and usually at a loss to the taxpayer). This largely explains the current crop of land use conflicts between farmers and others (mainly conservationists), e.g. Baddesley Common (Hampshire), West Sedgemoor (Somerset), Romney Marsh (Kent), Berwyn Mountains (Wales). The National Farmers' Union have responded by pointing out that an area the size of the Isle of Wight has been lost to British farming in the last 20 years and that 650,000 people are directly employed in agriculture and some 2 million (8 per cent of the labour force) rely for their jobs on British farms and their produce.

SPORTING ACTIVITIES

In the countryside the fostering of sporting interests has played a large part in shaping the fine detail of the landscape. In addition to the numerous fox coverts in areas like Leicestershire, many plant communities were established in both upland and lowland districts specifically for game purposes, e.g. grouse moors, pheasant and partidge coverts. In some remote areas the revenue from sporting activities plays a significant part in the local rural economy. A recent survey of countryside sports shows that 4 million people go fishing, nearly 600,000 go shooting or stalking and 214,000 follow hounds, either on foot or in the saddle, at some time during the year – and these people come from a complete cross-section of the British public. These activities account for about £1,000 million of direct annual expenditure and employ either directly or indirectly about 90,000 people. However, not all look favourably on these activities. The 'blood' sports (involving hounds) are now under great threat. The Co-operative Wholesale Society, the largest independent farmland owner in Britain, has now banned fox hunting over its 12,000 ha and so have certain Labour-controlled councils. Such a ban was also included in the recent Labour Party election manifesto despite claims that hunting actually ensures the preservation of many woodland plots and the hunted species. Thus, as in most areas of land use conflict, a political factor must be recognized.

Catering for certain types of sporting interest is largely in private hands and takes place on private land. Good examples are grouse shooting and salmon fishing. However, there is an increasing development of sporting facilities for the general public and this is having its impact on our countryside. Today, large sums are spent on sporting equipment and there has been a great diversification in sporting pursuits. Although some activities, such as water-skiing or hang-gliding, appeal only to a minority, the clashes with other countryside users which will result from their expansion will soon have to be resolved. The creation of special countryside areas just for active sport is a legitimate use since these activities may do much to relieve the tensions and aggressiveness generated by sedentary or monotonous urban employment.

TRANSPORTATION

For the foreseeable future we are mainly thinking in terms of road networks when transportation in the countryside is considered. In rural areas which attract many tourists the road network soon becomes heavily congested, having been established initially to meet only local needs. Any major improvement in the network may only result in more people flooding into the region. The better access then available may cause a greater strain on many of the other facilities in the district. So, to a certain extent, these 'improvements' can be self-defeating. For example, motorways recently constructed in southern England now give quick access from the populous south-east to the popular south-west and there is a danger of 'killing the goose that lays the golden egg'. A similar fear has been expressed for the Lake District in Cullingworth's review of these problems.

The construction of major road systems means a loss of much land already in other uses. Gradient considerations dictate that this will often be the better quality agricultural land. By their very nature these major roads will usually follow the most open and obvious routes into a region and will be visible from many vantage points. While improving access from outside, they may cause local resentment and a breakdown of established patterns of social association for those living permanently within the region. The controversy over the proposed Aire Valley Trunk Road in Yorkshire stems from such local resentment. Here, a major planning enquiry was repeatedly disrupted by local pressure groups and finally had to be abandoned after costs of about £30,000 had been incurred.

Improvement of minor roads and trackways provides further access for tourists but may also bring benefits to agriculture. As Jones points out, one of the most serious physical handicaps to upland pasture improvement is the lack of good access roads. Millions of hectares are not fully developed because vehicles carrying lime and other fertilizers just cannot reach these sites.

Another major threat to rural areas is the construction of large airports. With their voracious appetite for land, their very specific site requirements and their attraction of numerous ancillary facilities and activities, they represent a major source of land use conflict. One has only to think of Maplin Sands and Stansted to be immediately reminded of these conflicts.

Having seen something of the problems that face the environmental planner we may now turn to a consideration of conservation, bearing in mind that each activity above causes special difficulties for the conservationist. When several are combined they may make good conservation practice virtually impossible: a neat exemplification of this is contained in the two well-illustrated papers recently produced under the general title 'Alarm call for the Broads', (Moss; O'Riordan).

CONSERVATION PROBLEMS

TYPES OF CONSERVATION

A full inventory of our biota and the key sites considered to be of international, national and regional importance is a basic need for a proper consideration of conservation in this country. Ratcliffe has provided this in his two-volume survey which also includes a discussion of the criteria to be used in site selection. Margules and Usher have also reviewed similar conservation criteria and Peterken has written on conservation priorities for British woodlands.

O'Connor has listed four main types of conservation:

1. *Species conservation.* A concern for the protection of rare, interesting or beautiful species which are under some form of threat. Often this is a straightforward problem but not always so as the present controversy over the very rare thistle broomrape (*Orobanche reticulata*) illustrates. The 1981 Wildlife and Countryside Act gives complete protection to this plant with fines of £500 for every one killed. However, the creeping thistle (*Cirsium arvense*), upon which this parasitic broomrape grows, is an 'injurious weed' under the 1959 Weeds Act, with fines of £75 for farmers who fail to destroy it!

2. *Habitat conservation.* The aim is to maintain representative habitats over a wide ecological range which will provide reference points against which changes resulting from activities by man can be measured. These sites will allow important research into the functioning of natural ecosystems to proceed – work that is still in its infancy.

3. *Conservation as an attitude to land use.* Here the conservationist plays a positive role in planning and management of land which is subject to great pressures from various competing forms of land use. The aim is to strike a balance and so prevent disruption or destruction of natural ecosystems.

4. *Creative conservation.* More and more landscapes are now the result of large-scale human intervention (e.g. motorways, Rutland Water Reservoir, the Cotswold extractive mineral workings). The conservational opportunities afforded by these new sites should not be lost. Bradshaw has advocated 'active' conservation in these situations. This involves the *restoration* of disturbed areas, the *re-introduction* of suitable species (e.g. fish into cleared rivers; primary woodland ground flora species into secondary woodlands), and the *creation* of new conservation areas at totally degraded sites. Interest in *urban ecology and conservation* has greatly increased in recent years as witnessed by the appointment of a senior ecologist by the Greater London Council to study and monitor the many sites within the 610 square miles of the city (encompassing derelict land, disused railway lines, canals, dumps, reservoirs, woodland patches, riverside meadows, etc.).

The aims of conservation have moved a long way since the early days when there was a preoccupation with rare species – a concept of conservation still frequently held by the general public. We now think more in terms of whole landscapes, the resources they contain and the position of man in this complex. One of the major difficulties still remains the evaluation of an area for conservation purposes. Although a scoring system can be devised for assessing the worth of a feature, habitat or landscape, the values given will largely be the result of a subjective impression. We have not yet arrived at universally acceptable criteria and O'Connor notes the many elements of uncertainty in ecological evaluation.

An important consideration in conservation is the question of diversity. As Goldsmith emphasizes, diverse habitats are usually more interesting and are generally to be encouraged. These areas will usually support richer floras and faunas. There are areas, however, which are valued because they support extensive tracts of only one habitat, e.g. Pennine moorlands. Human activity in such areas may increase the diversity through trampling and disturbance, but this would not normally be welcomed. He concludes that there is ambiguity over the meaning of diversity and whether or not its increase or decrease is a desirable change. Each conservation area must be given a particular consideration in this respect.

The conservationist is sometimes popularly viewed as someone intent on 'fossilizing' habitats, eager to 'let well alone' by preventing other forms of land use. There are certainly cases where this approach is appropriate but in many situations an active role is called for rather than a passive one. Sometimes the continuation of a human activity is essential for the good conservation of a habitat or species. For example, the coppicing system led to a relatively rich ground flora and the vernal aspects of many woods are an attractive and much valued feature. The system has unfortunately declined, although several conservation groups are now attempting to re-introduce it. Likewise, Welch and Rawes have queried earlier contentions that sheep grazing on British uplands leads to extensive uninteresting grasslands with impoverished floras. On the contrary, they show that in the Northern Pennines while some species suffer elimination by grazing, others, especially some of the rarest Pennine plants, survive in heavily grazed places because grazing restricts the growth of their potential competitors.

We now know that if herds of large herbivores are to be conserved in good condition man must play the role of predators now rare or extinct. The case of the Kaibab deer has already been quoted (page 111) but the situation is no different for herds of red deer in Scotland. A regular cull is essential for the health of the population. Cutting the number of common gulls is essential on some offshore islands if rarer bird species (e.g. roseate tern) are to be encouraged. Ironically, the Royal Society for the Protection of Birds does this work, using poisons, and the Nature Conservancy also regularly kill gulls on the Isle of May in the Firth of Forth to protect vegetation and other birds. Active measures must also be taken against some plant communities if undesirable developments are to be halted. Fig. 10.8 illustrates this nicely. The Castor Hanglands Nature Reserve consists of three main plant communities: open grassy heaths, thick impenetrable thorn scrubs and mixed deciduous wood-

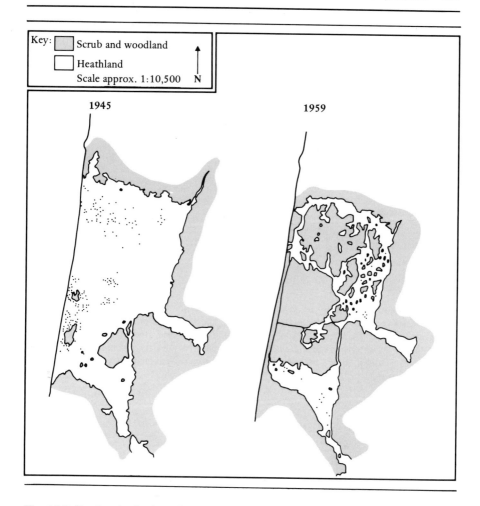

Fig. 10.8 Scrub colonization of grassy heathland at Castor Hanglands Nature Reserve, near Peterborough. Based on air photographic coverage for the years shown. (From Gent, 1966.)

land. The reserve is of importance in an arable landscape largely devoid of other areas of natural and semi-natural vegetation. It will become even more important with the planned expansion of nearby Peterborough to a population of 160,000 (previously to 250,000 but now revised). Unfortunately, the open grassy heaths are being rapidly invaded by scrub (Fig. 10.9). If this continues the floristic and faunal diversity of the reserve will greatly decline as will suitable areas for recreation. In this particular case experiments with sheep grazing and rotary cutters have been used to determine the most effective method of preventing scrub encroachment.

Fig. 10.9 Massive invasion of grassy heaths by scrub species at Castor Hanglands Nature Reserve, near Peterborough. Main grass species are: *Deschampsia flexuosa* (wavy hair-grass), *Brachypodium pinnatum* (false brome grass) and *Arrhenatherum elatius* (oat grass). Main scrub species are: *Crataegus monogyna* (hawthorn), *Prunus spinosa* (blackthorn), *Rosa canina* (dog rose) and *Viburnum lantana* (Wayfaring tree). In the background, mature scrub overtopped by oak and ash.

'ISLAND' BIOGEOGRAPHY

Biogeographers have long had a particular interest in the geography of remote islands. Such islands often have a unique fauna and flora, pose problems in terms of their isolation and how biota dispersed to reach them, and can act as natural laboratories where the boundaries are easily drawn. Indeed, studies of island biotas had great impact on the thinking of eminent scientists such as Darwin and Wallace. However, it was not until the 1960s that these investigations were given a sound theoretical base by the work of MacArthur and Wilson.

Their studies showed that the relationship between island area and species diversity is essentially linear, diversity being almost wholly a reflection of island area (though, of course, other factors are involved such as the number of habitat niches on an island). In their *equilibrium model* (Fig. 10.10a) for the biota of a single island, the initial rate of colonization will be high as the island is found quickly by species adept at dispersal and as vacant niches are occupied. This rate must drop sharply as chances increase that new arrivals will belong to species already established. With increasing competition between species and only a limited genetic stock established, the rate of extinction will climb steeply. The intersection point of the two curves gives the equilibrium number of species for that island. Next, MacArthur and Wilson considered the interplay between island isolation (near or far) and island area (small or large) in

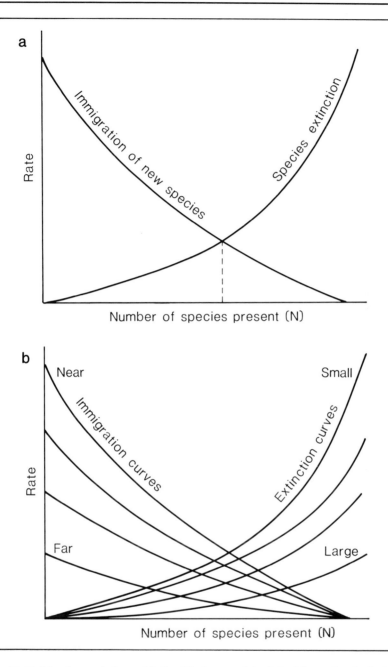

Fig. 10.10 Island populations: (a) Equilibrium model of the biota of a single island. The equilibrium number of species, indicated by the dashed line, is reached where the curve of rate of immigration of new species, not already on the island, intersects with the curve of extinction of species that are on the island. (b) Equilibrium models showing the inter-relationship between isolation and area of islands. An increase in distance (near to far) from source of colonists lowers the immigration rate, while an increase in island area (small to large) lowers the extinction rate. (Modified from MacArthur and Wilson, 1967.)

determining the shift of the equilibrium point (Fig. 10.10b). For example, increasing island size will tend to lower the extinction rate whilst increased distance from source will lower the rate of colonization, and vice versa.

These relationships can be expressed mathematically and one prediction is that quite high rates of extinction are the usual pattern (though some studies have queried this point). Turnover rates of species on islands can also be worked out. Of course, not all islands are initially virgin sites: those created by post-glacial sea-level rise are likely to be super-saturated with species at the start, gradually declining towards an equilibrium point rather than building up to it. Diamond and May have taken these ideas further by working out mathematically the *'incidence function'* for particular island species, i.e. their survival prospects. They have also developed *'assembly rules'* which indicate the likelihood of a species presence on an island, i.e. all possible combinations of species are considered but only certain mixtures exist in nature, thus allowing probability values to be assigned to each assemblage. By combining various forms of quantitative information, the basis exists for a 'predictive science of environmental management', according to the advocates of these ideas. A number of field studies have supported this theoretical framework (e.g. Wilcox; Simberloff and Wilson) but Gilbert has reviewed criticisms that lead him to reject largely the main thrust of the equilibrium theory.

Many of the ideas above on island biogeography were originally developed in relation to real islands. But there are many other types of 'island' to which these ideas could apply (e.g. woodland remnants, mountain tops, or even nettle patches in large fields – any isolated habitat in this sense qualifies). Most nature reserves in Britain also fall into this category. This leads to an aspect of conservation which naturally comes within the geographer's field of interest, namely, the question of the *size, shape* and *distribution* of our nature reserves. Streeter notes that with land values at a premium, information on the ideal size of nature reserves has become a vital economic as well as ecological concern for conservation. He defines the ideal size as that which will support a viable population of all those species regarded as characteristic of the ecosystem concerned. This size is determined by considering the area needed for occupation by the species in the system with the largest territorial requirements. These species are usually those at the end of a food chain. Table 10.3 gives these sizes for some forest species and on this approach the minimum size for a nature reserve would often need to be in excess of 40 ha (100 acres). Over half of our woodland National Nature Reserves fall below this size. Even as reserves for plant species they are often woefully inadequate: as Ratcliffe pointed out, woodland is the climatic climax of the British Isles, but it is now the most fragmented of all our habitats. The effects of this fragmentation and of size are considered by Helliwell in terms of the conservation value of British woodland sites.

Streeter also draws attention to the importance of shape. In long narrow woodlands much of the wood is edge and little is true woodland. Edge zones represent transitional environments, especially in terms of microclimate. Some species are favoured by a high proportion of edge, others are not. Game has argued that for some species (e.g. birds) linear shapes are better than circular.

Table 10.3 Territory sizes of some forest vertebrates (From Streeter, 1974)

Species	Territory – acres (ha)	
Sparrowhawk	96–1,280 (38·8–600)	
Tawny Owl	20–30 (8–12)	
Woodcock	15–50 (6–20)	
Pine Marten	220–640 (89–259)	
Red Deer	1 deer per 120 (48·5)	} Forestry Commission figures of
Fallow Deer	1 deer per 70 (28·3)	} 'acceptable' density

A fear often expressed by conservationists is that reserves will not only be too small but will become too isolated from each other. This is particularly likely in lowlands where much of the land use is already inimical to wildlife. The reserves thus appear as 'islands' in a 'sea of hostility'. If they are too widely spaced, movement of animals between reserves becomes hazardous or impossible. This could lead to inbreeding or failure to find a mate. In this context the hedgerow may well provide a relatively safe 'corridor' for movement or act as an important reserve itself for smaller animals. During the last 24 years about 80,000 km of hedgerow has been cut down in England and Wales and wholesale removal like this is to be deplored. Fortunately, according to a recent government survey, some farmers now realize that they were over-hasty in removing hedges and now see the error of their ways.

Several strategies have been presented showing the alternatives available for designing nature reserves in terms of their size, spacing and location within a region. Figure 10.11 shows two such schemes but, unfortunately, in a country like Britain nature reserves can seldom be designed from scratch – all too often one starts from a position where many problems already exist.

CARRYING CAPACITY

The major threat to conservation is the recreational use of the countryside. In their approach to the study of this impact on ecosystems, planners have made use of the concept of *recreational carrying capacity*, hoping to determine the permissible numbers that should be allowed to participate in activities within a particular environment. It is a concept very relevant to the conservationist's interests because if carrying capacities are exceeded then his work is seriously jeopardized. Attractive although this idea is, it is fraught with difficulties of definition and application. The Countryside Recreation Research Advisory Group recognize four different types of carrying capacity:

1. *Physical* – a term normally applied to man-made structures in the recreational environment and representing the maximum number of people or activities that can be accommodated for the purposes for which the facility was designed.

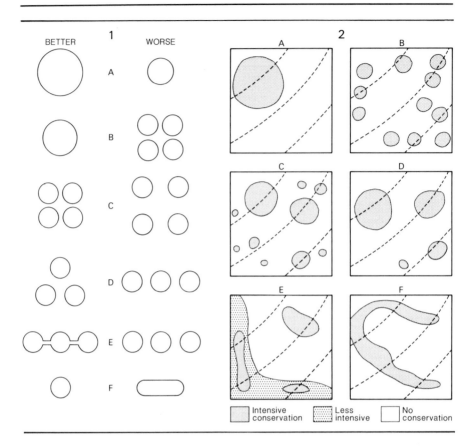

Fig. 10.11 The design of nature reserves, based on studies of island biogeography. (1) Scheme proposed by Diamond, 1975 where large is better than small, single is better than multiple, compact is better than scattered, linked is better than unconnected and circular is better than linear. (2) Scheme proposed by Helliwell, 1976, showing alternative patterns of land-use in a region or country with four major ecological 'zones' (separated by dashed lines). A: single large reserve overlapping two 'zones'. B: several small conservation areas covering all 'zones'. C: mixture of reserve sizes. D: fewer, larger sized reserves than in C. E: similar to D, but with areas of less intensive conservation linking areas of intensive conservation. F: single intensive conservation area straddling all ecological 'zones'. Solutions become increasingly satisfactory from A through to F. (Modified from sources quoted above.)

2. *Ecological* – the maximum level of recreational use acceptable before a decline in the ecological value of the site sets in.

3. *Economic* – the maximum level of recreation use acceptable before it begins to affect the economic well-being of other local non-recreational activities.

4. *Perceptual* – the maximum level of recreational use acceptable before there is a decline in the recreational experience of the participants.

Apart from 1. above, all are very difficult to measure and depend heavily on subjective determinations. People may vary greatly in their perception of the same environment and an individual's perception itself can change with time or alteration of circumstances. Because of this, Barkham argues that it is at best unprofitable and at worst seriously misleading to generalize about capacities from a few studies. It is an area in which much more detailed research is needed.

The *'World Conservation Strategy'*, 1980, clearly revealed that man is increasingly exceeding the carrying capacity of many parts of the planet. Britain's response to this document, *'The Conservation and Development Programme for the U.K.'*, 1983, consists of seven reports covering most aspects of our environment. The Rural Report is particularly concerned with conservation of the countryside and has two main thrusts: to preserve representative examples of landscape and habitats, many of which are substantially unchanged over the last 200 years and have a distinct regional character; to create new landscapes which, because of attention to sound ecological principles, will stand a good chance of becoming equally cherished in the future. As in all this work, a major problem is accurate forecasting of future developments. Gordon has provided a cautionary note for those who are inclined to take either a very pessimistic or a very optimistic view of environmental issues in the future.

The considerations outlined in this chapter make the achievement of a correct policy for the use of countryside ecosystems a challenging and extremely difficult task. Priorities will have to be determined in the light of short-, medium- and long-term goals. Decision-makers will seldom, if ever, be able to please all sectional interests, but they have a duty to see that the changes taking place in our countryside are in the long-term public interest and, at the same time, ecologically acceptable. We have moved into an area where, as Coddington notes, the problem of implementing conservation is institutional and political. To quote Fraser Darling, 'the term natural resource is a commonplace: natural resources are the currency of conservation, but land, water, plants, and animals are not the sum of them. Human well-being is an immense resource which can be squandered or marvellously regenerated. Planning and politics are potent ecological factors which conservation must grasp.' As we have seen, the ecosystem is an integrating concept which gives due weight to the important role of man. The relationship of man to land, a factor now largely determining the well-being of the other ecosystem components, is one that is central to geography.

DISCUSSION SECTION

When we looked at the pest problem we saw that the economic considerations could be as important as the ecological. Is this equally true of land use in natural and seminatural environments?

With our high population and relatively small land area we should not use land inefficiently, although we sometimes do. Increasing demands for recreational facilities and the need to produce more water, timber and home-grown food must mean a careful reappraisal of land use in many areas. With so many potential uses involved it becomes very difficult to judge the merits or otherwise of each particular claim. The superiority of an economic approach to this problem is advocated by James: 'In seeking a criterion by which an objective comparison can be made between the claims of a variety of activities for the use of a given area of land, the economic implications of these alternative activities provide society with its best measuring rod. This is not to imply that social, strategic or political issues are irrelevant or unimportant, but rather to suggest that these factors are ancillary to economic efficiency of resource use and that, in consequence, their influence should be subordinated to that of economic performance.' He advances two reasons for this. First, economic analysis of differing types of land use 'enables society to make valid comparisons between *quantifiable objectives* whose end-products are heterogeneous'. It also allows an evaluation of some *non-quantifiable objectives*. These are often subjective in character and defy direct evaluation, e.g. the amenity value of a woodland. As James shows, we can determine the difference between the income coming from the trees in a woodland planted purely for timber production and one planted with a mixture of species to enhance the diversity value of the site. This difference may be taken as the cost of the amenity, a cost resulting from the pursuit of non-monetary objectives.

Secondly, land itself is an economic good which possesses value. Its use involves the employment of other goods such as labour and capital to produce a commodity (food, water, recreation) which, in turn, is economic in character, being related to supply and demand and possessing a monetary value. 'In view of the economic relationships which the use of land evokes, it seems both reasonable and logical to suggest that the assessment of the claims of alternative activities to the use of a given area of land should, at least initially, be made on the basis of their respective efficiencies in the use of the economic goods incurred.'

A major problem in the economic approach is, of course, the measurement of intangible benefits in the equation. Some are more difficult to evaluate than others, while a few may well be impossible to assess precisely. How, for example, does one measure the value of benefits to physical and mental health derived from recreational activities? James shows, however, that we can make some headway in this difficult field. This may be done by considering the price, variously measured (e.g. the cost of travel), people are prepared to pay for a commodity and the quantity of that commodity subsequently demanded.

For forestry, Raup called for great flexibility in use because of uncertainties in the long run – but would this not also apply to other forms of land use?

Yes, there is no ultimate land-use classification. As our demands change or our technological capability improves so areas once thought only suitable for a particular form of land use may be reclassified. Some activities, however, by their very nature, will tend to impose a form of land use on an area far into the foreseeable future. Large reservoirs, for example, will remain for as long as possible in the landscape. If other interests are not catered for when such sites are planned then we have lost this flexibility over a large area.

In regions which have long experienced a particular form of land use there is marked inertia towards change. Opposition by the 'sitting' interest to reclassification schemes is likely to be highly vocal and well organized. But according to James, the use of our uplands is not the prerogative of the farmer or the forester. Farmers have no sacrosanct right to the use of the land and it is simply because of historical reasons that they now use over 80 per cent of the total land acreage. One approach to the problem of trying to achieve multiplicity of land use can be seen in the Wildlife and Countryside Act, 1981. The Nature Conservancy Council can now declare sites to be of special scientific interest (SSSIs) and several thousand more of these are likely to come into being. Unfortunately, most sites involve farmland and agreement may have to be reached with almost 30,000 owners and tenants, since some sites cover several farms. Farmers can be compensated for profit lost by not undertaking agricultural operations, but naturalists feel that this provision will make conservation impossibly expensive. MacEwen and MacEwen state that 'the central principle of the Bill was, therefore, that the conservation authorities would have to buy the goodwill of the awkward farmer or landowner (if it could) by making conservation profitable to him'. They go on to describe the Act as 'a dead end, from which another government will have to retreat before it can advance by a different course. It leaves agriculture and conservation on a collision course, but provides no way of regulating the conflict except by pouring small amounts of money into a bottomless pit.'

One approach to land use is to assess the land potential or capability rather than to map the existing use. *Land capability maps*, largely referring to agricultural potential, now accompany the Soil Survey Memoirs. These ideas have been extended to cover all forms of competing land use. The scheme shown in Table 10.4 is one used in the United States where land is graded according to its primary use. The emphasis is on ecological constraints and within each *primary use* there may be important secondary uses. For example, some farm land may also have a *secondary function* as recreational or wildlife land. Similar approaches have now been used in this country. McVean and Lockie provide an example from the Western Cairngorms (Fig. 10.12). A detailed study by Statham of the North York Moors includes a series of maps showing existing land uses, agricultural capability, landscape zones, optimum use in an 'ideal' competitive situation, optimum use in an 'economic' situation, potential conflict and opportunity areas, and optimum uses for an 'amenity' situation. Valuable though these maps are, they can only point the way: in the final analysis it comes down to the *environment as perceived* by planners and decision-makers who, in turn, are influenced by the perception of others, which may not always approximate to the environment as it is.

Apart from the obviously bad effects of the few who are responsible for vandalism, fires and litter in the countryside, what harm can the mass of visitors possibly cause if all they do is simply walk short distances?

Table 10.4 An outline of the Land-Capability Classification used by the US Department of Agriculture

Land-use suitability	Land-Capability Class and degree of limitations for use	
Suited for cultivation	I.	Few limitations. Potentially very productive sites. Wide latitude for use
	II.	Good land, but some slight limitations for certain crops or uses
	III.	Moderately good land but severe limitations for some uses. Regular cultivation possible providing potential hazards are taken into account
	IV.	Very severe limitations. Careful soil conservation measures must be applied. Suitable for limited or occasional cultivation
Not suited for cultivation	V.	Few limitations for use as grazing land or for forestry but not suited for cultivation because of wetness, rock outcrops, periodic flooding, etc.
	VI.	Moderate limitations for use as grazing or forestry land but too steep, wet, dry or stony for cultivation
	VII.	Severe limitations for grazing or forestry because of very steep, dry, wet or rough nature
	VIII.	Not suitable for cultivation, grazing or forestry. Extremely rough, dry or wet land. Useful only for wildlife, recreation or watershed protection

The following land-capability *subclasses* are recognized, based on four specific kinds of limitations:

Subclass	Degree of limitation for use
e	Limited by severe hazards of water or wind erosion
w	Limited by high water tables, poor drainage or excess water
c	Limited by climate; temperature and/or precipitation regimes unsuitable
s	Limited by root-zone conditions; soils may be shallow, stony, saline or low in fertility or water-holding capacity

This is best answered by considering some examples. Superficially, people just walking for enjoyment appear to do little harm. This would be so if numbers participating were small. However, this is not the case today. By reducing the height of tall grasses, light trampling may have a beneficial effect and encourage other species to grow. But more and more areas are now subjected to very heavy trampling. In these areas a few species such as *Poa pratensis* and *Plantago coronopus* are highly resistant to foot pressure but most, including many attractive flowering plants, are quickly eliminated. Loss of plant cover may lead to soil erosion, while soil compaction through trampling pressures can cause permanent puddles to form. Where puddling occurs, people seek alternative adjacent routes until they, in turn, become waterlogged. This results in a rapidly widening band of unsightly terrain. This is well seen on the Lyke Wake Walk across the North York Moors where sponsored walks

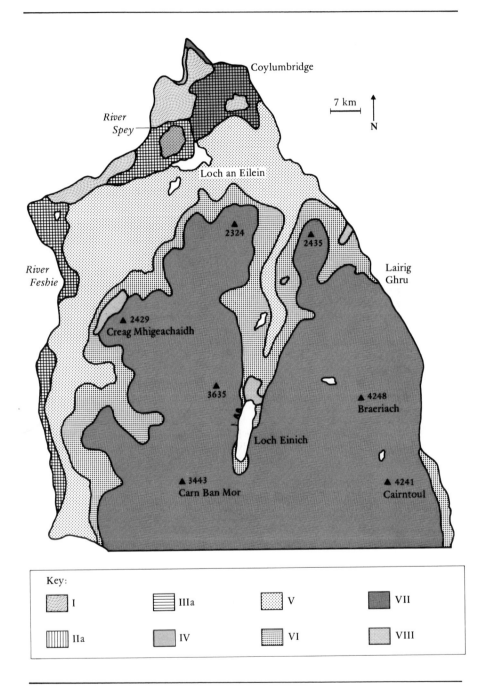

Fig. 10.12 Part of the Western Cairngorms: a division of the area into Land Capability Classes. See Table 10.4 for explanation. (Modified from McVean and Lockie, 1969.)

(one recently of 450 people in one huge phalanx) have widened the path by as much as 56 metres in poorly drained peaty areas.

Burden and Randerson have studied the effects of trampling on vegetation and soils in several semi-natural areas and have shown these to be quite significant. For example, an organized nature trail at Ranmore on the North Downs resulted in 7,729 people over a nine-day period in July 1969 following a route which included previously untrodden woodland and chalk grassland. Changes in path width and profile, soil compaction, litter and ground vegetation were clearly detectable; the path, for instance, increased in this short period from 0·5 m to 4 m wide.

Sand-dunes and salt-marshes are other areas frequently subjected to intense trampling. According to Schofield, a mature salt-marsh can suffer almost complete loss of plant cover by the movement across it of about 7,500 people per season. The elimination of key plants responsible for stabilizing the sand will result if similar numbers of people use dune communities for recreation. These numbers are not exceptional any longer and at many sites are regularly reached or surpassed. Slater and Agnew have also commented on the suscep- tibility of peat bog surfaces to public pressure, with particular reference to Borth Bog in North Wales. Once the damage is done, recovery may take many years and then only if people are rigorously excluded from the site. Bell and Bliss estimate that this time-span may be up to 1,000 years at some disturbed Alpine sites.

The presence of large numbers of people at a site is also disturbing to the animal life of the region. The species that people hope to see when they visit a site may be forced into hiding or driven to seek refuge elsewhere by the sheer pressure of hordes of *Homo sapiens*. Liddle, who has recently reviewed the ecological effects of human trampling, proposes a logical model that separates and examines the relationships involved. He also outlines the sequential management decisions required in areas subjected to this form of disturbance.

What part can the geographer play in conservation?

For effective conservation (and resource management) we need two types of basic knowledge. First, the detailed systematic knowledge of how each ecosystem component, including man, functions and relates to the other components. This is the field of the specialist, essentially engaged in analysis. The geographer can make a contribution here through his expertise in such fields as geomorphology, historical geography or land-use studies, as Coppock noted. Most of our information on this approach, however, will come from specialists in disciplines like botany, zoology and pedology. Because we are concerned with man, another range of specialists will make their contribution. These include economists, sociologists and psychologists.

Detailed research into these various aspects is still in its infancy. To quote O'Connor: 'The work of the International Biological Programme for the study of productivity and human welfare has involved the detailed investigation of a range of ecosystems in many parts of the world ... It is a salutary thought that some 40 or so people have now devoted about five years, equivalent to 200 man-years, to the study of a single hectare of mixed deciduous woodland in the English Lake District. It is now possible, for this wood, to put quantitative values . . . and we are within sight of a static, descriptive model of events in this wood. However, there is no basis yet for any statements of a predictive kind upon which a management policy for this or other woodlands could be based.' The wood in this study is Meathop Wood, already referred to several times in Chapter 8.

The second kind of knowledge we need is based on a synthesis of specialized studies. As Streeter points out, conservation is a synthesis of three separate disciplines: ecology, economics and sociology. Others would add more to his list. Now the geographer has always regarded himself as being in an important position, playing a key role in synthesizing material often initially provided by other disciplines. The holistic outlook of the geographer and his emphasis on spatial relationships should enable him to make a valuable contribution to environmental studies. Geography, by overlapping with a number of science and arts subjects, is in a unique position to integrate data and ideas from both. In theory at least, it should provide an excellent training for those concerned with problems of conservation and environmental management.

An area of environmental planning in which geographers would seem to have an obvious role to play is *Environmental Impact Assessment* (EIA), which developed in the United States. Bisset has reviewed the various schemes for EIA that have arisen. The basis of the approach is the *Environmental Impact Statement* (EIS) which Landy describes as a document required by the National Environmental Policy Act, 1969 (United States) for inclusion in every proposed project which is determined to be major and significant. The EIS is used in making decisions about the positive and negative effects of the undertaking on all aspects of the environment; it lists alternatives to the proposed action; discusses the relationship between local short-term uses of man's environment and the maintenance and enhancement of long-term productivity; and covers any irreversible and irretrievable commitment of resources which would be involved in the proposed action should it be implemented. In this country geographers have played a large part in setting up the Project Appraisal for Development Control (PADC) unit at Aberdeen University – a unit undertaking research and training in EIA and planning.

However, despite these developments, several writers have pointed out (e.g. Goodey, Coppock, and O'Riordan) that geographers have not yet made the expected impact on environmental studies that their philosophy and training would lead us to anticipate. A major weakness has been their failure to concern themselves with problem-oriented studies, particularly the community problems of environmental quality. Ecologists now recognize the necessity for a team approach for the full study of even relatively simple ecosystems. It therefore follows that in those ecosystems where the presence of man is an additional and vastly complicating factor, a team approach is even more essential. O'Riordan has drawn a similar conclusion, seeing the future role of the geographer interested in this area of study as not so much a specialist but more as a member of an interdisciplinary team concentrating on environmental problems.

REFERENCES

Anon, 1980. *World Conservation Strategy*, prepared by the International Union for Conservation of Nature and Natural Resources and others, Switzerland. (See also Allen, R., 1980. *How to Save the World*, Kogan Page, London.)

Anon, 1983. *Countryside Sports*. Standing Conference on Countryside Sports, College of Estate Management, Reading University.

Anon, 1983. *The Conservation and Development Programme for the U.K.*, Kogan Page, London.

Barkham, J. P., 1973. 'Recreational carrying capacity: a problem of perception', *Area*, 5 (3), 218–22.

Bell, K. L. and Bliss, L. C., 1973. 'Alpine disturbance studies: Olympic National Park, U.S.A.', *Biol. Conserv.*, 5, 25–32.

Bisset, R., 1983. 'A critical survey of methods for Environmental Impact Assessment', in *An annotated reader in Environmental Planning and Management* (eds O'Riordan, T. and Turner, R. K.), Pergamon Press, Oxford and New York, pp. 168–86.

Bonham-Carter, V., 1971. *Survival of the English Countryside*, Hodder and Stoughton, London.

Bowman, J. C., 1980. *Strategy for the UK forest industry*, Centre for Agricultural Strategy, Report No. 6, Reading University.

Bradshaw, A. D., 1977. 'Conservation problems in the future', in *Scientific aspects of Nature Conservation in Great Britain* (A Royal Society discussion organized by Smith, J. E., Clapham, A. R. and Ratcliffe, D. A.), The Royal Society, London, pp. 77–96.

Burden, R. F. and Randerson, P. F., 1972. 'Quantitative studies of the effects of human trampling on vegetation as an aid to the management of semi-natural areas', *J. appl. Ecol.*, 9, 439–58.

Coddington, A., 1974. 'The economics of conservation', Chapter 29 in *Conservation in Practice* (eds Warren, A. and Goldsmith, F. B.), Wiley, London, pp. 453–64.

Coombs, C. F. B., Isaacson, A. J., Murton, R. K., Thearle, R. J. P. and Westwood, N. J., 1981. 'Collared doves (*Streptopelia decaocto*) in urban habitats', *J. appl. Ecol.*, 18, 41–62.

Coppock, J. T., 1970. 'Geographers and conservation', *Area*, 2, 24–6.

Countryside Recreation Research Advisory Group, 1970. *Countryside Recreation Glossary*, Countryside Commission, London.

Cullingworth, J. B., 1974. *Town and Country Planning in England and Wales* (5th edn.) Allen and Unwin, London.

Darling, F. F., 1964. 'Conservation and ecological theory', *J. Ecol.*, 52 (Suppl.), 39–46.

Diamond, J. M., 1975. 'The Island Dilemma: lessons of modern biogeographic studies for the design of natural reserves', *Biol. Conserv.*, 7, 129–46.

Diamond, J. M., and May, R. M., 1981. 'Island Biogeography and the Design of Natural Reserves', in *Theoretical Ecology. Principles and Applications*, (2nd ed)., (ed. May, R. H.), Blackwell, Oxford and Boston, pp. 228–52.

Dower, M., 1964. 'Industrial Britain: the functions of open country', *J. Town Planning Inst.*, 50 (4), 132–41.

Gaffney, V., 1960. *The Lordship of Strathavon*, Third Spalding Club, Aberdeen.

Game, M., 1980. 'Best shape for nature reserves', *Nature*, London, 287, 630–32.

Gent, K. R., 1966. 'Vegetation development in an area of heathland, scrubland and woodland', B.Sc. Dissertation, Univ. of Leicester.

Gilbert, F. S., 1980. 'The equilibrium theory of island biogeography: fact or fiction', *J. Biogeography*, 7, 209–35.

Goldsmith, F. B., 1974. 'Ecological effects of visitors in the countryside', Chapter 14 in *Conservation in Practice* (eds Warren, A. and Goldsmith F. B), Wiley, London, pp. 217–31.

Goodey, B., 1970. 'Environmental studies and interdisciplinary research', *Area*, **2**, 16–18.

Gordon, L., 1983. 'Limits to the growth debate', in *An annotated reader in Environmental Planning and Management* (eds O'Riordan, T. and Turner, R. K.), Pergamon Press, Oxford and New York, pp. 362–72.

Helliwell, D. R., 1976. 'The effect of size and isolation on the conservation value of wooded sites in Britain', *J. Biogeography*, **3**, 407–16.

Helliwell, D. R., 1976. 'The extent and location of nature conservation areas', *Environmental Conservation*, **3**, 255–8.

James, G., 1974. 'Land use in upland Britain – an economist's viewpoint', Chapter 22 in *Conservation in Practice* (eds. Warren, A. and Goldsmith, F. B.), Wiley, London, pp. 337–60.

Jones, W. E., 1972. 'The future of the Uplands', in *The Dinas Conference*, Farming and Wildlife Advisory Group, Ministry of Agriculture, Fisheries and Food, Lampeter, pp. 4–10.

Landy, M., 1979. *Environmental Impact Statement Glossary*, IFI/Plenum, New York and London.

Lickorish, L. J. 1965. 'Planning for recreation and leisure', *J. Town Planning Inst.*, **51** (6), 243–7.

Liddle, M. J., 1975. 'A selective review of the ecological effects of human trampling on natural ecosystems', *Biol. Conserv.*, **7**, 17–36.

MacArthur, R. H. and Wilson, E. O., 1967. *The Theory of Island Biogeography*, Princeton University Press, Princeton.

MacEwan, A. and MacEwan, M., 1982. 'An unprincipled Act', *The Planner*, **68**, 69–71.

Margules, C. and Usher, M. B., 1981. 'Criteria used in assessing wildlife conservation potential: a review', *Biol. Conserv.*, **21**, 79–109.

McVean, D. N. and Lockie, J. D., 1969. *Ecology and land use in Upland Scotland*, Edinburgh University Press, Edinburgh.

Moss, B., 1979. 'Alarm call for the Broads': an ecosystem out of phase', *Geographical Magazine*, **LII**, 47–50.

O'Connor, F. B., 1974 'The ecological basis for conservation', Chapter 6 in *Conservation in Practice* (eds Warren, A. and Goldsmith, F. B.), Wiley, London, pp. 87–98.

O'Riordan, T., 1970. 'New conservation and geography', *Area*, **4**, 33–6.

O'Riordan, T., 1979. 'Alarm call for the Broads: signs of disaster and a policy for survival', *Geographical Magazine*, **LII**, 51–6.

O'Riordan, T. and Turner, R. K. (eds), 1981. *An annotated reader in Environmental Planning and Management*, Pergamon Press, Oxford and New York.

Peterken, G. F., 1977. 'Habitat conservation priorities in British and European woodlands', *Biol. Conserv.*, **11**, 223–36.

Ratcliffe, D. A. (ed.), 1977. *A Nature Conservation Review*, vol. I and II, Cambridge University Press, London and New York.

Raup, H. M., 1964. 'Some problems in ecological theory and their relation to conservation', *J. Ecol.*, **52** (*Suppl.*), 19–28.

Rooke, D. B. (ed.), 1974. *British Forestry*, Forestry Commission, HMSO, London.

Schofield, J. M., 1967. 'Human impact on the fauna, flora and natural features of Gibralter Point', in *The Biotic Effects of Public Pressures on the Environment* (ed. Duffey, E.), The Nature Conservancy, Monks Wood Symposium **3**, 106–11.

Simberloff, D. S. and Wilson, E. O., 1970. 'Experimental zoogeography of islands: a two-year record of colonization', *Ecology*, **51**, 934–7.

Simmons, I. G., 1969. 'Evidence of vegetation changes associated with Mesolithic man in Britain' in *The Domestication and Exploitation of Plants and Animals* (eds Ucko, P. J. and Dimbleby, G. W.), Duckworth, London, pp. 111–19.

Simmons, I. G., 1974. *The ecology of natural resources*, Edward Arnold, London.

Slater, F. M. and Agnew, A. D. Q., 1977. 'Observations on a peat bog's ability to withstand increasing public pressure', *Biol. Conserv.* **11**, 21–8.

Statham, D., 1972. 'Natural resources in the Uplands', *J. Town Planning Inst.*, **58** (10), 468–78.

Streeter, D. T., 1974. 'Ecological aspects of oak woodland conservation', in *The British Oak: its history and natural history* (eds Morris, M. G. and Perring, F. H.), Botanical Soc. Br. Isles, E. W. Classey, Faringdon, pp. 341–54.

Tivy, J. and Rees, J., 1978. 'Recreational impact on Scottish lochshore wetlands', *J. Biogeography*, 5, 93–108.

Warren, A. and Goldsmith, F. B.. (eds.), 1974. *Conservation in Practice*, Wiley, London.

Welch, D. and Rawes, M., 1964. 'The early effects of excluding sheep from high-level grasslands in the Northern Pennines', *J. appl. Ecol.*, **1**, 281–300.

Wilcox, B. A., 1978. 'Supersaturated island faunas: a species-age relationship for lizards on post-Pleistocene land-bridge islands', *Science*, **199**, 996–8.

General Index

Species Index